NUTRITIONAL DISORDERS IN CROP PLANTS

NUTRITIONAL DISORDERS IN CROP PLANTS

K. Balakrishnan
Professor of Crop Physiology
Agricultural College and Research Institute (TNAU), Madurai-625 104

A. Subbiah
Assistant Professor of Horticulture
Grape Research Station (TNAU), Theni Dt-625 526

J. Rajangam
Professor and Head
Department of Fruit Crops
Horticultural College and Research Institute, (TNAU) Periyakulam-625 604

P.P. Mahendran
Professor of Soil Science
Agricultural College and Research Institute, Madurai-625 104

M. Prakash
Professor
Faculty of agriculture,
Annamalai University, Chidambaram-608 003

SSPH

SATISH SERIAL PUBLISHING HOUSE
403, Express Tower, Commercial Complex
Azadpur, Delhi-110033 (India)
Phone : 011-27672852, Fax : 91-11-27672046
e-mail : info@satishserial.com, hkjain1975@yahoo.com
Website : www.satishserial.com

Published by :

SATISH SERIAL PUBLISHING HOUSE

403, Express Tower, Commercial Complex, Azadpur, Delhi-110033 (INDIA)
Phone : 011-27672852 Fax : 91-11-27672046
E-mail : info@satishserial.com, hkjain1975@yahoo.com

ISBN 978-93-86200-49-5

Composed, Designed & Printed in India

About the Authors

Dr. K. Balakrishnan is presently working as Professor of Crop Physiology at Agricultural College and Research Institute, Madurai. He was the recipient of ICAR Senior Fellowship during 1983 and obtained Ph.D in Crop Physiology in 1987 from Tamil Nadu Agricultural University, Coimbatore. He has also obtained Post Graduate Diploma in Journalism in 1988. He also worked in United Planter's Association of South India, Nilgris, Tamil Nadu. He has been teaching courses on Crop Physiology and Biochemistry to the Under Graduate and Post Graduate students in both Agriculture and Horticulture for the past 30 years. He served as Principle Investigator of Foliar diagnosis in Horticultural crops in plan scheme funded by Government of Tamil Nadu, Post Harvest Technology of Horticultural crops, Tropical fruits and Arid zone fruit crops funded by ICAR. He was the member of All India ICAR Syllabus Committee on basic Agricultural sciences during 2007-2008. He also served as a member of Board of studies, Tamil Nadu Agricultural University during 1997-2000. He has authored and co-authored 10 book chapters and 4 books and published 60 research papers in both National and International Journals. He has been the question setter and external examiner for M.Sc.(Ag.) and Ph.D thesis in several Universities. He was the Professor and Head, Department of Seed Science and Technology, Agricultural College and Research Institute, Madurai. during 2013-16.

Dr. A. Subbiah is presently working as Assistant Professor of Horticulture at Grapes Research Station, Theni, Tamil Nadu. He was the recipient of ICAR Junior Research Fellowship for his Master Degree during 1997 and completed his Ph.D., in 2004 from Tamil Nadu Agricultural University, Coimbatore. He also worked as Research Assistant at Cardamom Research Station, Pampadumpara, Idukki, Kerala (KAU), Senior Research Fellow at Department of Vegetable Crops, TNAU, Coimbatore and Technical Executive of Indian Spices Producers Company, Cochin, Kerala. He has also been involved in teaching courses on fundamentals of horticulture, production and postharvest technology of fruits to the Under Graduate students in Horticulture for the past 10 years. He also served as Co-Principle Investigator in the ICAR-NICRA Research Project on "Studies on Mango Flowering Phenology" and IDRC-CIFSF, Canada funded Project on "Enhanced Preservation of Fruits in Asia by using Nano Film

Techniques". Further, he is also In-Charge of Venture Capital Scheme "Mass multiplication of grape rootstocks and varieties. He was associated 10 various scientific societies involved in promotion of research and development in the field of fruits, spices and plantation crops as life member. He also served as a member of Board of Studies for UG Programme in Horticulture, Tamil Nadu Agricultural University during 2010-2013. He has authored and co-authored 4 book chapters and 6 books and published 23 research papers in reputed National and International Journals. He was the organizer of various training programmes related to production technology of mango, grapes and banana for the benefit of farming community. He is currently involved in research activities related crop improvement in grapes and standardization of canopy management and viticultural practices in grape.

Dr. J. Rajangam, Started his career as Asst. Professor (Horticulture) in the year 1995 at Tamil Nadu Agricultural University (TNAU), Coimbatore. As a Horticulturist, he has more than 20 years experience in various Horticultural crops. During his service, significant achievements were made pertaining to release of varieties Co-3 Chili, PKM-2 annual moringa and PKM 1 Davana and so many horticultural technologies were also standardized. He served as Professor and Head. HRS, Kodaikanal for a period of four years. During his tenure, he was officer in-charge of All India net work project on temperate fruit crops at HRS, Kodaikanal with CITH, Srinagar and PI of ICAR-AICRP (Fruits) at HC & RI, Periyakulam. He got rich experience in teaching of UG and PG for more than 16 years and guided 12 M.Sc (Hort.) students. He has published more than 30 papers in National and International research papers, 112 popular article 15 booklet and two books. He has visited Israel and Tanzania for post graduate certificate programme in protected horticulture and International Symposium on Moringa Cultivation and Utilization. Currently, he is Professor and Head, Department of Fruits, Horticultural College and Research Institute, Periyakulam (TNAU), Coimbatore.

Dr. P.P. Mahendran. B.Sc. (Ag) 1988, M.Sc. (Ag) 1991, Ph.D 2000; Senior Research Fellow 1992-1995, Assistant Professor 1995, Associate Professor 2005, Professor 2010 onwards, Tamil Nadu Agricultural University, Coimbatore; Specialization: Soil Fertility and Problem Soils Management; Experience: 25 Years of Teaching, Research and Extension. Published 54 Research Articles and three Text Books: Recipient of Young Scientist Fellowship by Tamil

Nadu State Council for Science and Technology, Chennai (2003), Best Researcher Award (2009), Best Teacher Award (2010), Best Extension Worker Award (2011) by TNAU and P.S.G. Ganga Naidu Medal for the Best Research in Sugarcane (2010) by MASU, TNAU.

Professor Dr. M. Prakash born in 1966, obtained his B.Sc.(Ag.) and M.Sc.(Ag.) from Tamil Nadu Agricultural University, Coimbatore, India in 1988 and 1990 respectively. He did his Ph.D from Annamalai University. His studious and scholarly aptitude during his studenthood fetched him prestigious student support awards like National Merit Scholarship, Government of India during 1981-1988 and ICAR Junior Fellowship during 1988-1990. Prof. M. Prakash is working on abiotic stress tolerance from his post graduation onwards. He started working on effect of antitranspirants on growth and physiology of brinjal and continuing his work on abiotic stress tolerance till date. He is having 24 years of teaching and research experience in the Department of Genetics and Plant Breeding wherein he has successfully guided 8 Ph. D. and 22 Masters Degree Programmes. He has also completed five funded research projects.

He is presently serving as Professor, Department of Genetics and Plant Breeding, Faculty of Agriculture, Annamalai University. He has earned the status of Departmental Research Support with Special Assistance Programme of University Grants Commission (UGC-SAP- DRS) to the Department with financial assistance of Rs. 50 lakhs at Phase I (2009-2014) and Rs. 1.025 Crores at phase II (2015-2020) level.

His credentials include publications of 85 refereed research papers, seven books published by National publishers, three books published by International publishers and one book chapter published in SESAME by globally reputed CRC Press. He has significantly contributed in knowledge sharing, through organization of 7 National Seminars, participation in 8 International Seminars, Conferences and Workshops and 18 National Seminars and Conferences. His expertise in abiotic stress tolerance is well recognised by invitations to serve in the Editorial Board/ Committees of several academic journals, awards like Crop Research Award in 2004 and Best Researcher Award in Annamalai University in 2010 and Fellowships like FELLOW of Indian Society of Plant Physiology, New Delhi in 2015 (FISPP), FELLOW of Association for the Advancement of Biodiversity Science, Bangalore in 2015 (FABSC) and FELLOW of National Academy of Biological Sciences (NABS), Chennai in 2016(FNABS).

Preface

The Indian Agriculture is known for its complexities and diversities in flora, fauna and soil types. It has made marvelous achievements as indicated by profound grain growth after the induction of green revolution in 1960's which helped the country to attain self-sufficiency in food grain production. With the introduction of high yielding varieties, the requirement for different essential nutrients is increasing. Nevertheless, the available nutrients in the soil reservoir are not sufficient to cater the needs of the newer efficient crop varieties. This causes the development of specific symptoms caused by the shortages of the particular element(s). Therefore, better understanding the availability in soils, functions, deficiency symptoms and their correction measures is of paramount importance to fulfill.

To satisfy the requirements, the book entitled "Nutritional Disorders in Crop Plants" has been brought out. A major objective in the preparation of this book is to bring together in one volume a comprehensive account of both principles of plant nutrition and nutritional disorders available in literature for many crops. This is absolutely warranted as the available information on various crops is scattered.

This book comprise of basic aspects of all seventeen essential elements, absorption of nutrients, rooting pattern, function, deficiency symptoms and their remedial measures to alleviate them. In this book, We do not pretend to claim credit for any original contribution on the subject. Rather, it should be viewed as a concise collection of various topics from different books, bulletins, journals, scientific papers etc., the names of which are indicated in the text itself or in the selected reference.

The book has been designed in such a way as to improve the knowledge on principles of plant nutrition essential for plant growth and the method by which such nutrient deficiencies can be diagnosed and effectively be remedied. The book has been written in simple language so as to understand not only by the scientists but also by students, the extension workers and farmers. The authors are thankful to Tmt. S. Vijiganapathy of Crop Physiology Unit, Agricultural College and Research Institute, Madurai for her neat typing of the manuscript of this book.

Agricultural College and
Research Institute, Madurai

K. Balakrishnan
A. Subbiah, J. Rajangam
P.P. Mahendran, M. Prakash

Contents

PART III: DIAGNOSIS OF NUTRITIONAL DISORDERS

Part I :

NUTRIENT ELEMENTS

Chapter 1

History of Plant Nutrition

The biggest challenge for India is its ever increasing population and to match it with food production. The population of India reached the 1.27 billions on 1st January 2014 and is projected to reach 1.63 billion around 2050 AD. Around 333 million tonnes of food grains are to feed this population. According to UNO, the world population is 7.6 billion in 2017 and it will be 9.8 billion in 2050 and 11.2 billion in 2100. In India, the arable land is limited and exceeded the carrying capacity of the land. Under these circumstances, there is a compelling need to increase agricultural productivity. Apart from the increasing population, the nutrient requirements of the people are also on the rise. It is estimated that in the next two generations, the world will consume twice as much food as has been consumed in the entire history of human kind. This situation unquestionably calls for a prudent approach for sustainable agricultural development as no other option would provide the answer.

The challenges of the next century are to be tackled within our limited resources. Specially, expansion in agricultural production is to be achieved within the available land, water and energy resources. Therefore, judicious use of resources, evolving technologies which are cost effective, resource saving are essential. The tasks are to be achieved within the frame work of protection of environment and preservation of natural resource base. This calls for a firm commitment to evolve new technologies for increasing the productivity and continuous use of available resources in the 21st century by all concerned with agricultural development.

Each crop has its own pattern of nutrient uptake. Under rice-rice-rice cropping system, 363 kg of N, 82 of P_2O5 and 400 kg of K_2O are removed from a hectare of soil annually. But farmers attach undue importance to nitrogen fertilizers than others. The ratio of the utilization of N, P and K fertilizers decreased from 1.0: 4.0: 0.14 during 1991 to 1.0: 0.29: 0.12 during 1995-96. Because of unbalanced fertilizer utilization pattern for the past three decades, the crop productivity decreased from 14 kg yield / kg applied fertilizers during 1970-71 to 9.30 kg during 1992-93. Such a skewed utilization of fertilizers has led to loss in soil productivity.

Moreover, the least importance attached to potassium fertilization led to explosion of myriad's of pest and disease problems and ultimate crop yield losses. Besides N, P, K, crop also needs nutrients like Ca, S, Zn, Fe, Cu Mn, B and Mo. In some cases, elements like Si and Cl are also essential for imparting special characters like pest and disease resistance. Though they are needed in micro level, their requirement is most essential for growth and full expression of their genetic potential. Although resistance and tolerance of crop plants to pest and diseases are genetically controlled, they are considerably influenced by nutrient. Mineral fertilizers not only affect the growth and composition of plants but also have profound effects on microbial activity in the soil rhizosphere and plant resistance and tolerance to root and shoot pathogens and pests indirectly. Usually a balanced nutrient supply that ensures optimum plant growth also considered optimum for plant resistance.

In recent times, the yield potential of crop varieties has already reached a plateau owing to several constraints. The most important among them is the deficiency of mineral nutrients noticed through the length and breadth of the country, where intensive cultivation has been practiced. The depletion of nutrients particularly micronutrients in the soil is increasing due to adoption of modern technologies in crop production such as raising high yielding crop varieties, multiple cropping, inter cropping etc., where the demand for nutrients is very high. The amount of nutrients depleted is not returned to the soil in the right proportion at right times leading to wide spread deficiency of nutrients especially micronutrients.

Fertility Status of Indian Soils

In the tropical regions, the organic matter of most soils of India is quite low (0.4 to 1.0%). Higher organic matter (2.0 – 3.0%) content in the humid region of north-eastern India has been noted because of perennial vegetation and acidic nature of soil. In India, sixty two per cent of the districts and union territories have low N content, 32.4%

have medium N content while only 4.9% belong to the high category. The deficiencies of nitrogen are mainly recorded in alluvial plains as well as black, red and laterite soils with the extreme occurring in the desert tract. On the whole, the low fertility status of N in the Indian soil is a matter of concern.

It has been observed that in 2.2% of the districts, the soils have the high level of P availability. These are located in Assam, Himachal Pradesh, and some arid regions of Rajasthan. On the other hand, 51% of the districts have a medium level of P and 46.8% belong to the low category. The deficiency of P is widely manifested in the alluvial plains of Uttar Pradesh, black soils of Madhya Pradesh and Maharashtra. In laterite soil, the availability is very limited. Under arid conditions, P deficiency is not very prevalent probably because of the limited depletion through crop removal.

A recent survey of Indian soils has revealed that 21% of the districts in the country are low in K. Nearly 44.2% of the districts have high K content, while the remaining 34.7% have medium K content. The need of K is likely to be felt with the growing intensity of cultivation. Alluvial soils of arid and semi-arid regions of north-western India (Rajasthan, Haryana, Punjab and Gujarat) are fairly rich in K but in the plains of Uttar Pradesh and Bihar, several districts come under low K content class. Red loams with lower base content account for medium available K in Tamil Nadu and Karnataka.

Deficiency of micronutrients is more likely in light textured and more calcareous soils. Among the micronutrients, Zn has attracted maximum attention in recent years. Zinc deficiency in maize is well-established on alluvial soils of Punjab, Haryana and eastern Uttar Pradesh, where the sandy texture and presence of $CaCO_3$ restrict the availability. Quite often, pH higher than 6.0 affects its availability. Response of crop yield to Zn has been marked in some black soils of Andhra Pradesh and red and yellow mixed soils of Karnataka. Some parts of Himachal Pradesh, Rajasthan, Madhya Pradesh, Bihar, Tamil Nadu and West Bengal have also recorded the good effect on Zn application. Similarly, the micronutrient responses, particularly Zn have been noticed in the alluvial soils of Punjab, Uttar Pradesh (East) and Haryana.

History of Plant Nutrition

The period of the development of the human race during which it began the cultivation of plants marks the dawn of agriculture. It is generally agreed that one area in the world that shows evidence of a very early civilization is Mesopotamia situated between the Tigris and Euphrates rivers in what is now Iraq. It is believed that human race invented agricultural practices 8,000-10,000 years ago in the area between Nile Valley in Egypt and Indus in India.

Theophrastus (327-287 B.C.) recommended the abundant manuring of the soils but suggested that rich soils be manured sparingly. Theophrastus also suggested that plants with high nutrient requirements also had a high water requirement. Not only did the Greek ancients recognize the merits of manure, but they also observed the effect that dead bodies had an increasing the growth of crops. The value of green-manure crops, particularly legumes, was also soon recognized. Theophrastus noted that a bean crop was ploughed under by the farmers of Thessaly and Macedonia. He observed that even when thickly sown and large amounts of seed were produced the crop enriched the soil. Virgil (70-19 B.C.) advocated the application of legumes.

Theophrastus suggested the mixing of different soils as a means of "remedying defects and adding heart to the soil." The addition of fertile soil to infertile soil could lead to increased fertility and the practice of mixing one soil with another may have provided better inoculation of legume seed on some fields. Again, the mixing of coarse textured soils with those of fine texture, or vice versa, may have caused an improvement in the water and air relations in the soils of the fields so treated.

The value of marl (a source of lime) was also recognized. The early dwellers of Aegina dug up marl and applied it to their land. Pliny (A.D. 62 - 113) stated that lime should be spread thinly on the ground and that one treatment was sufficient for many years. The age of the Greeks from 800 to 200 B.C. was indeed a Golden period for later centuries. Their writings, culture, and agriculture were copied by Romans, and the philosophy of many of the Greeks of this period dominated human thinking for more than 2000 years. Francis Bacon (1561-1624) suggested the principal nourishment of plants was water.

Willow tree experiment

Jan Baptiste **Van Helmont** (1577-1644), a reported that water was the sole nutrient of plants. He placed 200 lb of soil in an earthen container, moistened the soil, and planted a willow shoot weighing 5 lb. He carefully shielded the soil from dust, and only rain or distilled water was added. After five years, the experiment was terminated. The tree weighed 169 lb and about 3 oz. He could account for all but about 2 oz of the 200 lb of soil originally used. Because, he had added only water, his conclusion was that water was the sole nutrient of the plant. He attributed the 2 oz soil loss to experimental error.

About this same time, J.R. **Glauber** (1604-1668), a German chemist suggested that saltpeter and not water, was the 'principle of vegetation'. He further stated that because the animals ate forage, the saltpeter must have come originally from the plants.

Arthur Young (1741 - 1820), an English agriculturalist published a work entitled 'Annals of Agriculture', which was highly regarded and made a considerable impact on English agriculture.

The discovery of oxygen (O_2) by **Priestly** was the key to a number of other discoveries that went toward unlocking the mystery of plant life. **Jan Ingenhousz** (1730-1799) showed that air was purified in the presence of light. Coupled with this discovery was the statement by **Jean Senebier** (1742-1809), a Swiss natural philosopher and historian, that the increase in the weight of Van Helmont's willow tree was the result of air.

Justus Von Liebig (1803-1873), a German chemist, effectively destroyed the humus myth and **made the following statements:**

- Most of C in plants comes from CO_2 of atmosphere.
- H_2 and O_2 come from water.
- The alkaline metals needed for the neutralization of acids are formed by plants, as a result of their metabolic activities.
- P is necessary for seed formation.
- Plants absorb everything indiscriminately from soil and excrete materials from their roots that are non-essential.

The **law of the minimum**, stated by Liebig in 1862, is a simple but logical guide for prediction crop response to fertilization.

In 1843, **Lawes and Gilbert** tested the Liebig law in Rothamsted, England. They found some Limitations of Liebig's Law and settled the following points.

- Crops require both P and K but the composition of the plant ash is no measure of the amount of these constituents required by the plant.
- Non-legume crops require nitrogen. Without this, no growth will be obtained, regardless of the quantities of P and K present. The amount of NH_3 contributed by the atmosphere is insufficient for the needs of crops.
- Soil fertility can be maintained for some years by chemical fertilizers.
- The beneficial effect of fallow lies in the increase in available nitrogen compounds in the soil.

In 1886 German scientists, **Hellriegel** and **Wilfarth** concluded for the first time that bacteria in the nodules attached to legume roots assimilated gaseous N from the atmosphere and converted it to a form used by plants.

Criteria for Essentiality

An essential nutrient element is the one which is required for the normal life cycle of an organism and whose functions cannot be substituted by any other chemical compound. Besides, the chemical must be directly involved in nutrition as an essential metabolite or the element must be required for the action of an essential enzyme system. The criteria for essentiality of elements for plant growth have been proposed by Arnon and Stout in 1939and was refined by Arnon (1954):

i) A deficiency of the element makes it impossible for the plant to complete vegetative or reproductive growth stage of its life cycle.

ii) The deficiency is specific to the element in question and as such can be prevented or corrected only by supplying that particular nutrient element to the plant, and

iii) The nutrient must take part in metabolism as a metabolite or it must be required for action of an essential enzyme system.

Accordingly to Arnon's theory sixteen nutrient elements are found essential for plants. With the addition of nickel, now the number of essential elements has been enhanced to 17 (Table 1).

Table 1. Essential nutrients so far discovered

Elements	Essentiality established by	Year
Carbon	Since time immemorial	-
Hydrogen	Since time immemorial	-
Oxygen	Theodore de Saussure	1804
Nitrogen	Theodore de Saussure	1804
Phosphorus	C. Sprengel	1839
Potassium	C. Sprengel	1839
Magnesium	C. Sprengel	1839
Sulphur	Sachs J. Knop	1860
Calcium	C.Sprengel	1839
Iron	E. Gris	1843
Manganese	J. S. Mc Hargue	1922
Zinc	A.L. Sommer and C. B. Lipman	1926
Copper	A.L. Sommer, C. B. Lipman and C. Mc Kinney	1931
Molybdenum	D.L. Arnon and P.R. Stout	1939
Boron	K. Warington	1923
Chlorine	T.C.Broyer, A.B. Carlton, C.M.Johnson and R.P. Stout	1954
Nickel	H.P. Brown, M. Ross Welch and E. Earle Cary	1987

(Tisdale *et al.*, 1977)

Besides the seventeen elements, there are few more elements like rubidium, silicon, strontium, nickel, chromium, arsenic, sodium and vanadium etc. which are fund to stimulate the growth and to have beneficial effects at very low concentration. These nutrients are called as "beneficial elements" or potential microelements. These elements have been found to affect the uptake, translocation and utilization of essential elements, help in production of essential metabolite by activating enzymatic system/action and also counteract the toxic effects of some other elements or antimetabolites.

Classification of Essential Plant Nutrients

a) According to the quantity of nutrient present in plants

i) Basic Nutrients: C, H, O constitute about 96% of the total dry matter of plants.

ii) Macro/Major nutrients: Such elements are required in larger amount. N,P and K are Primary/Major nutrients. Ca, Mg, S are Secondary nutrients.

iii) Micronutrients: Required in small quantities (0.01-1000 ppm). Micronutrients are also called trace elements or Oligoelements, or Spurne elements. These are very efficient, and minute quantities produce optimum effects. On the othe hand even a slight deficiency or excess is harmful to the plants.

iv) Ultra micronutrients: Concentration required in 1 ppb (parts per billion). Nicholas (1963) put forward this concept giving the example of Mo and Co but it is faulty classification and is not acceptable today because the quantity of nutrients absorbed depends upon the types of plant.

v) Beneficial nutrients or Potential micronutrients: Such elements at very low concentrations and often under specific conditions have been shown to stimulate the growth of certain plants Se, Al, Rb, Sr, Si and As.

b) According to their Relative Requirements for Plant Growth

The essential plant nutrients can be classified based on their relative requirement by the plants:

a) **Major, Primary or Macro nutrients:** This group includes nitrogen, phosphorus and potassium which are required in maximum quantities by the plants. These nutrients are often deficient in almost all the soils because of their heavy depletion. The deficiency is corrected by application of fertilizers based on soil analysis and crop requirement. These elements are also referred as fertilizer elements. This group also includes C, H, O which are adequately present in atmosphere and need not be applied.

b) **Secondary nutrients:** This group includes Ca, Mg and S. They are required right from the beginning of the plant growth but in relatively lesser quantities.

c) **Micronutrients, trace elements:** This group includes Fe, Zn, Cu, Mn, B, Mo Cl and Ni. These elements are required in very minute quantities and they mostly act to stimulate, activate or accelerate the enzymatic activities in metabolic process of plants. These elements are required in traces and hence, they are also called as trace elements.

c) According to their Biochemical Behavior and Physiological Functions

From the physiological viewpoint, it is difficult to justify the classification of plant nutrients into macro and micronutrients depending upon their concentration in plant tissues. Based on their biochemical and physiological behavior, the plant nutrients may be classified into the following four groups (Table 2).

Table 2. Nutrient Classification based on their Biochemical Behavior and Physiological Functions

Nutrient elements	Uptake	Biochemical functions
Ist Group C, H, O, N, S	CO_2, HCO_3^-, H_2O, O_2, NO_3^-, NH_4^+, N_2, SO_4^{2-}, SO_2	Major constituent of organic matter, Essential elements of atomic groups which are involved in enzymatic processes. Assimilation by oxidation reduction reactions
IInd Group P, B, Si	As phosphates, boric acid or borate, silicate from soil solution.	Esterification with native alcohol groups in plants. The phosphate esters are involved in energy transfer reactions.
IIIrd Group K, Na, Mg	As ions from the soil solution	Non-specific functions establishing osmotic potentials. More specific reactions by which the conformation of the enzyme protein is brought into optimum status (enzyme activation). Bridging of reaction partners. Balancing diffusible and indiffusible anions.
IVth Group Fe, Cu, Zn, Mo	As ions or chelates	Incorporated in prosthetic groups mostly in chelated form. Enable electron transport by valancy change.

(Mengel and Kirkby, 2001)

d) Based on Functional basis along with their usual content

The plant nutrients can be classified based on the functional basis along with their usual contents present in plant tissues as a whole. It is mentioned as under:

Group I: The main structural element of plant tissues

i) Carbon : Constitutes nearly 45 per cent of the total tissues.

ii) Hydrogen : Constitutes nearly 6 per cent of the total tissues.

iii) Oxygen : Constitutes nearly 43 per cent of the total tissues.

Group II: Accessory structural elements of the more active and vital living tissues, especially proteins.

i) Nitrogen : 1-3 per cent, constitutes nearly 16-18 per cent of all the proteins.

ii) Phosphorus : 0.05 to 1 per cent. It is a constituent of nucleo proteins, phytin and lecithin.

iii) Sulphur : 0.05 to 1.5 per cent. It is a constituent of many sulphur containing protein.

Group III: Regulators and carriers for most plants

i) Potassium : 0.3 to 6.0 per cent of total tissues. It is not present as constituent but is found as salt and functions mainly in synthesis and translocation of carbohydrates.

ii) Calcium : 1 to 4 per cent of the total tissues. It is present as calcium pectate and as simple organic and inorganic salts.

iii) Magnesium : 0.05 to1 per cent of total tissues. It forms as essential constituent of chlorophyll and may function as a carrier of phosphorus.

Group IV: Catalyst and Activators for most plants

i) Iron : Present in 10 to 1000 ppm of total tissues. It functions in chlorophyll formation and probably in oxidation-reduction reactions. It may also exist in organic combinations.

ii) Manganese : Present in 5 to 500 ppm of total tissues. It is believed to function in chlorophyll formation and also in oxidation-reduction reactions.

iii) Zinc : It is present in 5 to 100 ppm of total tissues. It is believed to function as catalytic agents.

iv) Copper : It is present in 2 to 50 ppm of total tissues. It is believed to function as catalytic agents.

v) Boron : It is present in 2 to 75 ppm. Probably, it functions as a catalyst.

vi) Molybdenum : It is present in 0.01 to 10 ppm of total tissues. It is required in very minute amount for catalytic processes. Mo rich soils produce forage with about 100 ppm Mo which becomes toxic livestock.

vii) Chlorine : It is present in 100 to 250 ppm to several hundreds of total tissues plants need it in highest amount amongst all the micro-nutrients.

e) ICAR classification

In the handbook of manures and fertilizers (ICAR), another classification based on functions of elements is mentioned which is as under:

Group I Energy exchangers : H, O

Group II Energy stores : C, N, P and S

Group III Translocation regulators : K, Na, Ca and Mg

Group IV Oxidation-reduction regulators : Fe, Mn, Mo, Cu, B and Zn

f) Based on chemical nature

Cations - K, Ca, Mg, Fe, Mn, Zn, Cu

Anions - NO_3, H_2PO_4, SO_4

Metals - K, Ca, Mg, Fe, Mn, Zn, Cu

Non-metals - N, P, S, B, Mo, Cl

g) Based on forms of plant nutrients absorbed by plants

The essential elements are also be classified based on the forms of nutrients in which it is absorbed by plants (Table 3).

Table 3. Nutrient Classification based on forms of plant nutrients absorbed by plants

Nutrient Element	Ionic Form	Non-ionic form
Nitrogen	NO_3^-, NH_4^+	$CO(NH_2)_2$
Phosphorus	$H_2PO_4^-$, HPO_4^{2-},	Nucleic acid, Phytin
Potassium	K^+	-
Calcium	Ca^{2+}	-
Magnesium	Mg^{2+}	-
Sulphur	SO_4^{2-}	SO_2
Iron	Fe^{2+}, Fe^{3+}	$FeSO_4$ with EDTA
Manganese	Mn^{2+}	$MnSO_4$ with EDTA
Zinc	Zn^{2+}	$ZnSO_4$ with EDTA
Copper	Cu^{2+}	$CuSO_4$ with EDTA
Boron	$B_4O_7^{2-}$, HBO_3^-, HBO_3^{2-}	-
Molybdenum	MoO_4^{2-}	-
Chlorine	Cl^-	-
Nickle	Ni^{2+}	-

□□□

Chapter 2

Basics of Soil Fertility

The science dealing with soil as a natural resources on the surface of the earth, including pedology, physical, chemical, biological and fertility properties of soil and these properties in relation to their management for crop production. According to Buckman and Brady, soil may be defined as "A dynamic natural body on the surface of the earth, in which plants grow, composed of natural and organic materials and living forms".

Indian soils classification

Indian soils have been divided into eight major groups. They are by far the largest and the most important soil group of India.

Black Soil

These soils have been formed due to the solidification of lava spread over large areas during volcanic activity in the Deccan Plateau, thousands of years ago. They are black due to compounds of iron and aluminium. Mainly found in Deccan Plateau, Maharashtra, Gujarat, Madhya Pradesh, Karnataka, Andhra Pradesh, and Tamil Nadu. Apart from cotton cultivation, these fertile soils are suitable for growing cereals, oilseeds, citrus and vegetables, tobacco and sugarcane. They have high moisture retention level and lack in phosphorus, nitrogen and organic matter. It is called Regur and is ideal for cotton crop.

Red Soil

They are mainly formed due to the decomposition of ancient crystalline rocks like granites and gneisses and from rock types rich in minerals such as iron and magnesium. The term 'red soil' is due to the wide diffusion of iron oxides through the materials of the soil. Covers almost the whole of Tamil Nadu, Karnataka, Andhra Pradesh, Maharashtra, and Chhattisgarh, parts of Orissa, Jharkhand and Bundelkhand. They are generally deficient in nitrogen, humus and phosphorus, but rich in potash. These soils are suitable for rice, millets, tobacco and vegetables.

Laterite Soil

It is found in conditions of heavy rainfall with alternate wet and dry periods. The alterations of wet and dry season leads to the leaching away of siliceous matter and lime of the rocks and a soil rich in oxides of iron and aluminium compounds. It occurs in parts of Western Ghats, Eastern Ghats, Rajmahal hills, Maharashtra, Karnataka, Kerala, Orissa, West Bengal, Assam, Tamil Nadu, etc. These are poor in nitrogen and minerals. Best for tea, coffee, rubber, cinchona and coconut cultivation.

Forest and Mountain Soils

Such soils are mainly found on the hill slopes covered by forests. The formation of these soils is mainly governed by the characteristic deposition of organic matter derived from forest growth. In the Himalayan region, such soils are mainly found in valley basins, depressions and less steeply inclined slopes. Apart from the Himalayan region, the forest soils occur in higher hills in south and the peninsular region. It is very rich in humus but are deficient in potassium, and phosphorous. Suitable for tea, coffee, spices and tropical fruits.

Arid and Desert Soils

It is found in arid and semi-arid region in Rajasthan and adjoining areas of Punjab and Haryana lying between the Indus and the Aravallis receiving less than 50 cm of annual rainfall. This area is covered by a mantle of sand which inhibits soil growth. The phosphate content of these soils is as high as in normal alluvial soils. Nitrogen is originally low but its deficiency is made up to some extent by the availability of nitrogen in the form of nitrates. Thus, the presence of phosphates and nitrates make them fertile soils wherever moisture is available.

Saline and Alkaline Soils

In the drier parts of Bihar, Haryana, Punjab, Rajasthan and Maharashtra are the salt-impregnated or alkaline soils. Some of the salts are transported in solution by the rivers and canals, which percolates in the sub-soils of the plains. The accumulation of salts makes the soil infertile and renders it unfit for agriculture.

Peaty and Marshy Soils

Originate in the humid regions as a result of accumulation of large amounts of organic matter in the soil. They contain considerable amounts of soluble salts and 10 - 40% of organic matter.Peaty soils are found in Kottayam and Alappuzha districts of Kerala, where it is called Kari.Marshy soils high in organic matter are found in northern Bihar, coastal parts of Orissa, Tamil Nadu and West Bengal and parts of Uttar Pradesh.

Phases of Soils

Soil is a heterogeneous, multiphase, disperse and porous system. It is the most common media for plant growth.Minerals soils consist of four major components: Mineral materials, organic matter, water and air in various proportions. Approximately 50% of the total volume of the surface horizon of many soils is made up of inorganic materials (mineral matter) and organic matter (5%) and the remaining volume is porspace between the soil particles. Water and air occupy these pore spaces in various proportions. The proportion of air and water varies from one season to another. At optimum moisture for plant growth, the 50% of pore space possessed is divided roughly in half 25% of water space and 25% or air. The soil may be described as the three phase system: Soil solid, Liquid and Gaseous phase.

i) Solid phase

The soil material less than 2 mm size constitutes the soil sample. It is broadly composed of inorganic and organic constitutes. The majority of the soils of India are mineral soils. It accounts for nearly 50% of the total volume and 95% without of the solid phase is made up of inorganic or mineral matter. The remaining 5% weight comprises of organic matter. Inorganic constituents consist of carbonates, soluble salts, free oxides of iron, aluminium and silicon. The humus and humus like fractions of the solid phase constitute the soil organic matter. Soil is the habitat for enormous number of living organisms like roots of higher plants (Soil Macro flora), bacteria, fungi, actinomycetes and algae (Soil Micro flora). A gram of fertile soil contains billions of these micro-organisms. The live weight of the micro-organisms may be about

4000 kg/ha may constitute about 0.01 to 0.4% of the total soil mass. Soil also consists of protozoa and nematodes.

ii) Liquid phase

About 50% of the bulk volume of the soil body is generally occupied by soil pores which may be completely or partially filled with water. A considerable part of the rain which falls on soil is absorbed by the soil and stored in it to be returned to the atmosphere by direct evaporation or by transpiration through plants. The soil acts as the reservoir for supplying water to plants for their growth. The soil water keeps salts in solution which act as plant nutrients. Thus, the liquid phase is an aqueous solution of nutrients, when water drains from soil pores are filled with air.

iii) Gaseous phase

The air filled pores constitutes the gaseous phase of soil system and dependent on the liquid phase. The nitrogen and oxygen contents of soil air are almost the atmospheric air but the concentration of CO_2 is much higher (8 – 10 times more) which may be toxic to plant roots. This phase supplies O_2 and thereby prevents CO_2 toxicity. The three phases of the soil system have definite roles to play. The solid phase provides mechanical support for and nutrients to the plants. The liquid phase supplies water and along with it dissolved nutrients to plant roots. The gaseous phase satisfies the aeration (O_2) need of plants.

Soil Fertility and Productivity

Soil fertility is the inherent capacity of the soil to supply nutrients to plants in adequate amounts and in suitable proportions. Soil productivity is the capacity of the soil to produce crops with specific systems of management and is expressed in terms of yields (Table 1). All productive soils are fertile but all fertile soils need not be productive due to some problems like water logging, alkali, saline, adverse climate etc. According to modern terminology, soils fertility is the capacity of soil to produce crops of economic value and to maintain health of the soil without deterioration.

Table 1. Difference between soil fertility and soil productivity

Soil Fertility	Soil Productivity
i) It is considered as an index of available nutrient to plants.	It is a broader term used to indicate crop yields.
ii) One of the factors for crop production; the others are water supply etc.	It is the interaction of all the factors.
iii) Can be analyzed in laboratory	Can be assessed in the field under particular climate conditions
iv) It is the potential status of the soil to produce crops	Resultant of various factors influencing soil management.
v) Depends upon physical chemical and biological factors of soil	Depends upon location, fertility, physical conditions etc.
vi) Soil fertility is the function of available nutrients of soil. Soil fertility - f (Nutrient status of Soil)	Soil productivity is the function of soil fertility, management and climate.Soil Productivity = f (Soil fertility + Management + Climate)
vii) It is an inherent property of soil	It is not an inherent property of soil.
viii) The fertility of a certain soil is same in all the climates	Soil productivity differs according to the variations in climate and location.
ix) All fertile soils are not productive	All productive soils are certainly fertile

Soil chemical test for assessing fertility status

i) Soil pH

Soil reaction is the most important single chemical characteristic influencing many physical and chemical properties of soil. Soil reaction is measured by pH of a suspension of soil in water. The pH scale is a series of numbers and these numbers express the degree of acidity, alkalinity or neutrality of a solution. Soils are classified based on soil pH as follows.

pH	Class
< 4.5	Strongly acidic
4.5-5.5	Moderately acidic
5.5-6.5	Slightly acidic
6.5-7.5	Neutral
7.5-8.5	Slightly alkaline
8.5-9.5	Moderately alkaline
> 9.5	Strongly alkaline

ii) Electrical conductivity

Soils contain varying amounts of water soluble salts and differ in their total salt content. These soluble salts when present in optimum concentration serve as nutrients to the growing plants, but in excessive amounts are harmful to plant growth. Based on the crop tolerance to salinity, the soils are classified as follows.

EC (dS/m)	Soil salinity and crop tolerance
< 1.00	Normal soil, no deleterious effect on crops
1.00-2.00	Critical for germination
2.00-4.00	Critical for the growth of sensitive crops
>4.00	Injurious to most crop

iii) Soil organic matter

The role of soil organic matter in relation to soil fertility and physical conditions are widely recognized. Organic matter is the source of plant nutrients. A major portion of N (95-99% of the total), P (33-67% of the total) and S (75% of the total) in soils occur in organic combinations which mineralize to release the nutrients in inorganic forms to be used by plants. It serves as a reservoir of plant nutrients, promotes water storage and regulates microbial activity. Based on the organic carbon, the rating is given below

% Organic carbon	Rating
< 0.5	Low
0.5 to 0.75	Medium
> 0.75	High

iv) Available nitrogen (Alkaline permanganate method)

The available nitrogen in soil refers to a fraction of total nitrogen which is converted into forms accessible to the plant. This constitutes on an average only 0.5 to 2.5% of the total N in a soil at any given time. The soil rating based on the available nitrogen is given below.

Available N (Kg/ha)	Rating
< 280	Low
280 to 560	Medium
> 560	High

(v) Available phosphorus (Sodium bicarbonate method)

The term available phosphorus refers to the inorganic form occurring in the soil solution which is almost exclusively "orthophosphate". This orthophosphate occurs in several forms and combinations. Only a small fraction of the total amount present may be available to plants which is of direct relevance in assessing the phosphorus status. Depending on the available phosphorus, the soil is classified as follows

Rating	P(Kg/ha)	P_2O_5 Kg/ha
Low	< 9.8	< 22.5
Medium	9.8 to 24	22.5 to 55
High	> 24	> 55

vi) Available Potassium (Ammonium acetate method)

The available potassium is water soluble and exchangeable K. The neutral normal ammonium acetate extract contains both water soluble and exchangeable K. Potassium extracted by this extractant is considered as a suitable index of K availability in most soils. The three different ratings are given based on the available potassium.

Available Potassium (Kg/ha)		Rating
K	K_2O	
< 125	< 150	Low
125-300	150 - 360	Medium
> 300	> 360	High

vii) Available Sulphur (Monocalcium phosphate method)

The sulphate sulphur and part of organic sulphur forms the available sulphur in soils. Different reagents extract different quantities of sulphur from soil. The best reagent is one which correlates the amount of sulphur extracted with crop uptake. Among the various extractants, monocalcium phosphate (0.01 M) or 0.15% $CaCl_2 2H_2O$ is the best extractant for extraction of available sulphur.

Available Sulphur (ppm)	Rating
< 10	Low
10-20	Medium
> 20	High

viii) Exchangeable Ca and Mg (Ammonium acetate method)

Calcium and magnesium are the two most abundant alkaline earth cations in soils. Exchangeable magnesium is found in very small quantities in calcareous soils. The critical limits for exchangeable Ca and Mg is given below

Ca → less than 2.00 meq/100 g (deficient)

Mg → less than 0.5 meq/100 g (deficient)

or

Ca → less than 25% of CEC - deficient

Mg → less than 4 to 15% of CEC - deficient

ix) Available Cu, Zn, Fe and Mn (DTPA)

Micronutrient exist in soils either in water soluble, exchangeable or chelated forms and finally associated with primary and secondary minerals. Most of the micronutrients are in combination with soil organic matter and plants absorb them in ionic forms and chelated forms. The method commonly used for determining the available micronutrients in soil sample is given by Lindsay and Norvell (1978). DTPA method extracts all the four micronutrient cations viz, Zn, Cu, Fe and Mn from the soils simultaneously and particularly suited for neutral, alkaline and calcareous soils. DTPA as a chelating agent combines with free metal ions in the solution to form soluble complexes. Stability constants for the simultaneous complexing of Zn, Cu, Mn and Fe shows DTPA as a most suitable extractant. The critical concentration of the cationic micronutrients viz., Zn, Cu, Mn and Fe are given below (Table 2).

Table 2. Classification of soils into different categories based on the DTPA-extractable micronutrient cations

DTPA-extractable contents (mg kg^{-1} soil)				Class
Zinc	**Iron**	**Manganese**	**Copper**	
<0.3	<3.5	<2.0	<0.2	Acutely deficient
0.3-0.6	3.5-5.5	2.0-4.0	0.2-0.4	Deficient
0.6-0.9	5.5-7.5	4.0-6.0	0.4-0.6	Marginally deficient
0.9-1.2	7.5-9.5	6.0-8.0	0.6-0.8	Adequate
1.2-1.8	9.5-11.5	8.0-10.0	0.8-1.0	High
>1.8	>11.5	>10.0	>1.0	Very high

(Shukla *et al.*, 2016)

x) Available Boron (Hot water method)

Boron is unique in soils by its narrow range between deficiency and toxicity. Less than 1 ppm may lead to deficiency while 3 ppm may lead to toxicity. Saline soils are generally much higher in soluble B than non-saline soils as sodium salt of B are highly soluble. The critical limits of boron for different soil are furnished below

Soil texture	Critical limits (ppm)
Light textured soil	0.30
Heavy textured soil	0.60
Alkali soil	0.10
Acid soil	0.40

Available Molybdenum (Ammonium oxalate method)

Molybdenum deficiency is usually associated with soils below pH 5.00. The critical level of Mo was found to be 0.20 ppm at pH 5.00. Liming of acid soils results in an increase in available Mo. Molybdenum from the soil is extracted by ammonium oxalate $[(NH_4)_2 C_2O_4]$ solution adjusted to pH 3.3.

Soil Organic Matter

Role of Soil Organic matter

a) Improvement in soil Physical properties

i) Bulk density	:	Addition of organic matter to soil lowers the bulk density of the soil and makes the soil lighter.
ii) Aggregation	:	Organic matter improves the aggregate formation in soil by acting as binding agent. The stability of the aggregates is also improved thereby reducing the soil erosion.
iii) Pore space	:	Because of increased aggregation, the pore space of the soil get increased.
iv) Soil aeration	:	Because of better aggregation and porespace, the soil aeration get improved favouring respiration and growth of plant roots.
v) Soil temperature	:	Higher organic matter content of the soil helps to keep the soil warmer during winter and cooler during summer, because of its lower heatconductivity.
vi) Soil Colour	:	Imports darker colour to the soil
vii) Water holding	:	Soils with high organic matter imbibe and retain more water and help to capacity increase the water use efficiency/ conservation of soil moisture.

b) Improvement in Chemical properties

i) **Buffering Capacity of soils**: Addition of organic matter reduces the degree of acidity/alkalinity of soil and increases the buffering capacity of soils.

ii) **CEC of soils:** Soil organic matter increases the CEC of soils, thereby help in the adsorption/retention and availability of nutrients in soil. Soil humus is having the highest CEC value of about 300 m.e.q/100g. The loss of applied nutrients is minimised. It acts as a store house of nutrients in soil.

iii) **Solubility of soil nutrients**: The organic acids released during organic matter decomposition help for the solubilisation of soil mineral nutrients making them more available to crops.

iv) **Source of nutrients for plants:** Organic matter/humus contains largely C, N, P, S which forms a major source of nutrients for plants. In addition, it releases other secondary and micro nutrients into the soil during decomposition.

v) **Chelation:** Organic matter helps in the formation of mineral/ chelates with the plant nutrients more particularly metallic cations like Cu, Zn, Mn, Fe making them more available to crops and prevent them from precipitation/oxidation reactions.

vi) **Transformation of soil nutrients:** Organic matter helps for the activity of soil microbes which are responsible for the transformation and availability of various nutrients viz; N,P,S and other micronutrients.

vii) **Sink for heavy metal pollutants:** Organic matter complexes the heavy metal pollutants received from the various industrial wastes / effluents thereby prevent their toxicities to crop plants.

viii) **Metabolism of soil applied pesticides:** Combines with the organic molecules of soil applied pesticides and affects their bio activity persistence and biodegradation.

ix) **Soil Formation:** Soil organic matter plays an important role in the soil formation by the process called "Biogeochemical processes" which includes the process like podzolisation, laterisation, melanisation etc.

c) Biological properties

i) **Growth and activity of microbes:** Soil organic matter serves as the energy/substrate for the growth and activity of soil microbes which are involved in the biological N fixation, mineralization and immobilization of nutrients, degradation of crop residues, pesticides etc.

ii) **Growth and activity of earth worms :** Soil organic matter favours the activity of earthworms in soil, which helps for the improvement of soil aggregates, aeration, water retention and movement etc.

iii) **Soil Enzyme activity:** The enzyme concentration and activity will be higher in soils having higher organic matter due to increased microbial activity. This helps for the transformation of applied nutrients in soil viz; urea hydrolysis.

Soil Microorganisms

A soil can be called as a living system only when it has sufficient quantity of soil organic matter as well as microbial activity. Soil organic matter influences the physical and chemical properties of the soil to a greater extent though the proportion it makes in the soil is very less (<5.0%). It commonly accounts for atleast half the CEC of surface soil and is responsible perhaps more than any other single factor for the stability of the soil aggregates besides supplying energy for soil microbial growth and activity. The fundamental characteristic of soil is that it has its capacity to produce green plants. Plant and their products after being utilized by man and animal are thrown away as waster material to the soil. Fortunately, microorganisms act upon and convert them into useful or harmful substances. In the words of noble laureate Waksman "Without microbes, the life on this planet would come to a standstill". The different kinds of microbes such as bacteria, fungi, Actinomycetes, algae, protozoa and viruses present in soil bring about several changes.

Bacteria

These are the most dominant group of soil biota, their population ranges from hundred thousand to several hundred million per gram of soil depending upon physical, chemical and biological conditions of soil. They live in soil as cocci, bacilli or spirilli. Some of the bacilli persist in unfavourable conditions and great resistance to prolonged desiccations and also to high temperature as endospores

Most of the soil bacteria are heterotrophs and saprophytic in nature. They help in decomposing organic matter and involve in weathering and decomposition of rocks and minerals which help in soil formation processes. Most of the bacteria secrete extra polymeric substances which help in binding soil particles, improve soil structure and texture for improved soil porosity for better exchange and infiltration of gases. Bacteria present in soil also involves various geo-bio-chemical cycles like nitrogen, sulphur and carbon cycle etc.

Many beneficial bacteria present in soil has the capacity to use atmospheric nitrogen as biological nitrogen fixers. Rhizobium, Azobactor, Beijerinchia, Azospirillum are the common bacteria present in this category. Some of the bacteria like Bacillus sp., Pseudomonas sp., and fungi like Aspergillus, Penicillium, Fusarium present in soil help in solubalization of unavailable or fixed form of phosphorus present in soil.

Actinomycetes

These are numerous and widely distributed in soil. They are usually regarded as intermediate group between bacteria and fungi and produce hyphae or conidia or sporangia like fungi but the width of the mycelium is almost similar to the width of the bacterial cell. They differ from fungi in the composition of their cell wall. They do not have chitin and cellulose, which are commonly found in the cell wall of fungi.

Actinomycetes are usually second to bacteria in abundance. They are estimated to be ranging from 10^4 to 10^8 per gram of soil. The population of Actinomycetes increased in the presence of decomposing organic matter. They are intolerant to acidity and their number decline at pH 5. The most conducive range of pH is between 6.5 and 8.0. Waterlogging soil is unfavourable for the growth of Actinomycetes; whereas desertic soils of arid and semi-arid zones sustain sizeable populations, probably due to resistance of spores to desiccation. The percentage of Actinomycetes in the total microbial population increases with the depth of the soil and these can be isolated in sufficient number even from soil samples obtained from the C horizon of soil profile. Temperatures between 25°C and 30°C are conducive for the growth of Actinomycetes although thermophillic species growing at 55°C and 65°C are common in compost heaps where they are numerically extensive and mostly belong to genera Thermoactinomyces and Streptomyces. Many species of Streptomyces are known to produce antibiotics. Actinomycetes also play a role in the decomposition of resistance components of organic matter and in the formation of humus. Thermophillic Actinomycetes dominate and actively participate in the transformation taking place in the compost pits, manure pits and hay.

Fungi

These are present in soil as mycelia bits or rhizomorphs or various types of spore forms. Their number in the soil varies from few thousand to few million per gram. A wide range of soils under various vegetation types and geographical areas have been examined for the occurrence of fungi. Aspergillus, Mucor, Penicillium, Trichoderma, Cladosporium, Verticillium, Rhizopus, Fusarium, Cephalosporium, Botrytis etc. are the most commonly encountered genera of fungi in the soil.

Fungi, being aerobic are more numerous in surface layers of well aerated cultivated soil. The quality and quantity of organic matter present in soil has direct bearing on fungal population, since majority of specialized cells known as heterocyst which are implicated in nitrogen fixation. The waterlogged rice soils provide an ideal environment for the growth and multiplication of blue green algae. Blue green algae also add lots of organic matter to soils and improve soil aggregation.

They also grow on eroded and abandoned soils and also volcanic deposits when the required moisture is available. The mucilage of these algae has the properties of absorbing water quickly and retaining it for long time. Thus, the growth of these algae provides a substratum for the growth of spores, grass seeds etc., and thus help in building soil fertility. The thick growth of such algae on the surface of such soil acts as surface binders and prevents soil erosion.

Protozoans

These are also common in soil where the population ranges from 10,000 to 100,000 per gram of soil. They are characterized by a cyst stage in their life cycle which can help the species to withstand adverse soil conditions. Most of the soils contain protozoans like flagellates, amoebae or ciliates. Flagellated protozoan are predominated as compared to Amoebae and Ciliata. They are abundant in upper layer of soil. Application of organic manures increases the population of protozoans which is a reflection on the corresponding increase in the bacterial flora due to the addition of organic matter. Soil moisture level is another important factor affecting soil protozoa but they could survive even during the absence of moisture by forming cysts especially in dry condition.

Algae

These are found in different soils. They are abundant in places with enough moisture and sunlight. Their number, usually ranges from 100 to 10,000 per gram of soil. They form a scum on the surface of the soil visible to the naked eye. They are unicellular or filamentous or colonial. green algae, blue green algae and diatoms are numerous in soil. By virtue of the presence of photosynthetic pigments of their cells, algae are phototrophic and use carbon dioxide from atmosphere and give out oxygen. Green algae are dominant algal flora in acid soils but still numerous in neutral and alkaline soils. Chlorella, Chlamydomonas, Chrocococcum, Oedogonium are the most common green algae found in the soil. Blue green algae are pigmented having phycocyanin in addition to chlorophyll. They do not posses flagella and do not produce sexually. Some of the blue green algae produce commercial crops. Azolla and blue green algae are commonly used as biofertilizers in paddy cultivation. Phosphorus solubalizers are being generally applied to all crops irrespective of monocot or dicot plants. Similarly, mycorrhizal fungi are also isolated and used for phosphorus nutrition in different crops. Arbuscular mycorrhizal fungi application is most suited to nursery crops. But the quality of the produce is improved due to secretion of growth promoting hormones like Indole acetic acid, Gibberellic acid, Cytokinins and Vitamins that help in deterring root borne pathogens.

Nematodes

Incredible nematodes live in soil and participate in the food webs at several trophic levels. Soil inhabiting nematodes are classified as free living bacterial feeders, fungal feeders and predators. Free living nematodes are saprophytic in nature and feed on organic matter which helps in decomposition and indirectly participating in soil fertility but not so significant like other soil invertebrates such as earthworms, microorganisms and microarthropods.

Earthworms

Earthworms are the important soil invertebrates have an widespread distribution and show a high degree of adoptability to the prevailing soil conditions. Their involvement in the soil ecological process contributing to the soil ecological process contributing to the soil fertility was first recognized by Charles Darwin (1809-1884) and called the "Earthworms as farmers friend".

Globally more than 700 species of earthworms has been distributed in all the habitats except cold and desert regions. Ecologically they are classified as i) Epigeic earthworms confining to the topsoil feeding on the organic matter at surface feeding zones. ii) Endogeic earthworms are completely soil dwelling feeding on the organic matter at subterranean feeding zone. iii) Anecic earthworms are deep soil inhabitants making horizontal and vertical tunnels for feeding activities. Vermiculture is the art of maintaining and multiplication of earthworms in semi-arid conditions for the purpose of vermicomposing.

Soil Microarthropods

The microarthropods generally found in soil measuring less than 10 mm in length can be considered as microarthropods including proturans, diplurans, collembolans, mites, isopods, pauropods, symphylans etc. they are distributed in all kinds of soil including organically rich grasslands, pasture lands, garden land, cultivated, uncultivated, barren land etc. The population of these microarthropods varies from place to place depending upon many factors such as soil moisture, organic matter content, cultivation practices etc. Prolonged inundation adversely affect the soil microarthropods population and the population increases with precipitation of organically rich soils.

Soil Macroarthropods

The soil arthropods measuring more than 10 mm in length can be considered as macroarthropods including centipedes, millipedes, termites, ants, large number of ground dwelling beetles etc.

Millipedes and centipedes

Millipedes and centipedes are important soil macroarthropods found in arable and garden soil and are quite abundant during rainy season. Millipedes are generally recognized as feeders on dead and disintegrating plant material. They are also acting as agents for decomposition and enrich the soil fertility. They are comparatively longer in size (5-6 cm) and can be easily recognized by the presence of well developed dorsolateral expansions on all body segments. Millipedes are active burrowers in forest litter, compost pits and farm yard manure. Millipedes ingest the detritus along with soil and thereby mix the organic and inorganic components. Centipedes are distributed in moist habitats of tropical soils. They are often found under leaf litter, bark, rocks, cracks and crevices etc. They are carnivorous feeding on smaller insects which are associated in the leaf litter and smaller plants.

Termites

Termites are primarily cellulose feeders and are pests of many dry land crops and wood work of the buildings. Large population will be involved in burrowing the soil by moving in subterranean galleries to more than 100 meters in search of food. Termites after the soil properties through construction of mounds, nests, galleries and earthen runways on the ground. Termites wherever present play a significant role in soil modification and organic matter break down and their effects may be positive or negative for crop production. Termites activity may drastically influence the fertility process in natural ecosystems to the extent that the physical and chemical properties of soils may be largely determined by the digestive activities of the animals and modifications of the environment caused by them.

Ants

Ants are found everywhere on the earth marching in large numbers on the ground. Soil inhabiting ants exhibit wide variety of feeding habits such as omnivorous, carnivorous, fungivorous, granivorus, detrivorous etc. Ants exhibit territoriality, good communication system, alarm and assembly, polymorphism, division of labour etc.

They build crater like nests on the ground by excavating the soils from deeper regions and as a result, water percolation to the plants is found to be faster. Excavation of nests and deposition of heaps of soil ants hasten the decomposition of organic matter and also act as scavengers in removing dead insects from their habitats.

Beetles

Many beetles such as ground beetles, darkling beetles, tiger beetles and dung beetles are known to live in the soil during their immature stages. Among these, ground beetles and tiger beetles are predators, darkling and dung beetles are nature's scavengers. Dung beetles mainly carry large quantity of dung under the ground and this precious organic manure is being preserved by these beetles naturally to enrich the soil. The larvae and adults are making tunnels by burrowing the soil and organically enrich the soil by improving the soil fertility in many cultivated crops.

Cation Exchange Capacity

The cation Exchange Capacity (CEC) is a term used to describe the holding capacity of a particular soil for positively charged elements (cations). It may also be described as the capacity for a soil to exchange cations for another. In simple terms, the higher the clay content, the higher the CEC since clay particles, have the greatest surface area per unit volume of soil and, therefore, can hold the most cations. Cation exchange capacity (CEC) is a useful indicator of soil fertility because it shows the soil's ability to supply three important plant nutrients: calcium, magnesium and potassium. Cations are positively charged elements, the positive charge indicated by a + sign after the element symbol. The number of + signs indicates the amount of charge the element possesses. Cations are held by negatively charged particles of clay and humus called **colloids**. Colloids consist of thin, flat plates, and for their size have a comparatively large surface area. For this reason, they are capable of holding enormous quantities of cations. They act as a store house of nutrients for plant roots. Organic matter can have a 4 to 50 times higher CEC per given weight than clay. The source of negative charge in organic matter is different from that of clay minerals; the dissociation (separation into smaller units) of organic acids causes a net negative charge in soil organic matter and again this negative charge is balanced by cations in the soil. Because organic acid dissociation depends on the soil pH, the CEC associated with soil organic matter is called pH dependent CEC. This means that the actual CEC of the soil will depend on the pH of the soil. Given the same amount and type of organic matter, a neutral soil (pH ~ 7) will have a higher CEC than a soil with e.g. pH 5, or in other words, the CEC of a soil with pH-dependent charge will increase with an increase in pH. The CEC of a soil is expressed in cmol(+)kg (centimol positive charge per kg of soil) or meq/100 g (milli-equivalents per 100 grams of soil). Both expressions are numerically identical (10cmol(+)kg = 10 meq/100g).

Importance of CEC

The CEC is important because it provides a reservoir of nutrients to replenish those removed from the soil water by plant uptake. Similarly, cations in the soil water that are leached below the rootzone by excess rainfall or irrigation water are replaced by cations formerly bound to the CEC. Not only does 'buffer capacity' exist for cation nutrients, but also for soil pH. The higher the CEC, the more it takes to change soil pH. This principle applies to increasing pH with lime or high bicarbonate irrigation water as well as to decreasing pH with nitrogen fertilizers or elemental sulphur. The CEC of a soil is a good indicator of the nutrient holding and buffer capacity of the soil, but in itself is not particularly useful for managing soil properties. However, the concept of base saturation, which is a function of CEC, can be used to manage soil properties and nutrient availability.

Forms of nutrients absorbed by plants

All the nutrients contained in any organic source remain unavailable for crop use until the material is decomposed (mineralized), releasing its nutrients to the inorganic soil pool. Crops take up (absorb) and use almost exclusively inorganic nutrient forms. Plants take up nutrients as ions. Ions are electrically charged forms of each nutrient or element. Some are cations (positively charged) and others are anions (negatively charged) (Table 4).

Table 4. Forms of nutrients absorbed by plants and the relative mobility or elements in the plant body.

Element	Forms absorbed
Nitrogen	NO_3^-, NH_4^+
Phosphorus	$H_2PO_4^-$, HPO_4^{2-}, PO_4^{2-}
Potassium	K^+
Sulphur	SO_3^{2-}, SO_4^{2-}, SO_2
Calcium	Ca^{2+}
Magnesium	Mg^{2+}
Boron	BO_3^{2-}
Copper	Cu^{2+}
Iron	Fe^{2+}, Fe^{3+}
Manganese	Mn^{2+}, Mn^{4+}
Molybdenum	MoO_2^{2-}
Zinc	Zn^{2+}
Chlorine	Cl^-
Sodium	Na^+
Silicon	$So(OH)_4$
Nickle	Ni^{2+}

Nutrient Mobility

Bray (1954), defined mobility as the "overall process whereby nutrients reach the sorbing root surfaces, thereby making possible their sorption into the plant". This means that the mobility involves the solution or exchange of the nutrient as well as the movement to the root surface.

Anions (Cl^-, $BO_3^{3-}/HBO_3^{2-}/H_2BO_3^-$, HCO_3^-) are not adsorbed by the solid phase. These remain in solution and are carried away to longer distances in response to their adsorption by the plant roots. These are classified as the "mobile" nutrients. Cations (K^+, Ca^{2+}, Mg^{2+}, NH_4^+, Fe^{2+}/Fe^{3+}, Cu^{2+}, Ni^{2+}) are either held on the negatively charged surface of the colloidal clays and organic matter in the soil as exchangeable cations or are specifically adsorbed. Concentration of these cations in soil solution is controlled by i) kind and amount of the electrolytes in soil solution, ii) the amount of the different cations associated with solid phases including the exchangeable fraction, and iii) relative ease of their release.

Table 5. Mobile and immobile nutrients in soils and plants

Plant		**Soil**	
Mobile	**Immobile**	**Mobile**	**Immobile**
Cl	B	H_3BO_3, $H_2BO_3^-$	NH_4^+
Mg	Ca	Cl^-	Ca^{2+}
Mo	Cu	NO_3^-	Cu^{2+}
N	Fe	SO_4^{2-}	Fe^{2+}/Fe^{3+}
P	Mn		MoO_4^{2-}
K	Zn		$H_2PO_4^-$
Ni	S		HPO_4^{2-}
			Mg^{2+}
			Ni^{2+}
			K^+

Phosphate and molybdate ions are held in the soil by either anion exchange or ligand exchange or adsorbed or precipitated as sparingly soluble phosphates of Fe, Al, Ca, etc. Chemical equilibrium, dictated by different competing chemical reactions, determines the concentration of P or Mo in soil solution. Amounts associated with the solid phases, earlier in the exchangeable or labile forms, which are positionally closer to the soil surfaces, are readily available for absorption by the plant roots. Larger the distance of surfaces, more the time they take to reach the plant roots because of the restrictions imposed by the colloidal solid phases. Such nutrients are termed as the "immobile" nutrients. Classification of nutrients as mobile and immobile forms is given in Table 5. Interestingly N is both mobile (NO_3^-) and immobile (NH_4^+) in soil. Bray postulated that the roots explored the whole soil mass for absorbing the mobile nutrient. Soil rhizosphere, the root surface sorption zone, which constituted only a very small fraction of the soil volume, could be only explored for the immobile nutrients by the growing plants.

Nutrient Interactions

Interaction can be defined as the influence of an element upon another in relation to growth and crop yield. There may be positive or negative interaction of nutrients occurs either in soil or plant. The positive interaction of nutrients gives higher crop yield and such interactions should be exploited in increasing the crop production. Conversely, all negative interactions will lead to decline in crop yield and should be avoided in formulating agronomic packages for a crop.

There are mainly two types of interactions effect viz. antagonistic and synergistic effects. Antagonistic effect means an increase in concentration of any nutrient element will decrease the activity of another nutrient (negative effect). While synergistic effects means an increase of concentration of any one nutrient element will influence the activity of another nutrient element (Positive effect). The following antagonistic effects have been well established on the uptake of micronutrients by crops:

- Excess of P adversely affects utilization of Zn, Fe and Cu
- Excess of Fe adversely affects utilization of Zn and Mn
- Excess of Zn, Mn, and Cu induces Fe-deficiency in crops
- Excess of S and Cu induces Mo-deficiency in crops
- Excess of Lime induces deficiency of all micronutrients.
- Presence of carbonate and bicarbonate ions in soil due to sodicity or over liming reduces the availability of micronutrient cations to crops which suffer most iron deficiency.

- Lime X P, Lime X Mo, Mo X P, and Na X K are common negative interactions.
- Excess of Ca may induce P deficiency
- Excess of Ca and Mg depress K uptake
- Excess of Ca may reduce Mg uptake, if ratio is wider than 7:1
- Excess of K and NH_4^+ may reduce Mg uptake
- Excess of N, K and Ca may reduce B toxicity
- Excess of N, P, K may induces Cu deficiency
- Excess of NO_3^-N may cause Fe deficiency.

Nutrient deficiency in Soils

i) Nutrient deficiency in Indian soil

Nutrient balance sheets in most soils of India have been deficient and continue to be so. This is primarily because nutrient removals by crops far exceed the nutrient additions through manures and fertilisers. Indian soils have become deficient not only in major plant nutrients like nitrogen, phosphorus and in some cases, potash but also in secondary nutrients like sulphur, calcium and magnesium. Micronutrients such as zinc, boron and to a limited extent iron, manganese, copper and molybdenum have also been reported to be deficient (Table 6). Deficiency of micronutrients during the last three decades has grown in both, magnitude and extent because of increased use of high analysis fertilizers, use of high yielding crop varieties and increase in cropping intensity. This has become a major constraint to production and productivity crop plants.

Table 6. Nutrients deficiency in Indian soil

Nutrients	(Per cent)	Nutrients	(Per cent)
Nitrogen	90	Boron	33.0
Phosphorus	80	Iron	12.0
Potassium	50	Manganese	5.0
Zinc	49.0	Copper	3.0
Sulphur	41.0	Molybdenum	7.0

(Brajendra and Bhadana, 2011)

ii) Soil condition favourable for nutrient deficiency

Soil is a store house as well as source of nutrients needed by plants for plant growth. Intensive cultivation and management practices make soils less healthy and conducive for induction of nutrient deficiency (Table 7).

Table 7. Soil conditions inducing nutrient deficiency for crop plants

Nutrient	Conditions inducing deficiency
N	Excess leaching with heavy rainfall, low organic matter content of soils, burning the crop residue
P	Acidic condition, calcareous soils
K	Sandy, organic, leached and eroded soils; intensive cropping system without addition of fertilizer
Ca	Acidic, Alkali, or sodic soils
Mg	Similar to Calcium
S	Low organic matter content of soils; use of N and P fertilizers containing no sulphur, burning the crop residue
Fe	Calcareous silt and clays, high organic matter, calcareous soils
Zn	Highly leached acidic soils, calcareous soils, high levels of Ca, Mg, and P in the soils
Mn	Calcareous silt and clays, high organic matter, Calcareous soil
B	Sandy soils, naturally acidic leached soils, alkaline soils with free lime
Mo	Highly podzolized soils ; well drained calcareous soils

Problem soils

Saline and Alkali Soils

Salinity is one of the major abiotic stresses that limiting plant growth and productivity. Excess salt in the soil may adversely affect plant growth either through osmotic inhibition of water uptake by roots or specific ion effects. Specific ion effects may cause direct toxicity or, alternatively, the insolubility or competitive absorption of ions may affect plant nutritional balances and with NaCl, being the most prevalent. Both osmotic and specific ionic stresses from salinity can cause stunted growth and a reduced plant yield. Two different criteria are currently recognized as indices of salinity. These are the Sodium Adsorption Ratio (SAR) with a reported threshold of 12 and the Exchangeable Sodium Percentage (ESP) with a threshold of 15%. When the Chloride (Cl^-), Sulphate (SO_4^{2-}), Carbonate (CO_3^{2-}) and Bicarbonate (HCO_3^-) salts of Sodium (Na^+), Calcium (Ca^{2+}) and Magnesium (Mg^{2+}) are increased in soil, the soil becomes saline and alkaline. On the basis of amount of soluble salts, average quantity of exchangeable sodium and pH soils are classified as saline soil, alkali soil and saline alkali soil.

Exchangeable Sodium Percentage (ESP)

Exchangeable Sodium Percentage (ESP) is the relative amount of the sodium ion present on the soil surface expressed as a percentage of the total Cation Exchange Capacity (CEC).

$$ESP = \frac{\text{Exchangeable Sodium (in milli equivalent per 100g. of soil)}}{\text{Total Cation Exchange Capacity (in m.eq/100g soil)}} \text{ x } 100$$

ESP means Exchangeable Sodium Percentage and is also called Soluble Sodium Percentage i.e. SSP.

Sodium Absorption Ratio

Plants are detrimentally affected, both physically and chemically by excess salts in some soils and by high levels of exchangeable sodium. Soil with an accumulation of exchangeable sodium are often characterized by poor tilth and low permeability making them unfavourable for plant growth. Sodium adsorption ratio (SAR), along with pH, characterize salt-affected soils. It is an easily measured property that gives information on the comparative concentrations of Na^+, Ca^{2+}, and Mg^{2+} in soil solutions. The equation used to calculated SAR is given as follows:

$$\text{SAR (Sodium Adsorption Ratio)} = \frac{[Na^+]}{\sqrt{\frac{\left[Ca^{2+}\right]+\left[Mg^{2+}\right]}{2}}}$$

i) Saline soils (White Alkali)

Characteristics

Toxic concentration of soluble salts in the root zone viz., soluble salts are Cl^- and $SO4^{2-}$ of Na, Ca and Mg increases the salinity. Electrical conductivity (EC) of soil solution saturated extract is more than 4.0 millimhos/cm. [or decisiemens/metre] at 25°C. The Exchangeable sodium percentage (ESP) is less than 15% and pH is less than 8.5.

Two types of saline soils :

- Having substantial quantity of Ca and Mg.
- Having chiefly Na and damages the colloidal complex.

Formation of saline soils

Common in arid and semi-arid regions having annual rainfall less than 55cm. Lots of salts are deposited on soil surface in saline soils and saline soil layer shines white in dry-season.

Favourable Conditions

- High water table with a fairly high salt concentration.
- High temperature
- Low rainfall

The sources of salts are soil itself, ground water, irrigation water, canal and seawater.

Reclamation

Principle

Removal of excess salts to a desired level in the root zone. Leaching with water of good quality and adequate drainage are two essential component of any permanent solution of the salinity problem. Ponding of water is the most commonly used for leaching. Intermittent ponding is more efficient than continuous ponding.

Management practices

Management practices are water management and cropping system during initial years of reclamation.

ii) Sodic soils/ Black alkali/ Non-saline alkali soils

Characteristics

High exchangeable sodium percentage interferes with plant growth.

- EC < 4.0m.mhos/cm at 25°C (less than)
- ESP > 15%
- pH > 8.5 but generally 8.5-10
- Low infiltration rate.

Formation

Salt accumulation is the first stage in the sequence of processes and it is common to the family of salt affected soils; dominant in mean annual rainfall of 55-90cm and relatively low lying areas with insufficient drainage.

Reclamation

Basic principle

To replace exchangeable Na^+ by Ca^{2+} and thus released Na^+ salt be leached out of root zone. The amendments are as follows.

- Gypsum ($CaSO_4$ $2H_2O$)
- Iron pyrite (FeS_2)
- Several other amendments like calcium salts, acids, acid forming materials. The choice of amendment depends upon the nature and quantity of CO_3^{2-} present in soils.
- Bulky Organic manures, green manures, crop residues and other produce weak organic acids. These are applied in conjugation with Gypsum.
- Leaching with water of good quality.

Management practices

- Cropping with green manure
- Higher dose of N because of volatilization. Application of zinc in initial years of reclamation.
- Frequent irrigation with small quantities of water.

iii) Saline - alkali soils

Characteristics

- EC > 4 mmhos/cm at 25°C
- ESP > 15%
- PH < 8.5

Favourable factors are

- Aridity
- Poor external or internal drainage
- Irrigation by saltish water
- Rise in water table by excess irrigation
- Erratic use of irrigation water i.e. flooding followed by intense drought.

Saline-alkali soil is difficult to manage since its physical condition is very bad.

Detrimental effect of soil salinity and alkalinity

- Soil structure: in alkali condition, soil is dispersed and becomes compact.
- Less water permeability due to compactness.
- Low aeration
- Low microbial activities
- Unavailability of nutrients like P, Ca, N.
- Nutritional disturbances at pH 8.5 and more.
- Hinderance in water absorption.
- Effects of osmotic pressure: Increase in osmotic pressure of soil solution badly affects the plant's growth.
- Salt toxicity: In alkali soil, Na_2CO_3 is highly toxic.

The difference between saline and alkaline soils are given below (Table 8).

Table 8. Difference between saline and alkaline soils.

Saline	Alkaline
i) Mainly Cl^- and SO_4^{2-} of Na^+ but also Cl^-, SO_4^{2-} and HCO_3^- of Ca^{2+} and Mg^{2+} in small amount.	Mainly CO_3^{2-} of Na^+ but also CO_3^{2-} of K^+, Ca^{2+} and Mg^{2+} in small amount
ii) Soluble salt concentration is equal to or more than 0.1%	Soluble salt conc. <0.1%
iii) Exchangeable sodium % < 15%	ESP > 15%
iv) PH < 8.5	PH > 8.5
v) EC of saturated soluble extract at 25°C is more than or equal to 4 milli mho per cm.i.e EC > 4mmho/cm	EC < 4 mmhos/cm
vi) White/light grey colour hence called white alkali.	Black colour hence called black alkali.
vii) Flocculated soils therefore soil, aeration and permeability is normal	Dispersed and compact soil, aeration and permeability is low
viii) Easy to manage because physical condition of soil is good	Such soils can be managed because physical condition is not so good
ix) O.M. or humus is always found in soil.	Very less amount of O.M. or humus or even absent.
x) Can be reclaimed by mechanical methods upto some extent.	Use of amendments is must.
xi) In rainy season, some natural vegetation is grown.	No any natural vegetation except some grasses.

iv) Acid Soils

Soils having less than pH 5.5 are called acid soils. There are two criteria used to distinguish acid from non-acid soils.

a) The percent base saturation:

Percent Base Saturation (PBS)

$$\text{PBS} = \frac{\text{Exchangeable bases (Cmol/kg soil)}}{\text{CEC (Centimol/kg Soil)}} \times 100$$

Unit of PBS in Cmol/kg = m.eq./100g of soil

Humid region soils, dominated by silicate clays and humus, are acid if their per cent base saturation (PBS) is much below 80. A good cultivable soil has 60% bases occupied by calcium. It means Ca-dominant soil is fertile soil. At a given percent base saturation, the pH is normally highest in hydrous oxide clay minerals, intermediate in Kaolinitic and lowest in 2:1 type minerals. The greatest buffering occurs at about 50% base saturation. The buffering is uniform over pH range 4.5-6.5. It means same amount of lime is required to the soil pH from 5.0 to 5.5 as from 5.5 to 6.0.

b) pH: The major source of H^+ in most very acid soils is adsorbed aluminium.

$$Al^{3+} + H_2O \rightarrow Al(OH)^{2+} + H^+$$

Adsorbed Hydrogen is the second but minor source of H^+

Acid soils have relatively high amount of exchangeable H^+ and Al^{3+}.

Kaolinite and Illite clay minerals are dominant.

Processes involved in acid soil formation

- Laterisation of varying degrees.
- Podzolization with sub-temperate to temperate climate.
- Intense leaching of light alluvial soils.
- Marshy conditions with significant amount of partly decomposed organic matter.

Occurrence

Acid soils occur in almost all the major groups except black soils (vertisons). It occurs in northern-east, eastern and peninsular regions, coastal plains (Coastal alluvium, peaty soils, marshy soils) and Himalayan regions. Large area of acid soils are found in Assam, Manipur and Tripura.

Adverse effect on plant growth

a) Direct influences

i) Toxic effects of H^+ ions on root tissues.

ii) Influence of soil acidity on the permeability of the plant membranes for cations.

iii) Disturbance in the balance between basic and acidic constituents through the roots.

iv) Direct harmful effect by affecting the enzymic changes.

b) Indirect influences

i) Availability of various nutrients viz., P, Cu, Zn.

ii) High solubility and availability of elements like Al, Mn and Fe are in toxic amounts due to high soil acidity.

iii) Beneficial activities of soil micro-organisms are adversely affected e.g. Nitrification and Nitrogen fixation.

iv) Prevalence of Plant diseases.

v) Due to soil acidity, nutrients such as Ca and K may be deficient.

vi) Disturbances in the physical condition of soil. Deflocculated conditions of soil is due to high acidity.

Characteristics of acid soils

a) Physical

i) Light texture

ii) High permeability

iii) Poor water holding capacity

iv) Poor cation exchange capacity

v) Poor organic matter content.

vi) Mainly kaolinite and sometimes illite clay minerals.

b) Chemical

i) Base unsaturated soil, more anions than cations.

ii) Active and potential soil acidity.

iii) Availability of nutrients -

pH	Available nutrients	Remark
Low pH	Al, Fe, Mn, Zn, Cu, Co	All trace elements except B and Mo.
High pH	Mo, N, K, Ca, Mg, S	All major nutrients except 'P'
Neutral pH	P, B	Boron is available at pH 5.0-7.0

iv) Toxic effects on plants: Al^{3+} toxicity.

c) Biological

i) Population of Fungi more than that of bacteria.

ii) Rate of decomposition of biological materials and rate of mineralization and nitrification etc. are reduced when acidity is increased.

Management of acid soils

In two ways :

i) Either by growing crops suitable for particular soil pH or

ii) By ameliorating the soils through the application of amendments. Intensive and continuous cropping of arable crops in humid regions will aggravate soil acidity and use of acid producing fertilizers will further accentuate the process. Acid soils are made more suitable for agricultural use by liming which raises the soil pH. Liming is generally recommended for soils with pH less than 5.5. Dose of liming material depends on soil pH, soil texture, capacity and types of liming material. Generally at lower pH, higher amount of lime and at higher pH, lower amount of lime is required. For normal and 4.8 pH, 35 tonnes/ha and for pH 6-7, 4 tonnes/ha lime are required. On an average hydrated lime $Ca(OH)_2$ @ 5 tonnes/ha is applied in the field and is thoroughly mixed in soil. The gap between lime application and crop sowing is at least two to three months. Application of small quantity of lime every year gives good result, although lime can be applied once in two to three years for convenience.

Reactions in soil after application

$$CaCO_3 + H_2O + CO_2 \rightarrow Ca(HCO_3)_2$$

$$Ca(HCO_3)_2 \rightarrow Ca^{2+} + 2HCO_3^-$$

$$H\text{-}Soil + Ca^{2+} \rightarrow Ca\text{-}soil + 2H^+$$

$$H^+ + HCO_3^- \rightarrow H_2CO_3 \leftrightarrow H_2O + CO_2$$

Soil Physical Constraints

An optimum physical environment of soils is essential for better growth of plants, consequently for better yields. Based on soil physical properties viz, infiltration, bulk density, hydraulic conductivity, porosity (capillary and non capillary) aggregates etc. Soil physical constraints are identified as below:

i) Slow permeable soils
ii) Extensively permeable soils
iii) Sub soil hardening
iv) Surface crushing
v) Fluffy paddy soils
vi) Shallow soils

i) Slow permeable soils

Slow permeable soils are those having infiltration rates less than 6 cm/ day due to high clay content of the soil. Due to low infiltration rates, the amount of water entering the soil profile is reduced. Thus, increasing the run-off. Further, it encourages erosion of surface soil leading to nutrient removal in the running water. Moreover, due to heavy clay content, the capillary porosity is relatively high, resulting in impeded drainage and reduced soil conditions. This results in increase of some soil elements to the level of toxicity to the plants. It also induces nutrient fixation in the clay complex there, by making the nutrient becoming unavailable to the crop, eventually causing deficiency of nutrients.

Management

Slow permeable soils can be managed by adopting suitable practices viz.

- Provision of drainage facilities either through open or closed sub surface drains
- Forming contour and compartmental bunding to increase the infiltration rates of soils.
- Application of huge quantities of river sand or red soils of coarser texture to dilute heaviness of the soil.
- Application liberal doses of organic manures like Farm yard manure, compost, green manure, composted coir pitch, sewage waste, press mud etc.
- Adopting ridges and furrows, raised beds, broad bed and furrow systems.

ii) Excessively permeable soils

Excessively permeable soils are those having high amount of sand. Due to this, the soil is inert and unable to retain nutrients and water. These soils being devoid of finer particles and organic matter, the aggregates are weakly formed, the non-capillary pores dominating with very poor soil structure. Due to low retaining capacity of the soils, the fertiliser nutrients are also lost in the drainage water.

Management

The excessively permeable soils can be managed by adopting the techniques given below.

- Compacting the field with 400 kg stone roller (tar drum filled with 400 kg of sand or stones can also be used) 8-10 times at optimum moisture conditions.
- Application of clay soil upto a level 100 t ha^{-1} based on the severity of the problem and availability of clay materials.
- Application of organic materials like farm yard manure, compost, press mud, sugar factory slurry, composted coir pith, sewage sludge etc.
- Providing asphalt sheet, polythene sheet etc. below the soil surface to reduce the infiltration rate
- Crop rotation with green manure crops like sunnhemp and kolinchi.

iii) Sub soil hardening/hard pan

The sub soil hard pan in red soil is due to illuviation of clay to the sub soil horizon coupled with cementing action of oxides of Fe, Al, and Calcium carbonate, which increases the soil bulk density to more than 1.8 Mg m^{-3}. Further, the hard pan can also develop due to continuous cultivation of crops using heavy implements upto certain depth constantly. Besides, the higher exchangeable sodium content in black soils areas also result in compactness. All put together lowered in infiltration and percolation rates, nutrient movement and free air transport within the soil profile. It prevents root proliferation and limits the volume of soil available for nutrients uptake. Due to this, the contribution of sub soil fertility to crop growth is hampered.

Management

Soils with hard pan are managed by adopting following practices.

- Ploughing the soil with chisel plough at 0.5 m interval crisscross at 0.5 m depth once in 2-3 years.
- Application of organics to improve the aggregation and soil structure so as to prevent further movement of clay to the lower layers.
- Deep ploughing of the field during summer season to open up the sub soil.
- Cultivating deep rooted crops like tapioca, cotton so as to encourage natural breaking of the hard pan.
- Raising deep rooted semi perennial crops like mulberry, jasmine etc. can also help in opening up the sub surface hard pan.

Surface crusting

Surface crusting is due to presence of colloidal oxides of iron and aluminium in alfisols, which binds the soil particles under wet regimes. On drying, it forms a hard mass on the surface. The ill effects of surface crusting are:

- Prevents germination of seeds.
- Retards/inhibits root growth.
- Results in poor infiltration.

- Accelerates surface run off
- Creates poor aeration in the rhizosphere
- Affects nodules formation in leguminous crops.

Management

- When the soil is at optimum moisture regime, ploughing is to be given.
- Lime at 2 t ha^{-1} may be uniformly spread and another ploughing given for blending of amendment with the surface soil.
- Farm yard manure at 10 t ha^{-1} or composted coir pith at 12.5 t ha^{-1} or other organics may be applied to improve the physical properties of the soils, after preparation of land to optimum tilth.
- Scraping surface soil by tooth harrow will be useful
- Bold grained seeds may be used for sowing on the crusted soil.
- More number of seeds/hill may be adopted for small seeded crops.
- Sprinkling water at periodical intervals may be done wherever possible.
- Resistant crops like cowpea can be grown.

iv) Fluffy paddy soils

The traditional method of preparing the soil for transplanting rice consists of puddling, which substantially breaks soil aggregates into a uniform structure less moss. Under continuous flooding and submergence of soil for rice cultivation in a cropping sequence of rice-rice-rice, the soil particles are always in a state of flux and the mechanical strength is lost leading to the fluffiness of the soils. Impact of fluffiness is sinking of drought animals and labourers during pudding. This has been thus and invisible drain of finance for the farmers due to high pulling power needed for the bullocks and slow movement of labourers during the puddling operations. Further, fluffiness of the soil lead to very low bulk density and thereby leaching to very rapid, hydraulic conductivity and in turn the soil does not provide a good anchorage to the roots and the potential yield of crops is adversely affected.

Management

Following practices are need to be adopted to overcome this problem :

- The irrigation should be stopped 10 days before the harvest of rice crop
- After the harvest of rice, when the soil is under semi-dry condition, compact the field by passing 400 kg stone roller or an tar drum filled with 400 kg of sand for 8 times.
- The usual preparatory cultivation is carried out after compaction.

v) Shallow soils

The shallow soils are characterized by the presence of the parent rock immediately below the soil surface at about 15-20 cms depth. This restricts the root elongation and spreading. Hence, the crops grown in these soils necessarily be a shallow noted crops, which can exhaust the soil within 2-3 seasons. Therefore, frequent renewal of soil fertility is a must in these soils. These soils can be managed by growing crops which can withstand the hard rocky sub soils like mango, ber, fig, country goose berry, West Indian cherry, anona, cashew, tamarind etc.

❑❑❑

Chapter 3

Nitrogen

Nitrogen is widely distributed throughout the lithosphere, atmosphere, hydrosphere and biosphere. The total nitrogen content of surface mineral soils normally ranges between 0.05 and 0.02 per cent, corresponding to approximately 1750 to 7000 kg N ha^{-1} in the plough layer. Lower as well as higher amounts can be found depending on the various soil-forming processes. Of this total N content, in most cases less than five per cent is directly available to plants mainly as nitrate N (NO_3^- -N) and ammonium N(NH_4^+ -N). Organic N gradually becomes available through mineralization.

Forms of Nitrogen in the soil

In addition to nitrogen occurring as atmospheric dinitrogen gas in soil pore spaces, nitrogen occurs in both organic and inorganic forms in the soil. Soil organic matter exists as decomposing plant and animal residues. Organic matter formation and stability is largely related to long-term moisture and temperature trends. With higher average temperatures, soil organic matter decreases. As moisture increases, soil organic matter increases. Higher temperatures lead to more rapid and complete organic matter decomposition to soluble products which can leach from soil. Reduced tillage techniques in combination with legume rotations and judicious fertilizer use may help to increase organic matter levels with time.

Nitrite (NO_2^-) and nitrous oxide (N_2O) are present in soil in lesser quantities. Plants normally use nitrogen in both the ammonium and nitrate forms. Nitrite is actually toxic to plants. The nitrogen cycle, begins with nitrogen in its simplest stable form, dinitrogen (N_2), and follows it through the processes of fixation, mineralization, nitrification, leaching, plant assimilation, ammonia volatilization, denitrification and immobilization.

Biological Nitrogen Fixation

The biosphere is most dependent upon the only process that can remove N_2 from the air and convert it to a form usable by living beings. This process, called nitrogen fixation, is the beginning step in the synthesis of virtually all nitrogenous compounds. Nitrogen fixation is brought about primarily by nitrogen-fixing bacteria in soil and water, though a small amount is formed through nonliving processes involving lightning. Biological dinitrogen (N_2) fixation is one of the most important sources of N in agricultural system, since it has been estimated to be around 122 million tonnes per year. The most important N-fixing agents are the symbiotic associations between crop and forage/fodder legumes and bacteria of the genus *Rhizobium*. Nitrogen-fixing microbes have developed a unique enzyme system capable of breaking the triple bonds of the N_2 molecule and reducing the N atoms, an anaerobic process that requires the expenditure of considerable ATP (Figure 1). The primary product of nitrogen fixation is the ammonium ion, NH_4^+. The general equation for nitrogen fixation may be described as follows:

$$N_2 + 16ATP + 8H^+ + 8e^- \xrightarrow[\text{nitrogenase (Fe Mo)}]{} 2NH_3 + 16ADP + 16Pi + H_2$$

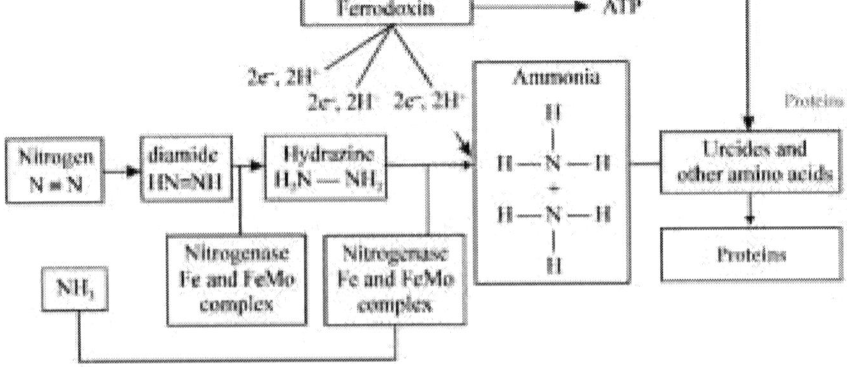

Figure 1. Simplified flow sheet of biochemical steps of Nitrogen fixation

Nitrogen fixers

The nitrogen fixing bacteria have been classified in to four groups viz.,

i) Symbiotic N fixers,

ii) Associative N fixers,

iii) Free-living N fixers, and

iv) Enophytic N fixers.

i) Symbiotic Nitrogen Fixers

Legumes comprise of one of the most important plant groups in agriculture. Nitrogen fixing members of this family include crops of economic importance like soybean, peas, beans and peanuts, and forage crops like lucerne and clover. The genus *Rhizobium* contains fast growing acid producing bacteria. The *Bradyrhizobium* contains slow growers, non acid producing bacteria and *Azorhizobium* nodulates on stem of *Sesbania rostrata. Rhizobium* maintains cross inoculation group (CIG) where CIG refers to the group of leguminous plants that will develop effective nodules when nodulated with Rhizobium obtained from the nodules from any member of that particular legume group.

Rhizobium leguminosarum bv., *viceae* CIG pea nodulated peas, lentil and Vicia; *Rhizobium leguminosarum* bv., *phaseoli*-CIG-bean nodulate *Rhizobium leguminosarum* bv.,*phaselous;* and *Rhizobium leguminosarum* bv., *trifoli*-CIG- clover nodulates *Trifolium* spp. The *Rhizobium meliloti*-CIG- alfalfa nodulate alfalfa, clover and fenugreek. The *Rhizobium loti*-CIG- lotus nodulate trifoli and lupine. The *Rhizobium fredii*-CIG-soybean nodulates soybean. *Rhizobium* spp. CIG-cowpea group nodulates Vigna, Arachis, Cajanus, Dolichus, Sesbania, Acacia, Prosopis, Green gram and black gram. *Rhizobium* spp.-CIG-chickpea nodulates the chickpea crop.

Bradyrhizobium japonicum nodulates soybean, but the *Bradyrhizobium* spp. also nodulate the cowpea group of legumes. *Photorhizobium* nodulates the *Aeschynomene* spp. Similarly *Sinorhizobium* is one of the fast growing soybean nodular. The non-*Rhizobium* strain *Anabaena azollae* lives in symbiotic association with fern Azolla suitable for rice aquatic environment and benefits the crop when incorporated as green manure crops (30-60 kg N ha^{-1}). Many environmental benefits also accrue from the incorporation of Azolla.

ii) Associative Nitrogen Fixers

The associative symbiotic N_2 fixers include *Azospirillum* and *Herbaspirillum*. The Azospirillum group includes six species namely, *Azospirillum lipoferum*, *Azospirillum brasiliense*, *Azospirillum amazonense*, *Azospirillum haloparefera*ns, *Azospirillum irakense* and *Azospirillum largomobilis*. All these are suitable for grasses, cereals, vegetable crops.

iii) Free-living Nitrogen Fixers

The aerobic free-living N_2 fixers include Azotobacter and Beijerinkia and the anaerobic fixers include Clostridium and Cyanobacteria (Blue green algae). The free-living microorganisms have frequently been referred to as the plant growth promoting bacteria (PGPB), where PGPB are defined as the free-living soil rhizosphere, rhizoplane and phylosphere bacteria that under same conditions are beneficial to plants in supplying N after their death and decomposition. These bacteria are capable of fixing atmospheric N_2, solubilize phosphorus and iron, and enhance production of plant hormones. It is well established that many of the soil plant associated bacteria are able to synthesize plant hormones to improve plant growth and development specifically root growth including length, density, volume and cation exchange capacity.

iv) Endophytic Nitrogen Fixers

The endophytic N_2 fixer includes *Gluconacetobacter diazotrophicus* fixes N_2 in sugarcane. It is also reported to possess different plant growth promoting characteristics. The estimate of the amount nitrogen fixed by various legumes are furnished in Table 1.

Table 1. Estimates of the amount of N_2 fixed by various legumes

Crops	Nitrogen fixed (kg N ha^{-1} $year^{-1}$)	
	Tropical/ subtropical	Temperate, cooler zone
Mung bean	70-88	201-302
Black gram	82-87	-
Soybean	60-79	82-168
Beans	40-70	68-79
Peas (field)	48-68	55-77
Horse gram	40-64	45-152
Groundnut	42-63	55-78
Cowpea	80-105	89-154
Lentil	98-106	88-114
Chick pea	103-109	110-118
Cloves	73-80	128-207
Alfalfa	84-97	229-290
Leucaena	67-79	74-584

(Mohanty and Singh, 2002)

Nitrogen Mineralization

Once nitrogen is fixed, it is subject to several chemical reactions which can convert it to different organic or inorganic forms. Mineralization occurs in soil as microorganisms convert organic nitrogen to inorganic forms. The first step of mineralization is called aminization. In which, microorganisms (primarily heterotrophs) break down complex proteins to simpler amino acids, amides, and amines. Heterotrophic microorganisms require preformed organic compounds as sources of carbon and energy. Autotrophic microorganisms can derive energy from the oxidation of inorganic elements or compounds such as iron, sulphur, ammonium, nitrite, or from radiant energy.

Nitrification

Microbial activity is also responsible for the two steps of nitrification. *Nitrosomonas* (obligate autotrophic bacteria) convert ammonium to nitrite. *Nitrobacter* species which coverts nitrite to nitrate. This step rapidly follows ammonium conversion to nitrite and consequently nitrite concentrations are normally low in soils.

$$\text{Organic Nitrogen} \xrightarrow{\text{Nitrosomonas}} \text{Nitrite} \xrightarrow{\text{Nitrobacter}} \text{Nitrate}$$

$$2NH_4^+ + 3O_2 \longrightarrow 2NO_2^- + 4H^+ + 2NO_2^- + O_2 \longrightarrow 2NO_3^-$$

Mineralization and nitrification are influenced by environmental factors that affect biological activity such as temperature, moisture, aeration, pH, and so forth. Nitrification, for example, occurs very slowly at cold temperatures and ceases once the temperature declines below freezing. The rate increases with increasing temperature until bacterial viability is reduced (around 95°F to 100°F), and then nitrification begins to decline as the temperature increases. Moisture and oxygen are necessary for microbial function in both the mineralization and nitrification processes. Excessive moisture limits oxygen availability, reducing mineralization and nitrification rates, which, perhaps, leads to anaerobic soil conditions. Rates of mineralization and nitrification proceed most rapidly at pH levels near 7 and decline as soils become excessively acid or alkaline.

Denitrification

Saturating soils with a ready supply of nitrate can cause denitrification. It is the conversion of nitrate to various gaseous forms of nitrogen which can be lost to the atmosphere (nitric oxide, nitrous oxide, dinitrogen). Denitrification occurs under oxygen-limiting conditions when anaerobic bacteria use nitrate in respiration in the presence of carbon sources such as organic matter.

Ammonia volatilization

Ammonia (NH_3) loss to the atmosphere is called ammonia volatilization. Technically, ammonia volatilization is different from gaseous loss of applied anhydrous ammonia, which is not retained in the soil. Instead, ammonia volatilization occurs when ammonium in the soil because of pH, it is converted to ammonia which is lost as a gas.

$$CO(NH_2)_2 + H^+ + 2H_2O \rightarrow 2NH_4 + HCO_3^-$$

Urea + Hydrogen + Water →Ammonia + Bicarbonate

$$NH_4^+ + OH^- \rightarrow NH_4OH + NH_3 + H_2O$$

Ammonia loss can be significant where applied fertilizers containing urea and does not incorporate them. This is particularly true if significant amounts of residue are present and conditions are warm and moist. The amount of total nitrogen loss from fertilizers containing urea due to ammonia volatilization can vary considerably from no loss to 50 percent or more of the applied nitrogen. The potential for ammonia volatilization is influenced by soil moisture, temperature, soil pH, soil buffering capacity, urease activity, residue cover, precipitation, wind and other factors.

Nitrogen Leaching

For leaching to occur, nitrogen must be in a water soluble, mobile form and abundant enough to transport nitrogen through the soil. Although urea and nitrite are mobile, neither exists in significant concentrations in soil. Nitrate susceptible to leaching. Nitrate leached below the root zone (four to six feet) for most agronomic crops will eventually leach downward until it reaches a saturated zone. The rate of nitrate movement downward depends on a variety of factors, including soil texture, precipitation and irrigation and crop uptake of water and nitrate.

Immobilization

Immobilization of inorganic nitrogen by soil microorganisms decomposing plant residues is not strictly a loss process. Immobilized nitrogen will be unavailable to plants for a time, but will eventually become available as residue decomposition proceeds and populations of microorganisms decline. Fertilizer nitrogen immobilization can be reduced by placing fertilizers below crop residues instead of incorporating fertilizer into the soil with residue.

The duration of the nitrate depression period during immobilization depends on environmental factors such as moisture and temperature and the carbon-to-nitrogen (C: N) ratio of the residue. Soil organic matter contains an average of about 50 percent carbon and 5 percent nitrogen. This ratio (10:1) is relatively constant for organic matter. Plant tissue low in nitrogen generally are more resistant to decomposition and require a longer time before the nitrogen is available to plants.

When a plant residue with a wide C: N ratio is incorporated into the soil, microbial decomposition starts. Microorganism populations increase greatly, as evidenced by increased release of CO_2 leaving the soil through respiration. The microorganisms take nitrogen from the soil for proteins. Consequently, for a time, the concentration of inorganic nitrogen in the soil declines, and may be deficient for plant growth. As residue decomposes, the C: N ratio narrows. At a ratio of approximately 17:1, nitrogen becomes available for plant use. Decomposition continues until the ratio is approximately 11:1 or 10:1.

Plant assimilation

Plants use nitrogen in primarily the nitrate or ammonium forms. If any preference exists, it is usually for ammonium early and nitrate late in the growing season. Research has shown that growth is optimized with a mixture of both ammonia and nitrate with ammonium used preferentially for synthesis of amino acids and proteins. Some plants also can directly use urea, although in most cases urea-nitrogen will hydrolyze to ammonium-nitrogen prior to uptake. For plants to take up nitrogen, it must move with water toward the root - a process called mass flow. Consequently, nitrate nitrogen that has moved below the root zone has potential to move up into the root zone as surface horizons of soil dry out and crops use water deeper in the profile. Conversely, plants may exhibit symptoms of nitrogen deficiency even though the soil contains adequate amounts of nitrogen, if moisture and consequently mass flow of nitrogen is limited.

Factors affecting nitrogen availability

Leaching Loss

The quantity of nitrate nitrogen lost in drainage water depends on the climate and cultural conditions. Loss is low in unirrigated and semiarid region and high in humid areas.

Rate of nitrification

Leaching loss of applied nitrogen depend on the rate of nitrification which is a biological transformation affected by substrate ammonium levels, presence of microorganisms etc.

pH

Neutral pH range (6.5 to 7.5) and temperature of 25 to 35^{0}C favours the activities of microorganisms.

Climate

Nitrate movement is closely related to water movement, the probability of leaching losses in different regions can be approximated by determining from climatic records whether or not precipitation, or more exactly percolation, exceeds evapo-transpiration over a given period.

Aeration

Denitrification is the dissimilatory reduction of nitrate (or nitrite) to gaseous products including NO, N_2O and N_2. Because denitrification is an anaerobic microbial transformation, it is influenced by the aeration, organic carbon content, temperature and pH of the soil. The main climatic variable influencing denitrification is rainfall, which influences the water content of the soil and thus soil aeration.

Effects of irrigation method on denitrification

Flood irrigation results in increased nitrogen losses via denitrification to a greater extent than does sprinkle irrigation practices.

Ammonia volatilization

Ammonia present in soils from manure and residue breakdown and from fertilizers that contain or produce it can be lost in significant quantities, especially from sandy soils and from alkaline/calcareous soils. This loss is most serious when the materials are applied to the soil surface. Surfaces applications not only minimize opportunities for the ammonia to react with soil colloids but temperature of surface soil are usually high, which enhances ammonia volatilization. Incorporating the manure and fertilizers into the top few centimeters of soil can reduce ammonia loss by 25-75%.

Source

Among the nitrogenous fertilizers, urea which is an increasingly important nitrogen source, is particularly susceptible to ammonia loss. The fertilizer nitrogen source and its management largely determine the potential for ammonia loss.

Flood water pH

The volatilization of ammonia is affected by a number of management, soil and environmental factors. Among these factors, flood water pH is an important determinant. The reduction in carbon dioxide causes a marked increase in flood water pH, levels above 9.0 at these values, ammonia is released from ammonium compounds and volatilizes directly to atmosphere.

Carbon Nitrogen Ratio

The C/N ratio in the organic matter (upper 15 cm) of arable soils commonly ranged from 8:1 to 15.:1, the medium being between 10.:1 and 12:1 C/N is lower of sub soils, in general, than for the corresponding surface layers. The C/N ratio in plant materials varies from 20:1 to 30:1 in legume and farm manure, to 100:1 in certain straw residues, and to as high as 400:1 sawdust. The heterotrophic florabacteria, fungi and actinomycetes become active and multiply rapidly, producing carbon dioxide in large quantities. Under these conditions, nitrate nitrogen essentially disappears from the soil because of the persistent microbial demand for this element to build microbe tissues. For the time being, then, little or no mineral nitrogen is available to higher plants. As decay occurs, the C/N ratio of the remaining plant material decreases because carbon is being lost and nitrogen conserved.

Acid Soils

Ammonium-N is naturally changed to the nitrate form by soil bacteria. Soil bacteria making the conversion of ammonium to nitrate are sensitive to soil pH. They do not function well when the soil pH is below 5.5. Logically, the conversion of ammonium to nitrate also slows when the soil pH is below 5.5.

Soil temperature

Nitrogen deficiency often occurs in winter or early spring because of low free N reserves, low soil temperatures and/or lack of root activity.

Nitrogen interaction with potassium

Potassium affects nitrate (NO_3^-) absorption and reduction. Rapid NO_3^- uptake depends on adequate K in the soil solution. Better N uptake and utilization with adequate K mean improved N use and higher yields.

Role of Nitrogen in Plants

Nitrogen is a major constituent of several of the most important plant substances. For example, nitrogen compounds comprise 40% to 50% of the dry matter of protoplasm and it is a constituent of amino acids, the building blocks of proteins. Most of the nitrogen taken up by plants is from the soil in the forms of NO_3^- , although in acid environments where nitrification is less likely to occur, ammonium NH_4^+ is more likely to be the dominating source of nitrogen. Amino acids and proteins can only be built from NH_4^+ so NO_3^- must be reduced.

Nitrogen enters the plant largely through the roots. A "pool" of soluble nitrogen accumulates. It's composition within a species varies widely depending on several factors, including day length, time of day, night temperatures, nutrients deficiencies, and nutrient imbalance. Short day length promotes asparagines formation; whereas glutamine is produced under long day regimes. Darkness favours protein breakdown accompanied by high asparagines accumulation. Night temperature modifies the effects due to night length and soluble nitrogen tends to accumulated owing to retarded synthesis and breakdown of proteins. Low night temperature conserves glutamine; high night temperature increases accumulation of asparagines. The pool of soluble nitrogen is much smaller than in well-nourished plants when N and P are deficient, since uptake of nitrate and further reduction and conversion of N to organic forms is restricted more than is protein synthesis. Before the nitrate taken up by the roots can be converted into organic nitrogen it must be reduced to ammonia.

Nitrate reduction is carried out in two steps. Nitrate is first reduced to nitrite (NO_2^-) in the cytosol by nitrate reductase using NADH or NADPH. Nitrite is then reduced to ammonia in the chloroplasts (plastids in roots) by a ferredoxin dependent nitrite reductase. In the chloroplasts, glutamine synthetase incorporates this ammonia as the amide group of glutamine using glutamate as a substrate. Glutamate synthase (Fd-GOGAT and NADH-GOGAT) transfer the amide group onto and 2-oxoglutarate molecule producing two glutamates. Further transaminations are carried out make other amino acids (most commonly aspargine) from glutamine. While the enzyme glutamate dehydrogenase (GDH) does not play a direct role in the assimilation, it protects the mitochondrial functions during periods of high nitrogen metabolism and takes part in nitrogen remobilization.

Once ammonia is available within the plant tissues, it is incorporated into amino acids and then into proteins and other organic combinations in a very orderly sequence of biochemical reactions. The key type reactions of assimilation include the reversible formations (amination of keto acids) of the two amino acids glutamate and aspartate and their amides, glutamine and asparagines. The other type reaction is the reversible transfer of ammonia (transamination) from one of these compounds to other appropriate keto acids to form the rest of the amino acids common to plants. Although in some cases, nitrates can be reduced by the roots, it is usually translocated in the anionic form through xylem tissue to the leaves and shoot apices. Here, the nitrate is reduced and incorporated first into amino acids and then into other structural and functional combinations.

Every nitrate ion reduced to ammonia produces one OH^- ion. To maintain a pH balance, the plant must either excrete it into the surrounding medium or neutralize it with organic acids. This results in the medium around the plants roots becoming alkaline when they take up nitrate. To maintain ionic balance, every NO_3^- taken into the root must be accompanied by either the uptake of a cation or the excretion of an anion. Plants like tomatoes take up metal ions like K^+, Na^+, Ca^{2+} and Mg^{2+} to exactly match every nitrate taken up and store these as the salts of organic acids like malate and oxalate. Other plants like the soybean balance most of their NO_3^- intake with the excretion of OH^- or HCO_3^-.

Plants that reduce nitrates in the shoots and excrete alkali from their roots need to transport the alkali in an inert form from the shoots to the roots. To achieve this, they synthesize malic acid in the leaves from neutral precursors like carbohydrates. The potassium ions brought to the leaves along with the nitrate in the xylem are then sent along with the malate to the roots via the phloem. In the roots, the malate is consumed. When malate is converted back to malic acid prior to use, an OH^- is released and excreted ($RCOO^- + H_2O \rightarrow RCOOH + OH^-$). The potassium ions are then recirculated up the xylem with fresh nitrate. Thus, the plants avoid having to absorb and store excess salts and also transport the OH^-.

Deficiency Symptoms

General symptoms are a uniform light green chlorosis of the leaves, and slow growth resulting in a delayed or sparse ground cover.

The development of nitrogen deficiency symptoms vary according to conditions experienced by the crop. When nitrogen is initially adequate during the establishment phase but becomes depleted during crop growth, plants may appear normal or near normal in colour and habit, except for yellowing and premature shedding of older leaves due to remobilisation of nitrogen from these tissues. In this case, the oldest leaves become uniformly yellow and slightly wilted. A light brown necrosis may spread from the tip or margins, but often the leaf is shed before it develops extensive necrosis. Necrotic tissue is supple rather than brittle.

Symptoms of chronic nitrogen deficiency include uniformly pale colour, reduced leaf size, loss of the normal sheen resulting in a dull appearance of the leaves, thin spindly vines and reduced activity of axillary buds leading to less branching. In severe cases, small purple-pigmented flecks or ringspots have been observed on the surface of older leaves of some cultivars.

Sources of Nitrogenous fertilizers

A large number of fertilizers containing nitrogen are available. These are given in Table 1.

Table 1. Nitrogenous fertilizers

Name of the fertilizers	Nitrogen (Per cent)
Sodium Nitrate - $NaNO_3$	16.5%
Potassium Nitrate - KNO_3	13.8%
Ammonium Sulphate - $(NH_4)_2SO_4$	21.2%
Ammonium Nitrate - NH_4NO_3	35.0%
Monoammonium Phosphate - NH_4HPO_4	12.2%
Diammonium Phosphate - $(NH_4)_2H_2PO_4$	21.2%
Urea - $CO(NH_2)_2$	46.6%
Anhydrous Ammonia - NH_3	82.2%
Potassium nitrate KNO_3	13
Sodium nitrate (nitrate of soda) - $NaNO_3$	16

□□□

Chapter 4

Phosphorus

Phosphorus is the tenth most abundant element in the earth's crust. About 200 minerals, occurring in nature have been reported to contain P with apatite group of minerals, being the most important. Indeed, fluorapatite and the related minerals act as the source material for bulk of the phosphatic fertilisers. The surface plough layer of soil (0-0.15m) contains P in the range of 200 to 2000 kg P ha^{-1}, the average being 1000 kg P ha^{-1}. Total P content of some representative Indian soils ranges from 580 to 2900 kg P ha^{-1}. Generally, P is present in soil predominantly in the inorganic forms, with organic P content varying from 20 to 80% of total P in the soil. The inorganic phosphates in soil comprise the i) Fe and Al phosphates (in acidic soils) and ii) Ca phosphates (in neutral and alkaline soils). The most important factors controlling the availability of P to plant roots are its concentration in the soil solution and the P-buffer capacity of the soil. The latter controls the rate at which P in the soil solution is replenished, i.e. the rate of desorption of P from the solid phase of the soil, which is faster in soils with a high buffer capacity. Also important are the size of the root system and the extent to which roots grow into the soil, and the efficiency with which roots take up P.

Phosphorus uptake

The concentration of phosphorus in the soil solution can range from 10^{-4} M, very high, to 10^{-6} M, deficient, to as low as 10^{-8} M in some very low-fertility tropical soils. These concentrations can be related to the amount of P in the soil solution and uptake of P. The minimum concentration of P to which the roots of soil-grown plants can deplete the external concentration of P in the rhizosphere soil solution is about 1 μm. The amount of nutrient transported by mass flow is related to the amount and rate of water movement to the root, the water use by the crop, and the concentration of the nutrient in the soil solution.

Diffusion is the main process by which P moves to the root surface. Diffusion involves the movement of ions along a concentration gradient, i.e. from a higher to a lower concentration. Thus, when plant roots remove nutrient ions from the soil solution and the concentration is lowered relative to that in the bulk solution, a concentration gradient develops and nutrient ions move down this gradient. The extent of depletion at the root surface depends on the balance between the supply from the soil and the demand by the plant. If the "absorbing power" of the root is large, this creates a sink to which nutrients diffuse. The root-absorbing power is not constant but depends on root metabolism and the nutrient status of the plant. The amount of P required at the root surface depends on the depletion profile that develops with time. The shape of this profile will depend on the balance between P uptake by roots, the rate at which P is replenished in the soil solution and the mobility of the phosphate ions by diffusion.

Plant roots can absorb P from soil solutions having very low P concentrations, in which P uptake is against a very steep P concentration gradient. This is because the P content of root cells and xylem sap is 100–1000 times larger than that of the soil solution. The transport of P across the cell membrane varies between plant species. Cultivars within the same species can differ in their capacity for active P uptake and these differences are probably largely genetically controlled.

Many plants have extensive root systems, which frequently have root hairs that extend out into the rhizosphere (the cylinder of soil surrounding the root), thereby increasing the effective surface area of the root system for the uptake of water and nutrients. Root hair formation is modified by environmental factors such as nutrient supply, especially that of N and P and it differs between species. In non-mycorrhizal plants, the extent of the zone of P depletion in the soil as a result of active P uptake by roots is often closely related to root hair length. Root hairs are more effective in absorbing P than is the root cylinder when the influx per unit area of each is compared because the smaller diameter and geometric arrangement of the root hairs maintain higher diffusion rates for P. In soils with little readily-available P, uptake by root hairs can account for up to 90 percent of total P uptake by the plant.

Factors affecting phosphorus availability

Soil pH

As the soil pH increases above about pH 7.0, soil P is increasingly "fixed" into less soluble/available forms by excess calcium. As the soil pH decreases below about pH 6.0, soil P is increasingly "fixed" into less soluble/available forms by excess soluble aluminum.

Soil Compaction

Phosphorous moves very little in the soil. Because of this, plant roots must be healthy and actively explore new areas of the soil in order to obtain adequate P nutrition. Anything that inhibits aggressive root growth is likely to reduce P uptake, even in high soil phosphorus.

Soil Aeration

Inadequate soil aeration is often related to soil clay content, soil drainage, and soil compaction. Most cultivated plants require adequate oxygen (O_2) in the soil atmosphere. A lack of adequate soil O_2 can reduce P uptake by as much as 50%.

Soil Moisture

As moisture stress increases, phosphorus availability and uptake decrease. Higher levels of soil phosphorus result in higher phosphorus uptake at all moisture levels. However, as soil moisture begins to exceed field capacity, the excess water excludes the needed oxygen from the soil and P uptake begins to suffer due to the lack of O_2 in the soil.

Soil Temperature

Cold soil reduces phosphorus uptake, as well as most other chemical and biological activity in the soil. As the summer begins to warm up, uptake efficiency of both the plants and the soil improve.

Soil Texture

Generally, low CEC soils require higher soil P to supply equivalent amounts of P to a crop. Such soils typically hold less water at any point in time, which slows P diffusion to the roots.

Soil Phosphorus

Soluble phosphorus, either from fertilizer or natural weathering, reacts with clay, iron and aluminum compounds in the soil and is converted readily to less available forms by the process of phosphorus fixation. Because of these fixation processes, phosphorus moves very little in most soils, stays close to its place of origin, and crops seldom absorb more than 20 percent of fertilizer phosphorus during the first cropping season after application. As a result, little soil phosphorus is lost by leaching. This fixed residual phosphorus remains in the rooting zone and will be slowly available to succeeding crops.

Organic Matter

Soils high in organic matter contain considerable amounts of organic phosphorus that are mineralized (similar to organic nitrogen), and provide available phosphorus for plant growth. In addition to supplying phosphorus, organic matter also acts as a chelating agent and combines with iron, thereby preventing the formation of insoluble iron phosphates. Heavy applications of organic materials such as manure, plant residues or green manure crops to soils with high pH values not only supply phosphorus, but upon decomposition, provide acidic compounds, which increase the availability of mineral forms of phosphorus in the soil.

Time of application

Fixation of soil phosphorus increases with time of contact between soluble phosphorus and soil particles. Consequently, more efficient utilization of fertilizer phosphorus is generally obtained by applying the fertilizer shortly before planting the crop. This practice is especially effective on soils with high phosphorus-fixing capacities.

Crop Residues

Incorporation of large amounts of crop residues can result in immobilization of available P by microbes. As they decompose the residue, they grow and reproduce, thus creating their own need for available P. During the decomposition of crop residue, soil microbes are effectively in competition with higher plants. Microbially immobilized P will gradually become available as decomposition is completed, the microbes die, and they are re-cycled.

Root Systems

Plant root geometry and morphology are important for maximizing P uptake, because root systems that have higher ratios of surface area to volume will more effectively explore a larger volume of soil. Soil P is essentially immobile, and the portion of soil P that is soluble and immediately available to plants is exceptionally small. Therefore, plant roots must constantly explore large volumes of soil to satisfy their need for P. There are significant differences between species in the relative size and effectiveness of their root systems. There can also be significant differences of this type between hybrids and varieties within the same species.

Mycorrhizae

Mycorrhizae are soil fungi that form a symbiotic association with plant roots. The thread-like hyphae of the fungus connect with plant roots and extend into the soil. The hyphae act like extensions of the plants root system by absorbing nutrients and transporting them back to the plant roots. A major benefit in this respect is an increase in P uptake. In exchange, the mycorrhizae receive sugars manufactured by the plant.

Role of Phosphorus in Plants

Phosphorus enters the plant through root hairs, root tips, and the outermost layers of root cells. Uptake is also facilitated by mycorrhizal fungi that grow in association with the roots of many crops. Phosphorus is taken up mostly as the primary orthophosphate ion ($H_2PO_4^-$), but some is also absorbed as secondary orthophosphate (HPO_4^{2-}), this latter form increasing as the soil pH increases. Phosphorus may be stored in the root or transported to the upper portions of the plant. Through various chemical reactions, it is incorporated into organic compounds, including nucleic acids (DNA and RNA), phosphoproteins, phospholipids, sugar phosphates, enzymes, and energy-rich phosphate compounds. viz., adenosine triphosphate (ATP). It is in these organic forms as well as the inorganic phosphate ion that P is moved throughout the plant, where it is available for further reactions.

Phosphorus is a vital component of the substances that are building blocks of genes and chromosomes. So, it is an essential part of the process of carrying the genetic code from one generation to the next, providing the "blueprint" for all aspects of plant growth and development. An adequate supply of P is essential to the development of new cells and to the transfer of the genetic code from one cell to another as new cells are formed. Large quantities of P are found in seeds and fruit where it is believed essential for seed formation and development. Phosphorus is also a component of phytin, a major storage form of P in seeds. About 50 percent of the total P in legume seeds and 60 to 70 percent in cereal grains is stored as phytin or closely related compounds.

Phosphorus deficiency

Plants lacking phosphorus, become small, tender, stunted shade plants, but have stronger stems. Phosphorus deficiency leads to poor lignifications. A dull dark to bluish-green colour is also characteristic phosphorus deficiency, particularly during the early stages. Since leaf growth is inhibited, but chlorophyll concentrations per unit leaf area are abnormally high, this applying rather to the blue-green chlorophyll A than to the more yellowish chlorophyll B. Severe phosphorus deficiency, causes leaves to grow pale, particularly in legumes.

Owing to the increased synthesis of anthocyans, reddish, reddish violet or deep purple discolouration develop, depending on the basic leaf colour. In cereals, this affects mainly the leaf sheath and stalk, but ears are sometimes also discoloured. Further symptoms accompanying prolonged deficiencies include the appearance of brown to dark brown necrotic spots on older leaves, dark brown necrotic lesions at the leaf margins and the death and falling of old leaves.

Sources of phosphate fertilizers

Phosphorus present in the phosphatic fertilizer is usually expressed as phosphorus pentoxide (P_2O_5) and the various sources of phosphatic fertilizers are given below (Table 1).

Table 1. Phosphatic fertilizers

Name of Fertilizers	P_2O_5 (Per cent)
Single super phosphate $Ca(H_2PO_4)_2$	16.0
Double super phosphate $2Ca(H_2PO_4)_2$	30.0-35.0
Triple super phosphate $3Ca(H_2PO_4)_2$	45.0-50.0
Basic slag $(CaO)_3P_2O_5SiO_2$	3.0-8.0
Dicalcium phosphate $CaHPO_4$	35.0-40.0
Rock phosphate $Ca_3(PO_4)_2CaF_2$	23.0-24.0
Bone meal $Ca_3(PO_4)_2CaF_2$	20.0-25.0

□□□

Chapter 5

Potassium

Potassium is one of the three major plant nutrients. It comprises 2.6% of the lithosphere. Most of the potassium in soil is held in the primary minerals like micas and potassic fieldspars and secondary minerals. Potassium occurs in soil solution as cations, held in mineral in three forms viz, unavailable, slowly available and readily available.

Unavailable Potassium

Depending on soil type, approximately 90-98% of total soil K is found in this form. Feldspars and micas are minerals that contain most of the K. Plants cannot use the K in this crystalline-insoluble form. Over long periods of time, these minerals weather (break down) and K is released. This process, however, is too slow to supply the full K needs of field crops. As these minerals weather, some K moves to the slowly available pool. Some also moves to the readily available pool

Slowly Available Potassium

This form of K is thought to be trapped between layers of clay minerals and is frequently referred to as being fixed. Growing plants cannot use much of the slowly available K during a single growing season. This slowly available K is not measured by the routine soil testing procedures. Slowly available K, can also serve as a reservoir for readily available K. While some slowly available K can be released for plant use during a growing season. Some of the readily available K can also be fixed between clay layers and thus converted into slowly available K.

Readily Available Potassium

Potassium that is dissolved in soil water (water soluble) plus that held on the exchange sites on clay particles (exchangeable K) is considered readily available for plant growth. The exchange sites are found on the surface of clay particles. This is the form of K measured by the routine soil testing procedure. Plants readily absorb the K dissolved in the soil water. As soon as the K concentration in soil water drops, more is released into this solution from the K attached to the clay minerals. The K attached to the exchange sites on the clay minerals is more readily available for plant growth than the K trapped between the layers of the clay minerals.

Factors affecting Potassium availability

i) Soil Factors

Cation Exchange capacity of the soil

The cation exchange capacity (CEC) of the soil reflects the soil's ability to hold K and other cations and store them in the soil for crop uptake. Clay minerals and soil organic matter are the two parts of soil that contribute to CEC. In general, the higher the CEC of the soil, the greater the storage capacity and supplying power for K.

The K fixation of the soil

Some soils have clay types that can fix large amounts of K from fertilizers or other sources. This reduces the availability of K to the crop. If dense layers develop in the subsoil, root penetration and rooting volumes are decreased, reducing the availability of K and other nutrients. Root systems are frequently shallow with roots concentrated in the upper layers where K supply may be adequate but where shortage of water can make it unavailable to plants.

Soil temperature

Low soil temperatures reduce K availability and uptake rate by crops.

Soil moisture

Moisture is needed for mass flow movement of K to the plant roots with water and for the diffusion of K to the roots to resupply that taken up by the roots. Drought stress or excess moisture reduces K availability and uptake by crops.

Soil tilth

Tilth is related to the friability and ability to get air into the soil. Air is needed for root respiration for K uptake. Tillage when soils are too wet leads to compaction.

ii) Plant Factors

The crop

Crops differ in their ability to take up K from a given soil. This is associated with the type of root system and surface area of the roots. Cereals, for example, have a much greater capacity to take up K from the plough layer than legumes does. Grasses have many more fibrous, branching roots, increasing the K absorbing surface.

Genotypes

Crop genetics come into play with the differences among varieties or hybrids of a given crop. They usually relate to the type of root system, root density and metabolic activity that affect K uptake.

Plant populations

As plant populations increase, yields of some crops are greater and demands on soil K are increased.

Crop yield

As crop yields increase, total K uptake also increases.

Role of Potassium in plants

Potassium is a univalent cation found in the largest concentration (100-200 mM) in plant cell sap and so it is called a 'master cation'. Potassium is ionic (K^+) free (not bound to any constituent) and highly mobile in plants. Enzymes serve as catalysts for chemical reactions, being utilized but not consumed in the process. They bring together other molecules in such a way that the chemical reaction can take place. Potassium "activates" at least 60 different enzymes involved in plant growth. The K changes the physical shape of the enzyme molecule, exposing the appropriate chemically active sites for reaction. Potassium also neutralizes various organic anions and other compounds within the plant, helping to stabilize pH between 7 and

8, which is optimum for most enzyme reactions. Plants depend upon K to regulate the opening and closing of stomata. Proper functioning of stomata is essential for photosynthesis, water and nutrient transport, and plant cooling. When K moves into the guard cells around the stomata, the cells accumulate water and swell, causing the pores to open and allowing gases to move freely in and out.

The role of K in photosynthesis is complex. The activation of enzymes by K and its involvement in adenosine triphosphate (ATP) production is probably more important in regulating the rate of photosynthesis than is the role of K in stomatal activity. When the sun's energy is used to combine CO_2 and water to form sugars, the initial high-energy product is ATP. The ATP is then used as the energy source for many other chemical reactions. The electrical charge balance at the site of ATP production is maintained with K ions.

Potassium also plays a major role in the transport of water and nutrients throughout the plant in the xylem. When K supply is reduced, translocation of nitrates, phosphates, calcium (Ca), magnesium (Mg), and amino acids are depressed. As with phloem transport systems, the role of K in xylem transport is often in conjunction with specific enzymes and plant growth hormones. An ample supply of K is essential to efficient operation of these systems. Potassium is required for every major step of protein synthesis. The "reading" of the genetic code in plant cells to produce proteins and enzymes that regulate all growth processes would be impossible without adequate K. When plants are deficient in K, proteins are not synthesized despite an abundance of available nitrogen (N). Instead, protein "raw materials" (precursors) such as amino acids, amides and nitrate accumulate.

The enzyme responsible for synthesis of starch (starch synthetase) is activated by K. Thus, with inadequate K, the level of starch declines while soluble carbohydrates and N compounds accumulate. Photosynthetic activity also affects the rate of sugar formation for ultimate starch production. Under high K levels, starch is efficiently moved from sites of production to storage organs. Potassium plays significant roles in enhancing crop quality. High levels of available K improve the physical quality, disease resistance, and shelf life of fruits and vegetables used for human consumption and the feeding value of grain and forage crops.

Photosynthetically active C_4 plants have lower potassium and protein contents than C_3 plants. The "potassium efficiency" of C_4 plants is better than that of C_3 plants. It was also reported that C_4 plants produced 68.5 kg of dry matter per kilogram of potassium, compared with only 42.7 kg for C_3 plants.

Potassium deficiency

Potassium deficiency slows down growth at first, but later the plants stop growing completely and become stunted. Plants lacking potassium have a stocky appearance with short internodes. Dicotyledons such as beets, celery, carrots, etc., have a rosette like appearance, whereas potatoes and field beans become rather bushy.

Due to growth inhibition, the younger leaves are smaller than those of plants that are adequately supplied with potassium and the leaf blades are often smaller. They usually have their normal green colour, but they are sometimes dark to bluish-green because the synthesis of organic substance is more strongly inhibited than that of chlorophyll. A metallic bronze shine has also been observed. The leaves are like an inverted cup, with downcurled edges because growth inhibition starts at the leaf margins and the intercostal leaf tissue is sometimes arched upwards so that the leaf has a wavy appearance. In fruit trees, the leaf blades sometimes curl forwards, particularly in the case of peaches, apricots, pears and cherries. Necrotic lesions at the margins of old leaves are accepted quite generally as a universal diagnostic feature of potassium deficiency. But these symptoms should not be mistaken for similar symptoms due to salt injury, spray damage or fungi attacks.

Sources of Potassium

The various sources of potassium fertilizers are given in Table 1.

Table 1. Potassium Fertilizers

Name of Fertilizers	K_2O (Per cent)
Muriate of potash KCl	60.0
Potassium sulphate K_2SO_4	48.0-52.0
Potassium magnesium sulphate K_2SO_4, $2MgSO_4$	22.0
Potassium schoenite K_2SO_4, $2MgSO_4 .6H_2O$	22.0-24.0

❑❑❑

Chapter 6

Calcium

Calcium is relatively abundant in soils and rarely limits crop production. It makes up about 3.6% of the earth's crust. It is present in soil minerals such as calcium silicates, aluminosilicates, carbonate minerals, sulphate minerals, phosphate minerals, fluoride minerals and deposited shells of molluscans and corals as $CaCO_3$. It is different from most plant nutrients in that it is only moved within the plant by the water moving from the roots through the leaves. Calcium is the predominant positively charged ion (Ca^{++}) held on soil clay and organic matter particles because it is held more tightly than magnesium (Mg^{++}), potassium (K^{+}), and other exchangeable cations.

Factors affecting calcium availability

Types of soil

Calcium deficiency, either in terms of plant need or for the maintenance of proper soil conditions, occurs under many circumstances. These are listed below (Table. 1).

Table 1. Types of soil which favours Ca deficiency

Types of soil	Calcium availability
Acid soils	Many crops will respond to calcium application where the degree of calcium saturation of the exchange complex is less than 25 percent
Sandy soils	Particularly those found in humid region where the rainfall is over 30 inches per year. Such soils are commonly derived from acid igneous rocks or siliceous sandstones.
Peat soils	Strongly acid peat soils are often low in calcium
Soils with the dominant clay montmorillonite	Calcium absorbed on kaolinite at 45 percent saturation is more available than it is on montmorillonie at the same degree of saturation.
Alkali soils	Where exchangeable sodium and pH are high. Plants may not be able to secure adequate calcium from such soils because of the extreme insolubility of calcium compounds, the low supply of exchangeable calcium, and the presence of a considerable excess of sodium over calcium.

Poor root growth

Factors inducing poor root growth like nematode damage, cold wet soils, compacted layer in soil, low pH, herbicide damage, etc are also contributes to Ca deficiency.

Soil pH

Acid soils have less Ca, and high pH soils normally have more calcium.

Soil CEC

Lower CEC soils hold less Ca and high CEC soils hold more.

Cations-calcium interactions

Calcium uptake is reduced by other major cations such as sodium (Na^+), potassium (K^+), magnesium (Mg^{++}), ammonium ($NH4^+$), iron (Fe^{++}) and aluminium (Al^{+++}).

Distribution of Calcium in Plants

Calcium moves toward roots by diffusion and mass flow in the soil. A number of calcium specific ion channels occur in the membranes of root cells, through which influx occurs, but these channels appear to be more involved in enabling rapid fluxes of calcium into the cytoplasm and organelles as part of signalling mechanisms. This calcium is moved into vacuoles, endoplasmic reticulum, or other organelles with movement occurring by means of calcium-specific transporters.

In general, calcicole (Ca loving plants) species contain high concentrations of intracellular calcium, and calcifuges species contain low concentrations of intracellular calcium. The different geographic distributions of these plants seem to be largely determined by a range of soil conditions other than just calcium concentration in the soil *per se*. In the calcareous soils favoured by calcicoles, in addition to high concentration of Ca^{2-}, pH is high, giving low solubility of heavy metal ions and high concentrations of nutrient and bicarbonate ions. In contrast, the acid soils favoured by calcifuges have low pH, high solubility of heavy metal ions, and low availability of nutrients.

The growth of calcicole species is related strongly to the concentration of calcium in the soil, but the inability of calcicole species to grow in acid soils is linked strongly to an inability to tolerate the high concentrations of nutrient ions viz., Al^{3+} Mn^{2+}, and Fe^{2+}. For calcifuge species, the difficulty in growing in a calcareous soil stems from an inability to absorb iron, although in some calcareous soils low availability of phosphate may also be a critical factor.

For calcifuge (Ca sensitive plants) species growing on calcareous soils, it seems as if the availability of iron is the most significant factor affecting plant growth, with chlorosis occurring due to iron deficiency. However, this deficiency is caused largely by immobilization of iron within the leaves, not necessarily a restricted absorption of iron. Calcicole species seem to make iron and phosphate available in calcareous soils by exudation of oxalic and citric acids from their roots. The high concentrations of bicarbonate ions in calcareous soils seem to be important in inhibition of root elongation of some calcifuge species.

The concentrations of calcium in plants are similar to the concentrations of potassium, in the range 1 to 50 mg Ca g^{-1} dry matter.

In general, monocotyledons contain much less calcium than dicotyledons. Calcium uptake by the plant is passive and does not require energy input. Calcium mobility in the plant takes places mainly in the xylem together with water. Therefore, calcium uptake is directly related to the plant transpiration rate. Conditions of high humidity, cold and a low transpiration rates may result in calcium deficiency. Salinity buildup might also cause calcium deficiency because it decreases the water uptake by the plant.

Roles of Calcium in Plants

Calcium has several distinct functions within higher plants. Calcium functions can be divided into four main areas: i) effects on membranes, ii) effects on enzymes, iii) effects on cell walls, and (iv) interactions of calcium with phytohormones. Calcium ions (Ca^{2+}) bridge phosphate and carboxylate groups of phospholipids and proteins at membrane surfaces helping to maintain membrane structure. Also, some effect occurs in the middle of the membrane, possibly through interaction of the calcium and proteins that are an integral part of membranes. Calcium is a key element in the structure of primary cell walls. In the primary cell wall, cellulose microfibrils are linked together by cross-linking glycans, usually xyloglucan polymers. Unlike K^+ and Mg^{2+}, Ca^{2+} does not activate many enzymes and its concentration in the cytoplasm is kept low. This calcium homeostasis is achieved by the action of membrane-bound, calcium-dependent ATPases that actively pump Ca^{2+} ions from the cytoplasm and into the vacuoles, the endoplasmic reticulum (ER), and the mitochondria. This process prevents the ion from competing with Mg^{2+}, thereby lowering activity of some enzymes; the action prevents Ca^{2+} from inhibiting cytoplasmic or chloroplastic enzymes such as phosphoenol pyruvate (PEP) carboxylase and prevents Ca^{2+} from precipitating inorganic phosphate.

Calcium appears to be involved in cell division and in cell elongation and is linked to the action of auxins. The loosening of cellulose microfibrils in the cell wall is controlled by auxins, giving rise to excretion of protons into the cell wall. Furthermore, auxin is involved in calcium transport in plants. Calcium is essential for the growth of pollen tubes. A gradient of cytoplasmic calcium concentration occurs along the pollen tube with the highest concentrations being found in the tip.

Deficiency symptoms

The primary symptom of calcium deficiency is the development of necrotic tissue on young leaves. The necrosis usually begins along the lateral margins and extends inward mainly in interveinal tissue. Necrotic tissue is mid to dark brown and brittle. Necrosis is not usually preceded by localized chlorosis, although the leaves may be uniformly paler than normal. The symptom is first observed on young expanding leaves, two or three leaves below the tip, but as the disorder intensifies, newly formed leaves are affected, and finally the apex may die. Necrotic leaves may be prematurely shed.

A second symptom may develop on mature leaves, which is apparently unrelated to the necrosis of young leaves. Necrotic spots appear, which are either clustered along the main veins from the petiole attachment point to about half way to the margin, or scattered uniformly across the interveinal areas. They are roughly circular with abrupt, irregular edges, and do not seem to accelerate leaf senescence. The necrotic tissue in this case is mid brown and is not brittle. Calcium deficient plants may produce soft, very small and misshapen storage roots depending on the varieties used.

Calcium appears to be transported from the soil solution to the upper plant parts via root tips. Any factor which prevents the growth of new roots (poor aeration low temperature etc) may be expected to prevent Ca uptake and thus induce deficiency.

Sources of calcium

Some of the calcium containing fertilizers are given in Table 1.

Table 1. Calcium fertilizers

Name of the fertilizers	Calcium content (Per cent)
Calcium chloride $CaCl_2$	36.0
Calcium nitrate - $Ca(NO_3)_2$	19.0
Calcium ammonium nitrate - $CaCO_3NH_4NO_3$	16.0
Dolomitic limestone $CaCO_3 + MgCO_3$	24-30
Single Superphosphate - $Ca(H_2PO_4)_2$	19.5
Lime stone	32.0
Gypsum - $CaSO_42H_2O$	29.0

❑❑❑

Chapter 7

Magnesium

The word 'magnesium' is derived from 'magnesia' for the Magnesia district in Greece where talc (magnesium stone) was first mined. However, there are other cities that are also named after the magnesium deposits in local regions. In 1808, Sir Humphry Davy discovered magnesium but named it magnium, because he considered magnesium to sound too much like manganese. Magnesium content in most agricultural soils ranges between 0.05 for sandy soils and 0.5% for clay soils. The sources of magnesium in soil are silicates, aluminosilicates, oxide minerals, hydroxide minerals, carbonate minerals and sulphate minerals. Magnesium is also present in soils as its various forms like water soluble, exchangeable and non-exchangeable. All these form is high as compared to water soluble and exchangeable forms. Exchangeable Mg in soils is usually ranged from 4 to 20% of the total Mg. Both fractions of Mg like exchangeable and water soluble are more important in plant nutrition.

Magnesium in plants

Magnesium is absorbed by plants from the soil solution as Mg^{2+} ion. Plant magnesium needs in most soils can be satisfied by the process of mass flow similar to that of Ca. Very little amount of Mg may reach to plant roots by interception. The amount of Mg taken up by plants, however, is usually less than that of Ca or K. In soils of temperate region, the amount of Mg concentration varies from 5-50 mg kg^{-1} in soil solution. The concentration of about 24 mg kg^{-1} in the soil is usually considered as an optimum level for the nutrition of most plants.

Factors affecting magnesium availability

Potassium

The interaction of K on reducing Mg concentration in plant tops is well documented. The K inhibits translocation of Mg from the root to the plant top.

Acid Soils

In acid soils, magnesium uptake by plants is reduced by K, Ca, Al, H, Mn and NH_4^+. The critical soil Mg concentration for maximum crop yield is higher in acid than in neutral soils. Soluble or exchangeable Al in soil interferes with uptake of Ca and Mg more than that of K, thereby increasing K/(Ca + Mg) values. Ammonium N is antagonistic to the uptake of Mg by plants, whereas NO_3^- fertilization often enhances Mg uptake.

Alkaline Soils

Magnesium activity in alkaline soils is influenced by the proportions of competing ions, viz., Ca, Na or K. Magnesium uptake in calcareous alkaline soils can be reduced by Ca, and sometimes by K. Gypsum applications reduce Mg concentrations in plants growing on alkaline soils.

Soil Organic Matter

Small amounts of magnesium are adsorbed to soil organic matter. Increasing soil organic matter concentrations increases the exchangeable cation capacity and improves the Mg supply available for plant uptake.

Soil pH

Low soil pH decreases Mg availability and high soil pH increases availability.

Soil Mg:Mn ratio

High available Mn can directly reduce Mg uptake. This may be independent of the acid conditions normally associated with excess available Mn in the soil.

Soil CEC

Low CEC soils hold less Mg, while high CEC soils can hold abundant Mg.

Cations Interactions

Cations viz., K^+, NH_4^+ affects the availability of Mg. The antagonism effect between Mg and K is well established. The ratio between potassium and magnesium (K: Mg) should be < 5.1 for field crops 3:1 for vegetable crops and 2:1 for fruit crops.

Phosphorus

Phosphorus uptake is often enhanced when applied with magnesium fertilizers.

Role of Magnesium in plants

In most commercially grown crops, magnesium concentrations average between 0.1 and 0.5% on a dry weight basis. However, total magnesium concentration may vary considerably between different plant families. The legumes can have nearly double the magnesium concentration as most cereal crops. Likewise, oil seed crops and root crops can also contain high concentrations of magnesium.

Magnesium has a number of key functions in plants. Metabolic processes and reactions that are influenced by Mg include: i) Photophosphorylation (such as ATP formation in chloroplasts), ii) photosynthetic carbon dioxide (CO_2) fixation, iii) protein synthesis, iv) chlorophyll formation, v) phloem loading, vi) partitioning and utilization of photoassimilates, vii) generation of reactive oxygen species, and viii) photooxidation in leaf tissues. Consequently, many critical physiological and biochemical processes in plants are adversely affected by Mg deficiency, leading to impairments in growth and yield. In most cases, the involvement of Mg in metabolic process relies on Mg activating numerous enzymes. An important Mg-activated enzymes is the ribulose-1,5-biphosphate (RuBP) carboxylase, which is a key enzyme in the photosynthesis process and the most abundant enzyme on earth. It is reported that up to 30% of the total Mg in plants is bound in chloroplasts. However, the appearance of Mg deficiency symptoms is highly dependent on light intensity. High light intensity increases the development of interveinal chlorosis, together with some reddish spots on the leaf blade. Therefore, the well-documented differences between plant species in the expression of visual Mg deficiency symptoms and also in critical deficiency concentrations of Mg in the leaf tissue may be related to the light intensity in a particular growth environment.

The other 70 to 85% of the magnesium in plants is associated with the role of magnesium as a cofactor in various enzymatic processes, the regulation of membrane channels and receptor proteins, and the structural role in stabilizing proteins and the configurations of DNA and RNA strands. Since magnesium is an integral component of the chlorophyll molecule and the enzymatic processes associated with photosynthesis and respiration, the assimilation of carbon and energy transformations will be affected directly by inadequate magnesium.

Over 300 enzymes require the presence of magnesium ions for their catalytic action, including enzymes utilizing or synthesizing ATP. Magnesium is an integral component of RNA, stabilizing the conformational structure of the negatively charged functional groups and also concurrently neutralizing the RNA molecule. In many cases, the role of the magnesium ion in the configurations and stabilities of many polynucleotides is not replaceable with other cations, since the ligand configurations are of a specific geometry that are capable of housing only magnesium ions. In addition, magnesium serves as a cofactor for enzymes that catalyze the hydrolysis and formation of phosphodiester bonds associated with the transcription, translation, and replication of nucleic acids.

Deficiency Symptoms

Magnesium is physiologically mobile within the plant. Therefore, if insufficient magnesium is available from the rhizosphere, magnesium can be reallocated from other plant parts and transported through the phloem to the actively growing sinks. Because of this mobility within the plant, symptoms of deficiency will first be expressed in the oldest leaves. Early symptoms of magnesium deficiency may be noted by fading and yellowing of the tips of old leaves, which progresses interveinally toward the base and midrib of leaves, giving a mottled or herringbone appearance.

In later stages of development, deficiency symptoms may be difficult to distinguish from those of potassium deficiency. Under mild deficiencies, a 'V'-patterned interveinal chlorosis develops in dicots as a result of magnesium dissociating from the chlorophyll, resulting in chlorophyll degradation. In some plants, a reddening of the leaves may occur, rather than chlorosis, as is the case for cotton, since other plant pigments may not break down as quickly as chlorophyll. The loss of protein from magnesium-deficient leaves, however, usually results in the loss of plastic pigments from most plants. On an individual leaf, as well as on a whole plant basis, deficiency symptoms may begin to appear only on the portions of a leaf or the plant that are exposed to the sun, with the shaded portions of leaves remaining green. Under severe deficiency symptoms, all lower leaves become necrotic and senescence with symptoms of interveinal yellowing progressing to younger leaves.

Light intensity can affect the expression of symptoms of magnesium deficiency. Partial shading of magnesium-deficient leaves has been shown to prevent or delay the development of chlorosis. Magnesium deficiency symptoms may begin to appear only on the portions of a leaf or plant that are exposed to the sun, with the shaded portions of leaves remaining green.

While magnesium is an essential element for all plants, these crops have been found to be especially responsive: lucerne, beet, broccoli, cabbage, cauliflower, celery, conifers, corn, cotton, cucumber, eggplant, lettuce, onion, pepper, potatoes, pumpkin, tobacco, tomato, and watermelon.

Sources of Magnesium

The commonly available magnesium fertilizers are given in Table 1.

Table 1. Magnesium fertilizers

Name of the Fertilizers	Magnesium Content (per cent)
Dolomitic lime $CaCO_3 + MgCO_3$	8-20.0
Epsom salts $MgSO_4.7H_2O$	10.0
Kieserite $MgSO_4.H_2O$	18.0
Potassium magnesium sulphate $K_2SO_4.2MgSO_4$	11.0

❏❏❏

Chapter 8

Sulphur

Sulphur is the 13th most abundant element in the earth's crust, averaging between 0.06 and 0.10%. The main S-bearing minerals in rocks and soils are gypsum, marcasite, epsomite, sulphur, pyrite, chalcopyrite ($CuFeS_2$) and cobaltite. Elemental S^0 occurs in volcanic deposits, and in deposits associated with calcite and gypsum. Minor amounts of sulphur occur as gases released during volcanic activity and as hydrogen sulphide (H_2S) of volcanic, hydrothermal and biological origin. Hydrogen sulphide, an important commercial source of S, is a contaminant in many natural gas fields.

Another source of soil sulphur is the atmosphere. In regions, where coal and other S-containing products are burned. SO_2 is released into the air and is later brought back to earth in precipitation. Plants may also absorb SO_2 by diffusion in to the leaves, which is metabolized by the plant.

Most soil sources of S are in the organic matter and are therefore concentrated in the topsoil. Elemental S and other forms as found in soil organic matter and some fertilizers, are not available to crops. They must be converted to the sulphate (SO_4^{2-}) form to become available to the crop. The sulphate form of S is an anion (negative charge), and therefore is leachable. As a rough rule-of-thumb, it can be considered to leach through the soil profile at about 50% as fast as nitrates (NO_3^-).

Different forms of sulphur

Sulphur can be classified into two major groups viz: i) organic and ii) inorganic forms. Most of the total S pool in soils is composed of organic fractions, primarily due to leaf litter inputs. Organic S in microorganisms constitutes less than 5 percent of total soil S, even in the microbially active zones. Transformations between major compartments of the S cycle result in a dynamic flux among organic and inorganic forms.

i) *Organic sulphur :* Organic sulphur in soil organic matter occurs in two primary forms: ester sulphates, which has direct C-S linkages. A few other organic forms also exist but they are of comparatively minor importance. Ester sulphates include compounds such as choline sulphate, phenolic sulphates, and sulphated polysaccharides. Carbon-bonded S is comprised principally of amino acids such as methionine cysteine and sulpholipids.

ii) *Inorganic sulphur :* In the atmosphere, sulphur occurs primarily as gaseous SO_2 and aerosol sulphate (SO_4^{2-}). Coal burning emissions are the principal source of atmospheric oxidized S with most of the S emitted as SO_2. Sulphur dioxide is oxidized at a conversion rate of about 0.1 percent per hour in the humid atmosphere to sulphuric acid (H_2SO_4).

Factors affecting sulphur availability

There are various factors which affect the sulphur transformation in soils.

Type and amount of clay

Higher amount of clay decreases the sulphur in soil such decrease may be due to higher retention of soluble sulphate sulphur by clay minerals usually through the adsorption mechanism.

Oxides and hydrous oxides of Fe and Al

These compounds in soil which can retain a substantial amount of sulphate sulphur on their surfaces rendering sulphur unavailable to the plant.

pH

In acid pH, the availability decreases due to higher retention through adsorption.

Organic matter

Soils containing an adequate amount of organic matter have greater ability to supply sulphur to the plant because of greater release of sulphate sulphur from the mineralization process.

Texture

Clay and clay loam soils are believed to contain a higher amount of sulphur than that of sand and sandy loam soils. Such variation is mainly due to textural differences and also partly from the association of organic matter with clay colloids.

Alternate wetting and drying

Sulphur availability in soils increases due to alternate wetting and drying under actual field conditions. Drying after soil saturation or submergence reduces the sulphur content in soils.

Nutrient ions

Various cations like H^+, K^+, Ca^{2+}, Mg^{2+} and Na^+ and anions like OH^-, $PO4^{3-}$, NO_3^-, Cl^- etc. affect the availability of sulphate sulphur in soils through adsorption mechanism.

Presence or absence of Vegetation

Higher amount of sulphur is mineralized in cropped than that of uncropped soils. Such increase in sulphate sulphur in soils may be ascribed to the "Rhizosphere effect" resulting from the excretion of amino acids and sugars by plant roots and the resultant stimulation of microbial activity.

Interactions of sulphur with other nutrients

The interaction between N and S has been found to be synergistic. Application of N increases the concentration and uptake of S and vice-versa.

High Response Crops

While sulphur is an essential element for all plants, these crops have been found to be especially responsive: viz., broccoli, cabbage,viz, cauliflower, maize, sugar beets, sugar cane and watermelon.

Microbial decomposition

Immobilization refers to the process of converting inorganic S (SO_4^{-2}) to organic S, and is essentially the reverse of mineralization. Microorganisms remove available S from solution and convert it into proteins and other organic compounds. Although, this process removes S from the available pool, the S is still in a reverse pool that could eventually become available to plants via mineralization. Since mineralization and immobilization are primarily biological processes, any factors affecting microbial growth will influence both of these S transformations. Important factors include soil temperature, water content, pH, C: S ratio, aeration and residue composition. The highest mineralization and immobilization rates will occur under aerated, warm, and moist conditions with near neutral pH levels, because these conditions are optimal for microbial activity.

The N:S ratio is relatively stable in organic matter, remaining near8:1, but the C:S ratio is more varied, and strongly affects the relative amounts of mineralization and immobilization. Residues with C:S ratios higher than about 400:1 result in net immobilization of S from the soil, while C:S ratios lower than 200:1 result in net mineralization of S into the soil.. Between ratios of 200:1 and 400:1, a combination of the two processes occurs. Each progressive decomposition cycle of soil residues results in C expelled as carbon dioxide, thereby lowering the C:S ratio. Under conditions favouring oxidation, for each 1% of the organic matter in the soil approximately 6kg of sulphur are released per annum.

Role of sulphar in plants

Sulphur is an essential element for growth and physiological functioning of plants. The total sulphur content in the vegetative parts of crops varies between 0.1 and 2% of the dry weight. The uptake and assimilation of sulphur and nitrogen by plants are strongly interrelated and dependent upon each other and at adequate levels of sulphur supply, the organic N/S ratio is around 20:1 on a molar basis. In most plant species, the major proportion of sulphur (up to 70% of the total S) is present in reduced form in the cysteine and methionine residues of proteins. Additionally, plants contain a large variety of other organic sulphur compounds such as thiols (glutathione; ~1 to 2% of the total S) and sulpholipids (<"1 to 2% of the total S); some species contain the so-called secondary sulphur compounds such as alliins and glucosinolates.

In general, plants utilize sulphate (S^{6+}) taken up by the roots as a sulphur source for growth. Sulphate is actively taken up across the plasma membrane of the root cells, subsequently loaded into the xylem vessels and transported to the shoot by the transpiration stream. In the chloroplasts of the shoot cells, sulphate is reduced to sulphide prior to its assimilation into organic sulphur compounds. Plants are also able to utilize foliarly absorbed sulphur gases.

The sulphur requirement varies strongly between species and it may fluctuate during plant growth. The sulphur requirement can be defined as 'the minimum rate of sulphur uptake and utilization that is sufficient to obtain the maximum yield, quality, and fitness,' which for crop plants is equivalent to 'the minimum content of sulphur in the plant associated with maximum yield' and is regularly expressed as $kgSha^{-1}$ in the harvested crop. In physiological terms, the sulphur requirement is equivalent to the rate of sulphur uptake, reduction, and metabolism needed per gram plant biomass produced over time and can be expressed as mol S g^{-1} plant day^{-1}.

The major proportion of the sulphate taken up by the roots is reduced to sulphide and subsequently incorporated into cysteine, the precursor and the reduced sulphur donor for the synthesis of most other organic sulphur compounds in plants. Even though root plastids contain all sulphate reduction enzymes, reduction predominantly takes place in the chloroplasts of the shoot. The reduction of sulphate to sulphide occurs in three steps. First, sulphate is activated to adenosine 5_-phosphosulphate (APS) prior to its reduction, a reaction catalyzed by ATP sulphurylase. Second, the activated sulphate (APS) is reduced by APS reductase to sulphite, a reaction where glutathione most likely functions as reductant. Third, sulphite is reduced to sulphide by sulphite reductase with reduced ferredoxin as reductant. Plants synthesize a great array of secondary metabolites as they are physically immobile, and the presence of secondary compounds may give either repellent or attractant properties.

Glucosinolates are characteristic compounds of at least 15 dicotyledonous families. Of these, the Brassicaceae are the most important agricultural crops. Glucosinolates act as attractants, repellents, insecticides, fungicides, and antimicrobial protectors. More than 80% of the total sulphur in *Allium* species is present in secondary compounds. *Allium* species contain four *S*-alk(en)yl-L-cysteine sulphoxides, namely *S*-1-propenyl-, *S*-2-propenyl-, *S*-methyl- and *S*-propyl-L-cysteine sulphoxides. Iso-alliin is the main form in onions, whereas alliin is the predominant form in garlic. Alliins supposedly contribute to the defense of plants against pests and diseases. Alliin acts as an antioxidant by activating glutathione enzymes and is regarded as having an anticarcinogenic and antimicrobial effect.

Deficiency symptoms

Sulphur deficiency leaves begin to develop chlorosis, which starts from one edge of the leaves and spreads over intercostal areas; however, the zones along the veins always remain green. The symptoms in cereals are less specific than in cruciferous crops. In early growth stages, plants remain smaller and stunted and show a lighter colour than plants without symptoms. The general chlorosis is often accompanied by light green stripes along the veins. Leaves become narrower and shorter than normal.

A characteristic secondary symptom of severe sulphur deficiency is a reddish-purple colour due to the enrichment of anthocyanins in the chlorotic parts of brassica leaves. Under field conditions, the formation of anthocyanins starts 4 to 7 days after chlorosis. The phenomenon is initialized by the enrichment of carbohydrates in the cells after the inhibition of protein metabolism. Plants detoxify the accumulated carbohydrates as anthocyanates, which result from the reaction with cell-borne flavonols to avoid physiological disorders. Many other nutrient deficiencies are also accompanied by formation of anthocyanins, which therefore, is a less specific indicator for sulphur deficiency.

In particular, leaves which are not fully expanded produce spoon-like deformations when struck by sulphur deficiency. The reason for this is a reduced cell growth rate in the chlorotic areas along the edge of the leaves, while normal cell growth continues in the green areas along the veins, so that sulphur-deficient leaves appear to be more succulent. The grade of the deformation is stronger, the less expanded the leaf is when the plant is struck by sulphur deficiency. Marbling, deformations, and anthocyanin accumulation can be detected up to the most recently developed small leaves inserted in forks of branches.

Sources of sulphur

Some of the sulphur containing fertilizers are given below (Table 1).

Table 1. Sulphur containing fertilizers

Name of the Fertilizers	Sulphur Content (per cent)
Ammonium Sulphate $(NH_4)_2SO_4$	23.0
Ammonium Thio Sulphate $NH_4S_2O_3$	26.0
Calcium Sulphate (Gypsum) $CaSO_4$, $2H_2O$	15-17.0
Epsom Salts $MgSO_4 \cdot 7H_2O$	14.0
Ferrous Sulphate $FeSO_4$, $7H_2O$	10.5
Potassium Sulphate K_2SO_4	17.5
Sulphuric Acid H_2SO_4	32.0

❑❑❑

Chapter 9

Zinc

Zinc minerals are $Zn(OH)_2$, ZnO, $ZnCO_3$ and Zn_2SiO_4. The total zinc content in soils varies from 3 to 770 mg kg^{-1} with the world average being 64 mg kg^{-1}. Zinc ions (Zn^{++}) are held on the surface of clay and organic matter particles. Soil organic matter holds zinc in a chelated form. Chelation is a process by which certain metals are held within the structure of large organic molecules. Because zinc is held on soil particles and by chelation, it does not move through the soil and is not leached under most conditions. Zinc enters plants primarily via root absorption of Zn^{2+} from the soil solution, which is a dynamic and complex process. Uptake depends on ion concentrations at the root surface, plant demand and root absorption capacity. Zn reaches the plant root surface through mass flow, diffusion and root interception mechanisms.

Mass flow is passive nutrient transport from the soil to the roots and is driven by transpiration. When the solution moving through the soil to the root contains a relatively large concentration of Zn, mass flow becomes the dominant mechanism bringing Zn to the root surface. When the Zn concentration is low, particularly in soils with low plant available Zn, diffusion plays important role in the transport of Zn and other nutrients, such as P, K, Cu, Fe and Mn, to the root surface because mass flow can only carry a small fraction of the nutrients required by the plants. In contrast to mass flow; diffusion operates only in the immediate volume of soil surrounding a root. The interception of nutrients by roots is an important uptake mechanism for soil immobile nutrients such as Zn. Thus, root interception (i.e., root growth and root surface area) is also an important factor in determining plant availability of Zn.

Factors affecting the zinc availability

i) Soil factors

- Zinc solubility is highly pH dependent and decreases 100 - fold for each unit increase in pH. The availability of soil Zn to plant decreases with increased soil pH. Its availability is high in soils with pH below 6. In alkaline soils, the availability is low and hence calcareous soils are frequently Zn deficient due to the alkaline reaction.
- Organic matter decomposition gives rise to certain chelating agent which contribute to Zn availability to plants. Low organic matter soils are prove to Zn deficiency.
- Coarse soil fractions contains and retain low Zn. Therefore, light textured soils are generally Zn deficient.
- High P availability in soils induces Zn deficiency. The antagonistic effect of high P on Zn availability is aggravated in calcareous soils.
- The highly mobile $ZnSO_4$ complex is an important species in soils and contribute significantly to total Zn in solution. Solubility and mobility of Zn^{2+} in soils are believed to be increased by the presence of SO_4^{2-} and subsequent formation of this complex.
- Liberal application of nitrogen fertilizers can stimulate plant growth and increase Zinc requirement beyond the available supply.
- The continuous flooding of the soil may induce Zn deficiency due to the deleterious effects of H_2S on plants under anaerobic conditions.
- Zinc deficiencies are common during cool and wet seasons, namely poor light as well as cool temperature and excessive moisture aggravate Zn deficiency. Increase in soil temperature have been shown to increase the availability of Zn to crops.
- Generally, zinc is relatively immobile in soils. Most of the zinc was retained in the surface of the soil and only little if any down ward movement occurred. Hence, any removal of top soil, either by erosion or in the process of land shaping for gravity increases Zn deficiency.
- Species and varieties of plant may vary markedly in their susceptibility to Zn deficiency. Genotype differences in sensitivity to Zn deficiency are known to occur in several other crops.

Zinc uptake and movement in plants

Roots take up Zn from the soil solution as a divalent cation (Zn^{2+}), and at high pH, Zn is absorbed as a monovalents ($ZnOH^{+}$) cation. Zn accumulation in plant roots exhibits biphasic kinetics, with initial rapid entry and binding within the root cell wall followed by a slower linear transport phase across the plasma membrane. Zn^{2+} movement from the external solution to the root cell wall free space occurs via diffusion, and Zn is subsequently transported across the plasma membrane via ion transport proteins. The cytoplasm of a plant cell is negatively charged, and there is an electrical gradient that draws cations, including Zn^{2+} into the cell. From the root surface, nutrients are transported into the root xylem through the epidermis, cortex and endodermis. Zinc may pass through the root to the xylem either through the extracellular spaces between root cells or through the cytoplasmic continuum of root cells linked by plasmodesmata (symplast). The apoplastic cation flux is largely determined by the cation exchange properties of the cell wall. Nutrients in the xylem move toward shoots in the transpiration stream. This driving force for transpiration arises from the gradient in water potential between stomatal cells and the atmosphere.

Role of Zinc in plants

In plants, zinc does not undergo valency changes and its predominant forms are: low molecular weight complexes, storage metalloproteins, free ions and insoluble forms associated with the cell walls. Zinc can become inactivated within cells by the formation of complexes with organic ligands or by complexation with phosphorus. Depending on the plant species, between 58% and 91% of the zinc in a plant can be in a water-soluble form (low molecular weight complexes and free ions). This water-soluble fraction is widely considered to be the most physiologically active and is regarded as a better indicator of plant zinc status than total zinc contents. The low molecular weight complexes are normally the most abundant soluble form of zinc and are probably the most active form of the metal.

Low molecular weight complexes of zinc

The low molecular weight complexed forms of zinc are probably physiologically active because they can be degraded and may also be involved in homeostatic mechanisms where they may act as a buffer system to bind excess free zinc ions.

Zinc in enzyme activity

In plants, zinc acts as a functional, structural or regulatory co-factor of a large number of enzymes. More than 70 metallo-enzymes containing zinc have been identified and these occur in all of the six classes of enzymes: Oxidoreductases, transferases, hydrolases, lyases, isomerases and ligases. The zinc atom is usually tightly bound to the apoenzyme and can only be removed with severe chemical treatment. It forms strong complexes with radicals of polar groups containing oxygen, nitrogen and sulphur. Zinc plays a vital role in many important enzyme systems including: Carbonic anhydrase (transport of CO_2 in photosynthesis), Several dehydrogenases: alcohol dehydrogenase Glutamic dehydrogenase. L-lactic dehydrogenase, malic dehydrogenase, D-glyceraldehyde, phosphate dehydrogenase, and D-lactate dehydrogenase, Aldolase, Carboxypeptidase, Alkaline phosphatase, Superoxide dismutase (converting superoxide radicals to hydrogen peroxide and water), RNA polymerase (protein synthesis), Ribulose bi-phosphate carboxylase (important role in starch formation) and Phospholipase.

Carbohydrate Metabolism

Zinc exerts an effect on carbohydrate metabolism through its effect on photosynthesis and sugar transformations.

a) **Photosynthesis**: A deficiency of zinc can cause a reduction in net photosynthesis by 50%-70% depending on the plant species and the severity of deficiency. This reduced efficiency of photosynthesis could be due to a reduction in the activity of the enzyme carbonic anhydrase. Zinc is a constituent of carbonic anhydrase, but the carbonic anhydrase in dicotyledons is a larger molecule and contains more zinc than carbonic anhydrase in monocotyledons (such as the cereals). A sharp decline in carbonic anhydrase activity occurs when plants suffer a zinc stress and this has effects on the carbon dioxide assimilation pathway. Carbonic anhydrase activity is closely related to zinc content but only has an effect on photosynthesis and dry matter production when the activity is very low. With extreme zinc deficiency, carbonic anhydrase activity is absent.

Zinc is a constituent of other enzymes involved in photosynthesis, including ribulose 1,5-biphosphate carboxylase (RuBPC) which has been found to catalyse the initial step of carbon dioxide fixation in photosynthesis. The reduction of photosynthesis in zinc deficient plants can also be due to decrease in chlorophyll content and the abnormal structure of chloroplasts.

b) **Sugar transformations**: Enzymes involved in the formation of sucrose, such as aldolase are adversely affected by zinc deficiency. A decline in the level of sucrose in plants are due to lower activity of sucrose synthetase activity. Zinc may play a role in the metabolism of starch because the starch content, activity of starch synthetase, and the number of starch grains are all depressed in zinc deficient plants.

Protein Metabolism

In general, the amount of protein in zinc deficient plants is greatly reduced. The mechanism by which zinc deficiency affects protein synthesis is considered to be due to a reduction in RNA and the deformation and reduction of ribosomes. Zinc is necessary for the activity of the enzyme RNA polymerase and it protects the ribosomal RNA from attack by the enzyme ribonuclease. The importance of zinc in protein synthesis suggests that relatively high zinc concentrations are required by meristematic tissue where cell division as well as synthesis of nucleic acid and protein is actively taking place.

Auxin Metabolism

Zinc is required for the synthesis of auxin (a growth regulating compound-indole acetic acid, IAA). Stunted growth and 'little leaf' are the most distinct visible symptoms of zinc deficiency and are the result of disturbances in the metabolism of auxins, especially IAA. Low levels of IAA may be the result of inhibited synthesis or enhanced degradation of IAA.

Reproduction

Flowering and seed production are severely depressed in zinc-deficient plants. Reduction in seed production in zinc-deficient plants may be due to i) increased formation of abscisic acid causing premature loss of leaves and flower buds, and ii) disruption of the development and physiology of anthers and pollen grains.

Deficiency symptoms

Visual symptoms of zinc deficiency in plants are fairly characteristic and are relatively easy to identify. These distinctive symptoms are useful for recognizing acute Zn deficiency and for indicating Zn responsive soils, but not the hidden or marginal deficiencies. The most common symptoms of Zn deficiency include: stunted growth, shortened internodes and petioles and small malformed leaves (little leaf) which results in the "rosette" symptom in the early growth stages of dicotyledons and "fan shaped" stems in monocotyledons. The deficiency symptoms first appear on young leaves as Zn is immobile under conditions of deficiency. These leaves remain small, cup upward and develop interveinal chlorosis and necrotic spots on the upper leaf surfaces which later join to each other to form brown necrotic and brittle patches. The necrosis is often more noticeable on middle aged leaves which eventually wilt, bend and collapse. Zinc deficiency is typically patchy, even within a single field and symptoms develop rapidly but depend greatly on the degree of stress. Zinc deficient plants are unthrifty, lack vigor; give patchy appearance with short and thin stems. In young plants, interveinal areas are with dark brown necrotic lesions.

Plants vary considerable in their requirements for zinc. Crops with high zinc requirements include maize, onion, and spinach. Those with medium requirements are barley, beans, beet, cucumber, lettuce, potato, radish, sorghum, soybean, tobacco, and tomato.

Sources of Zinc

Several zinc fertilizer materials are commercially available. Zinc sulphates and zinc chelates are probably the most common zinc fertilizer materials used (Table 1).

Table 1. Zinc fertilizers

Name of the Fertilizers	Zinc (Per cent)
Zinc oxide ZnO	70.0
Zinc sulphate heptahydrate $ZnSO_4$ $7H_2O$	21.0
Zinc sulphate monohydrate $ZnSO_4H_2O$	33.0
Zinc chelates ZnEDTA	12.0

❑❑❑

Chapter 10

Copper

Copper is one of the oldest known metals and is the 25th most abundant element in the Earth's crust. Copper containing minerals are $CuCO_3$, $Cu_3(OH)_2(CO_3)_2$, $Cu(OH)_2Cuo$ and $CuFe_2O_4$. The transition metals are noted for the variety of complexes they form with bases. In these complexes, Cu^{1+} and Cu^{2+} act as electron acceptors. Chelating bases are so named because they have two or more electron donor sites (often on O, S, or N atoms) that form a 'claw' around the copper ion. Such complexes are important in soil chemistry and in plant nutrition.

Copper in Soil

Copper concentration in the earth's crust averages about 50-70 ppm. Igneous rocks contain 10-100 ppm Cu, while sedimentary rocks contain 4-45 ppm Cu. Total Cu concentration in soils ranges from 1-40 ppm and averages about 9 ppm. Cu is supplied to plant roots by diffusion of organically bound, chelated Cu, similar to chelated Fe. Organic compounds in the soil solution are capable of chelating solution Cu^{+2}, which increases the solution Cu^{+2} concentration above the predicted by Cu mineral solubility. Copper occurs in the soil almost exclusively in the divalent form. The largest fraction of copper is usually present in the crystal lattices of primary and secondary minerals. The copper concentration of the soil solution is usually very low and it has been observed that more than 98% of soil solution copper is complexed with organic matter. Copper is, in fact, more strongly bound to organic matter than other micronutrient cations. Copper complexes play an important role in regulating copper mobility and availability in the soil. Since copper is strongly bound to soils, it is very immobile in the soil. The result is that copper added to the soil as a result of the use of copper containing fertilizers is thus largely restricted to the upper soil horizons.

Factors Affecting Cu Availability

Texture

The potential for Cu deficiency is greater in excessively leached, coarse textured soils.

Soil pH

Soil Cu decrease with increasing pH due to decreased mineral solubility and increased adsorption.

Interactions with Other Nutrients

High Zn, Fe, and P concentrations in soil solution can depress Cu absorption by roots and intensify Cu deficiency.

Plant Factors

Plants vary greatly in their sensitivity to Cu (Table 1). Genotypic differences are related to i) differences in Cu absorption rates, ii) greater soil exploration due to greater root mass and/or root hairs, iii) increased Cu solubility due to root exudate influence on soil pH or redox potential, iv) more efficient Cu transport from roots to shoots, and/or v) lower Cu requirement.

Table 1. Crop Sensitivity to Cu Deficiency

High Sensitivity		**Mild Sensitivity**		**Low sensitivity**	
Lucerne	Lettuce	Apples	Cucumber	Beans	Potato
Beets	Onion	Barley	Oats	Beans, snap	Rapeseed
	Rice	Broccoli	Radishes	Grasses	Rye
Carrots	Spinach	Cabbage	Strawberries	Grapes	Soybean
Citrus	Sudan grass	Cauliflower	Sweet corn		Turf
Flax	Wheat	Corn	Tomato	Peas	
			Turnip		

Role of copper in plants

The general requirement of copper by plants is relatively small, in the range of 2-20 ppm in the dry plant material. Once absorbed, Cu^{2+} is readily reduced to Cu^{+} and donates an electron to reduce O_2. Which enables it to function in many oxidation – reduction reactions in plants.

Photosynthesis and Respiration Both photosynthesis (reduction of CO_2 to carbohydrates) and respiration (oxidation of carbohydrates to CO_2) involve the transfer of electrons that requires Cu. Approximately 50% of Cu in the chloroplast is found in plastocyanin, a protein involved in energy transfer in photosynthesis reactions. Cu is part of the enzyme cytochrome oxidase that catalyze electron transfer in respiration.

Lignin Formation in Cell Walls Several enzymes (polyphenol oxidase and diamine oxidase) important to synthesis of lignin contain Cu.

Cellular Processes in higher plants

One of the Cu accumulation sites in higher plants are chloroplasts. This metal is directly involved as a component of plastocyanin in the photosynthetic electron transport chain. However, Cu content in chloroplasts, usually several times higher than that found in plastocyanin. Cu has been found also in several other proteins or their complexes Table 2. Some amount of Cu can also bind with chromosomes, maintaining probably their normal structure. Cu has a distinct regulatory role in electron transport between PS II and PS I as the constituent of plastocyanin.

Table 2. Copper content in the functional components of plants

Protein	**Cu atom/protein**
Polyphenol oxidase	1
Zn/Cu superoxide dismutase	2
Ascorbate oxidase	8
Cu amine oxidase	2
Galactose oxidase	1
Plantacyanin	1
Plastocyanin	1

Deficiency symptoms

Plants differ in their sensitivity to copper deficiency. Stunted growth, distortion of young leaves, necrosis of the apical meristem, bleaching of young leaves, and "summer dieback" in trees are typical visible symptoms of copper deficiency.

Young tissues show chlorosis, distortion, and necrosis (death). The death of the growing points often leads to excessive tillering in cereal crops and excessive branching in dicots (non-grass crops). Some vegetables show a blue-green colour before advancing to chlorosis. Excessive wilting, lodging and reduced disease resistance result from the weak cell walls caused by Cu deficiency. Reduced seed and fruit yield is caused mainly by male sterility. Copper deficiency often causes a complete failure to set flowers. Copper is found to be evenly distributed in the plant, but is relatively immobile. Therefore, a constant supply is needed throughout the growing season.

An increase in the number of auxiliary shoots in dicotyledons are secondary symptoms caused by necrosis of the apical meristem. Summer dieback has been described as the dying off of twigs and growing points on the terminal shoots of trees. Copper deficiency affects grain, seed, and fruit formation much more than vegetative growth. The main reason for the decrease in the formation of generative organs is the nonviability of the pollen from copper-deficient plants. When the copper supply is adequate, the anthers containing pollen and the ovaries have the highest copper content in the flowers and obviously the highest copper demand.

Impaired lignifications of cell walls is the most typical anatomical change induced by copper deficiency in higher plants. This gives rise to the characteristic distortion of young leaves, bending and twisting of stems and twigs and an increase in the lodging susceptibility of cereals, particularly with a high nitrogen supply. Nitrogen application accentuates copper deficiency; and when nitrogen supply is high, the application of copper fertilizers is required for maximum yield.

Sources of Copper

Copper fertilizers are available in various forms (Table 3). Copper can be broadcast or banded in soils or applied as a foliar spray (Table 3).

Table 3. Copper Fertilizers

Source	Copper (per cent)
Copper chelate CuEDTA	13.0
Copper sulphate $CuSO_4 \cdot 5H_2O$	24.0

❑❑❑

Chapter 11

Iron

Iron is the fourth abundant element, comprising about 5% of the earth crust. Which occurs as ferromagnesian silicates. Major part of the Fe released during weathering in precipitated as oxides or hydroxides and small portion is incorporated into secondary silicate minerals or complexed by soil organic matter. Soluble Fe^{+2} increases significantly when soils become waterlogged. The pH-dependent relationship for Fe^{3+} in soil solution is:

$$Fe(OH)_3 + 3H^+ \leftarrow Fe^{3+} + 3H_2O$$

The most abundant form of Fe in soils is ferric oxide (Fe_2O_3) or hematite, which is extremely insoluble and imparts a red colour to the soil. As pH increases by one unit, activity of Fe^{+++} decreases by 1000-fold due to the formation of insoluble Fe^{+++}hydroxide. Fe solubility increases under reduced condition.

Factors affecting the iron availability

Soils and plants are the most important factors affecting Fe availability to plants. The availability of Fe is particularly sensitive to changes in the soil environment. Part of the sensitivity to these changes is related directly to the performance of the root system in exploring the soil volume for this non mobile element, and part is related to the pool or bonding of the element in the soil.

Soil pH

Soil reaction is one of the most important factors affecting Fe availability to plants. The soil pH is known to influence the solubility of Fe in soil. As pH increases, Fe is converted to less soluble forms, principally to the oxide Fe_2O_3. The reaction responsible for the reduced solubility of Fe with increasing pH is well understood.

Oxidation – reduction

One of the soil properties which has a marked effect of Fe behaviour is the oxidation-reduction or redox potential. Oxidation-reduction factors affect the valence of Fe and thereby its uptake by plants. Oxidation-reduction reactions involve the transfer of electrons from one ion of molecular to another. Oxidation is the donation and reduction is the acceptance, of electrons from other substances. Oxidizing agents accept electrons from other substances and thereby undergo reduction. Reducing agents donate electrons to other substances. Iron (Fe^{2+}) loses an electron to form Fe^{3+}. The oxidation-reduction of Fe is described by the following equation:

$$Fe^{3+} + e^{-} \rightleftarrows Fe^{2+}$$

Oxidation-reduction processes affect the valence of Fe and thereby its uptake by plants. When Fe^{3+} is reduced to Fe^{2+}, its concentration increases and so does uptake. On the other hand, when Fe^{2+} is oxidized to Fe^{3+}, its concentration is reduced and uptake by plants is also reduced.

Organic matter

The presence of organic matter may promote the availability of Fe, presumably by supplying soluble complexing agents that interfere with its fixation. When Fe is added to a soil in chelated form, or when chelation occurs by soluble organic compounds, the concentration and gradient in soil solution is usually higher than the concentration and gradient of unchelated Fe.

Microorganism

There are five ways that microorganisms may affect the availability of Fe. These are: i) the release of inorganic Fe ions during the decomposition of organic material, ii) immobilization of Fe by incorporation into microbial tissue, iii) oxidation of Fe generally to a less available form, iv) reduction of an oxidized form of Fe under conditions where oxygen is limited, and v) indirect transformations; e.g. changes in pH or oxidation potential. Soil sterilization is known to increase Fe^{2+} in soil. It is well established that flooded soils are subject to a succession of Fe transformations from the Fe^{3+} to the Fe^{2+} state under reducing conditions. This is caused by a wide variety of facultative anaerobic soil bacteria.

Plant genetic variability

The total quantity of Fe in a soil is usually sufficient for plant growth but the use of this Fe by the plant is genetically controlled by an adaptive mechanism. This mechanism is activated in Fe-efficient plants in response to Fe stress but remains inactive if Fe is sufficient.

Iron-efficient plants release H^+ ions and reductants from their roots when they are under Fe stress. The pH at their root zone is lowered which favours Fe^{3+} solubility and reduction of Fe^{3+} to Fe^{2+}.

Interaction with other nutrients

A number of nutrients appear capable of reducing the availability, absorption and utilization of Fe in crop plants. Iron deficiency induced by heavy applications of P has been widely reported. Similarly, Fe deficiency is caused by the accumulation of excess Mn and Cu in the soil. High Mn in soils or plants may oxidize Fe to an inactive state.

Bicarbonate

Fe deficiency is most often observed on high calcareous soils in arid regions, but it can also occur in acid soils low in total Fe. Irrigation water and soils high in bicarbonate (HCO_3^-) may enhance Fe deficiency due to high pH associated with HCO_3^- accumulation. Calcareous soil pH ranges from 7.3 to 8.5, coinciding with the highest incidence of Fe deficiency and lowest solubility of soil Fe. HCO_3^- forms in calcareous soils by:

$$CaCO_3 + CO_2 + H_2O \rightarrow Ca^{2+} + 2HCO_3^-$$

Excessive water and poor aeration

Iron chlorosis is often associated with cool, rainy weather when soil moisture is high and soil aeration is poor. Also, root development and nutrient absorption are reduced under cool, wet conditions, which contribute to Fe stress. High pH or HCO_3^- induced chlorosis often disappears when these soils dry. Flooding and submergence of soils can improve Fe availability by increasing Fe^{+2} concentration.

Plant Factors

Although diffusion of both Fe^{+3} and Fe^{+2} to the root occurs, Fe^{+3} is reduced to Fe^{+2} before absorption. Plant genotypes differ in their ability to take up Fe and are classified according to their sensitivity or tolerance to low levels of available Fe (Table 1).

Table 1. General sensitivity of crops to Fe deficiency

Sensitive	Moderately Tolerant	Tolerant
	Lucerne	Amaranthus
Beans	Barley	Barley
Broccoli	Cabbage	Grasses
Cauliflower	Maize	Millet
Citrus	Cotton	Oat
Field beans	Forage legumes	Potato
Flax	Grasses	Rice
Forage sorghum	Ornamentals	Sugar beet
Fruit trees	Rice	Sunflower
Grain sorghum	Tomato	Wheat
Grapes	Turf grass	
Ornamentals		
Groundnut		
Soybean		
Strawberries		
Vegetables		
Walnut		

The ability of plants to absorb and translocate Fe appears to be a genetically controlled adaptive process that responds to Fe deficiency or stress. Roots of Fe-efficient plants alter their environment to improve Fe availability and uptake by i) excretion of H^+ and organic acids from roots, ii) excretion of chelating compounds from roots, and iii) enhanced rate of Fe^{+3} to Fe^{+2} reduction at root surface.

Fe Uptake and Transport

Iron arrives in the vicinity of the root as various chemical compounds or organic complexes, rarely as elemental Fe. Iron in the soil solution can be moved to the plant root as a component in the bulk soil pore solution moving toward the root as water is taken into the plant to replace that lost by transpiration or used in growth processes. Iron also can move to the root by diffusion from a region of high concentration to a lower one. Roots also can intercept Fe compounds in the soil as the roots grow and expand into additional soil volume. Root density and extension are very important factors in the plant's ability to obtain Fe.

Iron uptake by the plant is not as simple as with other essential elements. Iron is taken up by plant roots in greatest amounts in the zone of the root between cell elongation and maturation, about 1 to 4 cm behind the root tip. Uptake of Fe by the plant is an active process, that is, energy is expended by the plant to take in Fe. Iron uptake is dependent on the plant's ability to reduce Fe^{+++} to Fe^{++} and remove it from the complex or chelating compound. The chelated Fe in the soil solution moves to the root by mass flow or by diffusion. At the root, Fe is reduced and removed from the chelating molecule and moved across the cell membrane.

Iron is absorbed into the rhizodermal (epidermal) cells and endodermal cells in the root. From the endodermis, Fe is loaded into the pericycle cells and then to the xylem. Most of the Fe transported to the shoots ends up in the shoot apoplast. From here, it can be moved across the cell plasmalemma and into the cytoplasm and into organelles.

Strategies used by higher plants for Iron

Acquisition

Qualitative differences exist in the plant kingdom among genotypes in their iron deficiency - induced adaptation mechanisms and these differences can be classified according to two strategies.

In Strategy I-type, plants respond to iron deprivation with both morphological and physiological changes. Morphological modifications include enhanced development of lateral roots and differentiation of specialized transfer cells, increasing the surface area for reduction and transport of iron. Enhanced reduction of Fe(III) to Fe(II) under iron-deficiency stress seems to be the most typical feature of this strategy and is characterized by an increased acidification of the rhizosphere and the activation of plasma membrane bound reductases.

Strategy II- type systems are characterized by an iron deficiency induced release of specific Fe(III) chelating compounds (phytosiderophores) and the Fe(III) chelates as intact molecules.

The distribution of these two strategies in the plant kingdom shows a strong systematic order and it became evident that the chemotaxonomic boundary for the two strategies is not between dicotyledons and monocotyledons as suggested in the older literature but between most higher plants (Strategy I) and grasses (Gramineae, Strategy II). Besides, it is considered that Strategy II is more efficient than Strategy I for example, grasses can grow on calcareous soils that will not support the growth of dicotyledons. This can probably be related to the fact that Strategy II is less pH-dependent than Strategy I and it has been observed that the release of phytosiderophores and Fe(III) uptake are only slightly depressed at high soil pH.

Strategy I

Under iron deficient conditions, non graminaceous plants basically release reductants or chelators and enhance proton excretion in the rhizosphere and finally transport the generated Fe(II) through the plasma membrane. This implies that, at least, three different and concurrent steps (pH-lowering at the rhizosphere, Fe(III) reduction, and Fe(II) transport) are involved in this mechanism. In order to decrease the pH-value, increasing the amount of soluble Fe(III), plants extrude protons into the rhizosphere and it is usually admitted that these proton pumps are driven by plasma membrane ATP-ases. For every one unit drop in pH, iron becomes a thousand fold more soluble. The requirement of a reduction step in cellular iron uptake is clearly established and the existence of plasma membrane ferric reductase activity has been described in both prokaryotic and eukaryotic cells responsible for the plasma membrane Fe(III) chelate reductase activity in plants.

Strategy II

The Strategy II-plants use a chelation based strategy to acquire iron. In response to Fe-deficiency, Fe(III) binding non-protein amino acids termed phytosiderophores are synthesized and released. These plants also induce a high-affinity uptake system for the Fe(III) - phytosiderophore complexes that transport them into the root as intact complexes. One of the best characterized transporter families of this type is the YSL family. YSL proteins are believed to mediate the uptake of metals that are complexed with phytosiderophores or nicotianamine.

Role of iron in plants

Iron is absorbed by roots as Fe^{2+} and Fe^{3+}. Because, Fe can exist in two oxidation states, it accepts or donates an electron ($Fe^{3+} + e^- \rightleftarrows Fe^{2+}$), depending on oxidation potential. Transfer of electrons between organic molecules and Fe provides the electrochemical potential for many enzymatic transformations in plants. Several enzymes are involved in chlorophyll synthesis and when Fe is deficient, chlorophyll production is reduced, which result in the characteristic chlorosis symptoms of Fe stress. Fe is a structural component of porphyrin molecules; cytochromes, hemes, hematic, ferrichrome, and leghemoglobin. These substances are involved in oxidation-reduction reactions in respiration and photosynthesis. As much as 75% of cell iron is associated with chloroplasts, and up to 90% of iron in leaves occur with lipoprotein in chloroplast and mitochondria membranes.

Fe-containing cytochromes in the chloroplasts function in photosynthetic reduction processes where ferredoxin and Fe-S protein, is an electron acceptor. Ferredoxins are the first stable compound of the photosynthetic electron transport chain. Reduction of O_2 to H_2O during respiration is also a common function of Fe compounds. Fe is a constituent of nitrogenase, the enzyme essential for N_2 fixation by rhizobia and other microorganisms. Fe may also be capable of partial substitution for Mo involved in NO_3^- reductase in legumes. The heme proteins contain a characteristic heme iron–porphyrin complex, and this acts as a prosthetic group of the cytochromes. These are electron acceptors–donors in respiratory reactions. Other heme proteins include catalase, peroxidase, and leghemoglobin.

Catalase has a major role in the photorespiration reactions, as well as in the glycolate pathway and is involved in the protection of chloroplasts from free radicals produced during the water-splitting reaction of photosynthesis. The reaction sequence of peroxidase shown above includes cell wall peroxidases, which catalyze the polymerization of phenols to form lignin. Peroxidase activity is noticeably depressed in roots of iron-deficient plants, and inhibited cell wall formation and lignification, and accumulation of phenolic compounds have been reported in iron-deficient roots.

As well as being a constituent of the heme group, iron is required at two other stages in its manufacture. It activates the enzymes aminolevulinic acid synthetase and coproporphorinogen oxidase. The protoporphyrin synthesized as a precursor of heme is also a precursor of chlorophyll. Although iron is not a constituent of chlorophyll, it is required for the conversion of Mg protoporphyrin to protochlorophyllide, means that it is essential for chlorophyll biosynthesis.

Deficiency symptoms

The typical symptoms of iron deficiency in plants are chlorotic leaves. Often the veins remain green; whereas the laminae are yellow, and a fine reticulate pattern develops with the darker green veins contrasting markedly with a lighter green or yellow background. In cereals, this shows up as alternate yellow and green stripes. Iron deficiency causes marked changes in the ultra structure of chloroplasts, with thylakoid grana being absent under extreme deficiency and the chloroplasts being smaller. Plants suffering from iron deficiency show distinct symptoms; the most obvious is interveinal chlorosis. Chlorosis is a general term used to describe leaves that are abnormally yellow. Chlorosis caused by iron deficiency is usually restricted to the leaf tissue between the veins, while the veins themselves remain green.

The yellowing develops because iron is used in the production of chlorophyll. Iron doesn't move within plants, and even healthy plants cannot take iron from older leaves and send it to younger leaves. In iron deficient plants, the levels of iron in younger leaves are too low for normal chlorophyll production, and chlorosis first appears on new growth. As the deficiency progresses, it may also affect older leaves. In severe cases of iron deficiency, leaves will be smaller than normal, will turn pale yellow, and will develop angular brown spots between the veins. Leaf margins may also turn brown.

Sources of Iron

Different sources of iron containing fertilizers available in the market. Among these, Ferrous sulphate is commonly used (Table 1).

Table 1. Iron containing fertilizers

Name of the fertilizers	Iron (per cent)
Iron Frits	14 to 19
Basic Slag	10 to 13
Ferrous Sulphate - $FeSO_4 7H_2O$	31.0
Ferrous ammonium sulphate - $(NH_4)_2SO_4FeSO_4 6H_2O$	14.0
Iron chelates:Fe EDTA	9-12.0

❑❑❑

Chapter 12

Manganese

Manganese biogeochemistry in soils is complex, because it is present in several oxidation states (II, III, IV, VI and VII), while in biological systems it occurs preferably as II, III and IV. Divalent manganese (Mn II) is the most soluble species of Mn in soil; whereas the solubility of Mn III and Mn IV are very low. Pyrolusite is the most stable manganese containing mineral in the oxidative environments and controls the solubility of Mn^{2+} in the oxidised soils. Mn oxides can form co-precipitates with iron (Fe) oxides, exhibiting amphoteric behavior. In addition, Mn interacts both with cations and anions in oxidation-reduction reactions. These reactions are influenced by a varieties of physical, chemical and microbiological processes.

Manganese in Soils

Both pH and redox conditions influence Mn bioavailability in soils. In most acid soils at low pH (<5.5), an increased redox potential of Mn oxides, increasing the concentration of soluble Mn^{2+}, which is the predominant Mn form in the soil solution and the most available Mn form for plants. At higher soil pH (up to pH 8), chemical Mn^{2+} auto-oxidation is favoured over $Mn0_2$, Mn_20_3, Mn_30_4 and even Mn_20_7, which are not normally available to plants. Furthermore, high pH allows Mn adsorption into soil particles, decreasing their availability. A reducing environment can be produced when there is an excess of water, poor drainage or applications of organic material. Different organic molecules can dissolve solid Mn oxides through transfer of electrons, transforming them into an available Mn form for plants.

Environmental conditions also affect Mn soil contents. The highest concentrations of soluble and exchangeable Mn are found after hot, dry summers and under warm waterlogged conditions in acid soils. This is probably due to the inhibition of Mn-oxidizing organisms, thereby allowing the chemical reduction of Mn oxides in these soils.

A mobilization of Mn^{2+} is produced by the rhizosphere acidification due to the release of H^{+}or low molecular weight organic acids from plants (Fig 1). Organic acids released in anion forms from roots can chelate Mn^{2+} released from the MnOx (Mn oxides). Mobilization of micronutrients (including Mn) into the rhizosphere is due mainly by its acidification and complexation with the organic acids (citrate) in various plant species. Soil microorganisms can also help Mn mobilization and immobilization, depending on soil conditions. In aerated soils, microorganisms may mobilize Mn through MnOx reduction favoured by H^{+} root excretion. In contrast, Mn-oxidizing bacteria can decrease Mn availability in aerated and calcareous soils or in poorly aerated and/or submerged soils. Another key factor in the Mn dynamics in soil is organic matter (OM). Given that OM is negatively charged, it has a great Mn adsorption capacity, forming Mn complexes which decrease the amount of exchangeable Mn. However, the Mn adsorbed by OM can be exchanged by the H^{+} released from the roots.

Reduced Mn (Mn^{2+}) form is the only available metal form for plants. It can be taken up via an active transport system in epidermal root cells and transported as divalent cation Mn^{2+} into the plants. Manganese uptake by roots is characterized as a biphasic process. The initial and rapid uptake phase is reversible and non-metabolic, with Mn^{2+} and Ca^{2+} or other cations being freely exchanged in the rhizosphere. In this phase, Mn^{2+} appears to be adsorbed by the negatively charged cell wall constituents of the root-cell apoplastic spaces. The second phase is slow, with Mn^{2+} being less readily exchanged. Its uptake into the symplast is dependent on plant metabolism.

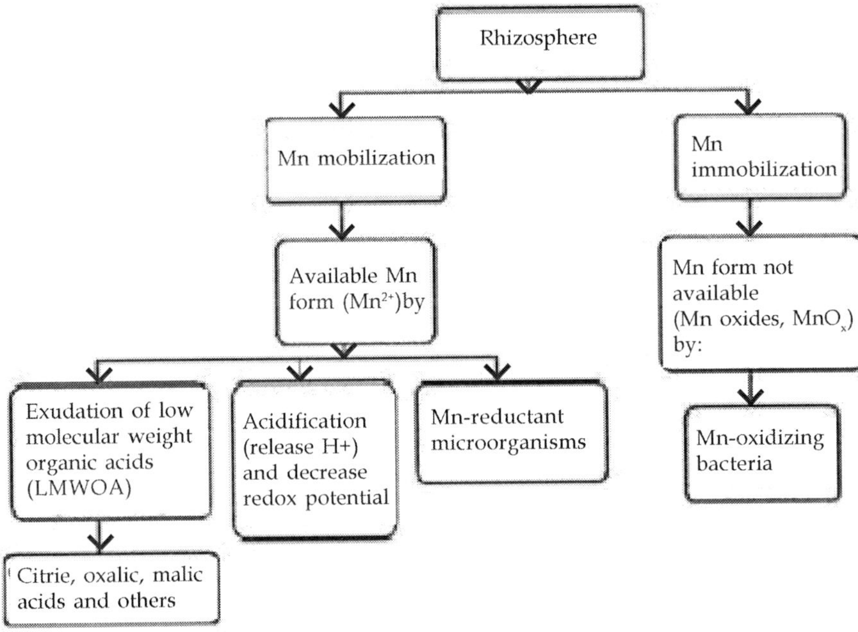

Figure 1. The role of soil rhizosphere in the mobilization and immobilization of Mn in soils.

Factors Affecting Mn Availability

Soil pH

Management practices that influence soil pH affect Mn^{2+} availability and uptake. Liming acid soils decreases solution and exchangeable Mn^{2+} by precipitation as MnO_2. On the other hand, low Mn availability in high-pH and calcareous soils and in overlimed, poorly buffered, coarse-textured soils can be overcome by acidification through the use of acid forming N or S materials. High pH also favours the formation of less available organic complexes of Mn.

Excessive Water and Poor Aeration

Waterlogged soils exhibit reduced O_2 and lower redox potential, which increases soluble Mn^{2+}, especially in acid soils. Mn availability can be increased by poor aeration in compact soils and by local accumulations of CO_2 around roots and other soil microsites. The resulting low-redox conditions will render Mn more available without appreciably affecting the redox potential or pH of the bulk soil.

Soil organic matter

Low Mn^{2+} availability in high organic matter soils is attributed to the formation of unavailable chelated Mn^{2+} compounds. In contrast, addition of natural organic materials will increase solution and exchangeable Mn^{2+} in mineral soils. High rates of green manure crops can cause reducing conditions that increases Mn^{2+}.

Weather Effects

Under dry soil conditions, Mn availability is generally reduced because of lower diffusion rates. In cold, wet conditions, Mn availability can also be reduced through decreased organic matter mineralization and reduced root growth. In saturated or waterlogged soils, soluble Mn^{2+} can increase to excessive levels due to chemical reduction of MnO_2 to soluble Mn^{2+}.

Interaction with other Nutrients

High levels of Cu, Fe, or Zn can reduce Mn^{2+} uptake by plants. Addition of acid forming NH_4^+ fertilizers will enhance Mn uptake by decreasing soil pH.

Plant factors

Several plant species exhibit differences in sensitivity to Mn deficiency caused by differences in plant metabolism (Table 1).

Table 1. Crop Sensitivity to Mn Deficiency

High Sensitivity		Mild Sensitivity		Low sensitivity	
Apples	Peas	Alfalfa	Parsnip	Asparagus	Rice
Beans	Potato	Barley	Potato	Corn	Turfgrass
Citrus	Radish	Broccoli	Rice	Cotton	
Cucumber	Sorghum	Cabbage	Tomato	Field beans	
Fruit trees	Soybean	Carrot	Turnip		
Grapes	Spinach	Cauliflower			
Lettuce	Strawberries	Celery			
Oats	Sugar beet				
Onion	Tomato				
Peaches	Wheat				

Role of Manganese in Plants

Plants absorb Mn^{+2} and low-molecular weight organically complexed Mn. Mn concentration in plants typically ranges from 20 to 500 ppm, while Mn-deficient plants contain < 15-20ppm Mn. Mn^{+4} must be reduced to Mn^{+2} for absorption by root by:

$$MnO_2 + 4H^+ + 2e^- \rightarrow Mn^{+2} + 2H_2O$$

The main Mn role in photosynthesis is its involvement in the water-splitting system of photosystem II (PSII), which provides electrons necessary for photosynthetic electron transport. In water photolysis, a group of four Mn atoms (Mn cluster) is associated with the oxygen evolving complex (OEC) bound to the reaction center protein of PSII. The Mn cluster in PSII accumulates four positive charges, which oxidize two water molecules, releasing one 0_2 molecule and four protons. Therefore, this metal cluster is considered a catalyst compound of water oxidation, where Mn ions are close to a redox-active tyrosine residue.

Manganese also plays a role in ATP synthesis in RuBP carboxylase reactions and the biosynthesis of fatty acids, acyl lipids and proteins . In addition, Mn plays a primary role in the activation and as cofactor of various enzymes in plants (~35), such as: Mn-superoxide dismutase, Mn-catalase, pyruvate carboxylase and phospho-enolpyruvate carboxykinase . Manganese is also essential for the biosynthesis of chlorophyll (through the activation of specific enzymes), aromatic amino acids (tyrosine), secondary products, like lignin and flavonoids. It also participates in the biosynthetic pathway of isoprenoids and assimilation of nitrate.

As a cofactor of superoxide dismutase (SOD), manganese participates in the plant's defense against oxidative stress, produced by elevated levels of activated forms of oxygen and free radicals (reactive oxygen species, ROS), which are harmful to plants.

Manganese SOD (MnSOD) belongs to the group of metal-containing SOD enzymes which are classified according to their metal cofactor: iron SOD (Fe-SOD), localized in the chloroplast; copper-zinc SOD (Cu/Zn-SOD), located in the chloroplast, cytosol, and possibly in the extracellular space. Manganese SOD is found mainly in mitochondria and peroxisomes. MnSOD may play an important role in enhancing salt tolerance.

Deficiency Symptoms

On young, expanding leaves, small areas between minor veins become pale and sunken, and eventually develop into necrotic spots. While all interveinal zones of the leaf are affected, they do not develop at the same rate. Indeed, it is common to see a peppering of brighter, cleared spots across the leaf blade and necrotic pits develop. Initially, the leaves most affected are those which are rapidly expanding, a few nodes back from the tip. But as the symptom intensifies, both older and younger leaves are affected. Pits in older leaves tend to be darker and are often concentrated near the base of the midrib and main veins. The young leaves become pale, thickened and brittle, and may curl under or buckle as they expand. When pitting occurs in very young leaves, their subsequent expansion causes the pits to develop into larger holes. In severe cases, only a skeleton of the leaf remains. Severely-affected leaves become necrotic, the necrosis spreading from the tip or lateral margins. The growing points at the shoot tip and in leaf axils remain active. Axillary shoots are generally less affected than the subtending leaf.

In cultivars where the leaf blade is deeply divided into lobes, the above symptoms may be poorly developed. Instead, small, chlorotic, thickened and deformed young leaves may be produced. The leaf surface may be finely pitted especially in the vicinity of main veins.

Sources of Manganese

Manganese sulphate ($MnSO_4.3H_2O$) and chelated manganese (MnEDTA) are the most common fertilizer sources of manganese. Both are very soluble and are effective as foliar treatments (Table 1).

Table 1. Manganese Fertilizers

Name of the fertilizers	Manganese (per cent)
Manganese dioxide - MnO_2	55-65.0
Manganese sulphate - $MnSO_4 3H_2O$	24.0
Manganese chelate - Mn EDTA	12.0

❑❑❑

Chapter 13

Boron

Boron (B) is a member of the subgroup III of metalloids and has intermediate properties between metals and nonmetals. Despite its low abundance in the nature, B is widely distributed in both lithosphere and hydrosphere. Tourmaline is a stable mineral which control the boron solubility in soils. Boron availability in soil and irrigation water is an important determinant of agricultural production. In soil solution, B exists primarily as boric acid [$B(OH)_3$], which can be easily leached under high rainfall conditions leading to deficiencies in plants that grow there. On the contrary, under low rainfall conditions, B cannot be sufficiently leached and therefore may accumulate to levels that become toxic to plant growth. This is very often in arid and semiarid regions with high-boron groundwater, where the accumulation of B in topsoil due to the evaporation of groundwater reaches toxic levels that reduce crop yields.

Boron in Plants

Boron absorbed by plants is predominantly undissociated boric acid ($H_3BO_3^o$). The anion forms ($H_2BO_3^-$, HBO_3^{-2}, BO_3^{-3}, and $B_2O_7^{-2}$) exist when soil pH > 7, although plants absorb these less readily than $H_2BO_3^o$. Active B uptake (against a concentration gradient) across the plasma membrane requires co-absorption of H^+. As inside cell, pH is greater than outside cell pH, H^+ readily moves across the membrane, sometimes accompanied by $H_2BO_3^o$. This relationship helps explain why B uptake is reduced in alkaline soils where the H^+ gradient is smaller.

Once inside root cells, $H_2BO_3^o$ is readily transported in the xylem to leaves, where much of it occurs in cell walls. B translocation in the phloem from leaves to other plant parts is restricted, thus B accumulates, especially in older leaves. This explains why B toxicity symptoms first appear in older leaf tips. This is consistent with B-deficiency symptoms first appearing in apical meristems and other plant parts receiving water and nutrients through the phloem.

Factors affecting Boron availability

Soil organic matter

Most of the available boron in soils is found in soil organic matter. Organic matter complexes with B to remove it from soil solution when levels are high after B fertilization. Soil organic matter decomposes to resupply the soil solution to maintain adequate B levels when solution B is removed by crop uptake or leaching. Soils with low organic matter content have reduced B-supplying capacity and will usually require more frequent B fertilization at lower application rates. Soil organic matter must be decomposed to release complexed B, so conditions such as cool, wet weather or hot, dry weather which decrease organic matter breakdown will reduce available B in soils.

Soil texture

The texture of soil is an important factor affecting the availability of boron. Fine-textured soils generally require more boron than do the coarse-textured soils to produce similar boron concentrations in plants.

Soil Salinity

An antagonistic relationship existed between soil boron application levels and sodium adsorption ratio (SAR) of irrigation waters. Increasing soil salinity levels decreased the boron concentration in soils.

Soil pH and liming

Boron availability to plants decreases with increasing soil pH, especially above pH 6.5. However, strongly acid soils (pH less than 5.0) also tend to be low in available B because of B sorption to iron and aluminum oxide surfaces of soil minerals.

Soil microbial activity

Microorganisms break down soil organic matter, so plant-available B is released from organic complexes. Conditions favouring improved microbial activity are warm, moist soils with adequate aeration. Soil conditions which hinder optimum microbial activity are drought conditions, cold and wet soils, and poor soil tilth (poor aeration).

Soil fertility

Balanced soil fertility generally results in improved B uptake by plants. It results in improved plant vigour and root growth allows greater uptake of B and other nutrients.

Drought conditions

During periods of drought, the top soil dries out so plant roots are unable to feed in the uppermost soil layer where most of the available B occurs. Dry weather also limits the availability of B, because it restricts water flow, which transports available B in solution. Crops are more likely to become B deficient during drought periods in soils with low levels of available boron.

Tillage

Boron is more available to plant roots when the surface soil is tilled. Tillage allows soil mixing and improves aeration and drainage. These conditions are optimum for organic matter breakdown which releases available boron.

Interactions with other ions

When Ca availability is high, plants can tolerate higher B availability. Under low Ca supply, many crops exhibit lower B tolerance. Greater Ca^{2+} supply in alkaline and recently overlimed soils restricts B availability; thus, high solution Ca^{2+} protects crops from excess B. The Ca:B ratio in leaf tissues has been used to assess B status of crops, where B deficiency for most crops is likely when Ca:B ratio is greater than 1, 200:1.

Soil Moisture

Boron deficiency is often associated with dry weather, where low soil moisture reduces B release from organic matter and B uptake through reduced B transport (diffusion and mass flow) to absorbing root surfaces.

Environmental Factors

One of the chief environmental factors affecting the response of plants to the availability of nutrients is the intensity of light. The faster the plant grows, for example, under high light conditions, the faster it will develop boron deficiency symptoms in a particular growth period.

Boron deficiencies are generally found in dry soils where summer or winter drought is severe. Relative humidity also affects boron accumulation, for example, an increase in percent relative humidity from 30 to 95 resulted in a decrease from 16.5 to 9.9 mg B per plant.

Plant Factors

Crop species exhibit differences in sensitivity to Boron deficiency (Table 1).

Table 1. Crop Sensitivity to B Deficiency

High Sensitivity		Mild Sensitivity		Low sensitivity	
Alfalfa	Cotton	Apple	Peaches	Asparagus	Peppermint
Almond	Olive	Carrot	Pears	Barley	Potato
Apple	Peanut	Clover	Radish	Bean	Rice
Apricot	Plum	Grapes	Spinach	Citrus	Rye
Broccoli	Roses	Lettuce	Tomato	Cucumber	Sorghum
Cabbage	Sunflower		Strawberries	Maize	Soybean
Canola	Turnip			Grasses	Wheat
Carnation	Walnut			Oat	
Cauliflower				Pea	
Celery					
Cherry					
Conifers					

Crop species differ in their capacity to take up B, even when they are grown in the same growth medium. These differences generally reflect different B requirements for growth. In general, dicots (cotton and leguminous plants) have 4-7 times higher B requirement (20-70 mg B kg^{-1}) than monocots 5-10 mg B kg^{-1}. As the most important functions of B in plants are thought to be its structural role in cell wall development and stimulation or inhibition of specific metabolism pathways, thereby, differences in the B demand of graminaceous and dicotyledonous species are probably related to the differences in their cell wall composition.

Role of boron in plants

Boron concentration in monocots and dicots varies between 6 and 18 ppm and 20 and 60 ppm, respectively. B deficiency often occurs with <20 ppm in mature leaf tissue of most crops. B toxicity to plants is uncommon in most arable soils unless it has been added in excessive amounts in fertilizers or urban compost. In arid regions, however, B toxicity may occur naturally or may develop because of a high B content in irrigation waters. B toxicity in most plant occurs at > 200 ppm B.

Boron is involved in the metabolism of carbohydrates. It affects the activity of key enzymes of the oxidative phase of the pentose phosphate pathway (PPP), viz., glucose-6-phosphate dehydrogenase and 6- phosphogluconate dehydrogenase. Low activities of phenolases in boron deficient plants also contribute to enhanced accumulation of phenolics. The phenolic compounds, particularly caffeic acid and quinones, are highly reactive. Their excessive production enhances the generation of superoxide ions (O_2^-) causing peroxidative damage to cellular constituents. Boron involvement in ascorbate metabolism may also expose boron deficient plants to oxidative injury. Boron enhances the activity of membrane localized NADH oxidase (semidehydroascorbate reductase), which catalyzes transfer of electrons to ascorbate free radical. Boron is important for nodule development as well as nitrogen fixation by leguminous plants. Boron plays an important role in plant reproductive development. Often, the effect of boron deficiency is more pronounced on reproductive yield than on biomass production. The reproductive parts of flowers anthers, ovary, stigma possess a relatively higher concentration of boron than in other plant parts and show aberrations when plants are not adequately supplied with boron.

Deficiency symptoms

In addition to cell wall structure, B is essential for normal transport of water, nutrients, and photosynthetic sugars to rapidly developing meristematic (growing) tissues, such as root tips, leaves, buds, and storage tissues. Thus, B deficiency commonly appears as a structural deformity in actively growing regions. For example, in legumes, rosetting of new leaves is a common B-deficiency symptom caused by decreased cell division in apical regions. Boron is also needed for normal development of legume root nodules. B deficiency affects reproductive growth more than vegetative growth. Adequate B increases flower production and retention, and seed and fruit development.

In plants where B is immobile, B translocation from older to actively growing tissues is reduced. The first visual deficiency symptom is cessation of terminal bud growth, followed by death of young leaves. In B-deficient plants, the youngest leaves become pale green, losing more colour at the base than at the tip. The basal tissues break down, and if growth continues, leaves have a twisted appearance. Plants that exhibit B mobility or phloem transport are those species that also produce higher levels of polyols (i.e., sorbitol, mannitol), including apple, pear, plum, peach, olive, and carrot.

B-deficiency symptoms vary widely with crop species but commonly affect both root and top growth. B-deficiency symptoms often appear in the form of thickened, wilted or curled leaves; a thickened, cracked, or water-soaked condition of petioles and stems; and a discolouration, cracking, or rotting of fruits, tubers, or roots. Internal cork of apple is caused by B deficiency. Low B in citrus fruits results in uneven thickness of the peel, lumpy fruit, and gummy deposits in the fruit. The breakdown of internal tissues in root crops gives rise to darkened area referred to as "brown heart" or "black heart". For example, hollow-heart in ground nut occurs when low B limits Ca translocation that inhibits cell wall development and cell division.

Sources of Boron

Boric acid and sodium tetraborate are the most commonly used boron fertilizers (Table 1).

Table 1. Boron Fertilisers

Name of the fertilizers	Boron (per cent)
Borax - $Na_2B_4O_7 10H_2O$	10.5
Boric acid - (H_3BO_3)	17.5
Solubor - $(Na_2B_4O_7.5H_2O+Na_2B_{10}O_{16}.10H_2O)$	19.0

❑❑❑

Chapter 14

Molybdenum

It's oxidation state in soils varies from II to VI; whereas Mo is predicted to be in the IV oxidation state under anoxic conditions. Adsorption on soil particles increases with decreasing pH from 7.75 to 4.45. Therefore, more Mo is available for plants at higher soil pH values. Large amounts of Al and Fe oxides enhance soils adsorption and reduce the amount of water-soluble Mo.

Factors affecting Mo availability

Soil pH

MoO_4^{-2} availability increase approximately tenfold per unit increase in soil pH. Liming to increase soil pH increases Mo availability and prevent Mo deficiency. Alternatively, Mo availability is decreased by application of acid forming fertilizers.

Soil texture

In sandy, low organic matter soils, MoO_4^{-2} retention is low, increasing the probability of Mo deficiency, especially with high Mo requiring crops (i.e., crucifers, legumes).

Fe/Al Oxides

Soils high in Fe/Al oxides tend to be low in available Mo.

Interaction with other Nutrients

Phosphorus enhances Mo absorption by plants, probably due to exchange of adsorbed MoO_4^{-2}. In contrast, high levels of solution SO_4^{-2} depress MoO_4^{-2} uptake by plants. The antagonism of SO_4^{-2} on MoO_4^{-2} uptake is due to their similar transport mechanism across the cell wall. On soils with marginal Mo levels, application of S fertilizers may induce Mo deficiency.

Both Cu and Mn can also reduce Mo uptake; however, magnesium encourages Mo absorption. NO_3^- encourages Mo uptake, while NH_4^+ sources reduce Mo uptake. The effect of NO_3^- uptake may be related to release of OH^- by roots that would increase Mo uptake. The effect of NO_3^- uptake may be related to release of OH^- by roots that would increase Mo solubility.

Copper-molybdenum antagonism is well known and toxicity arising from excess Mo in herbage is effectively prevented by applying Cu to the soil. Soils high in Fe_2O_3 are frequently deficient in Mo.

Soil moisture

Mo deficiency is accentuated under dry conditions due to reduced mass flow or diffusion. Mo availability increases with soil water, where plants grown on poorly drained soils can accumulate excess Mo.

Plant Factors

Crops vary in their sensitivity to molybdenum (Table 1).

Table.1 Crop sensitivity to Mo deficiency

High Sensitivity	Mild Sensitivity	Low sensitivity	
Lucerne	Beet root	Apple	Grapes
Broccoli	Cabbage	Barley	Peach
Cauliflower	Citrus	Beans	Potato
Legumes	Oats	Carrot	Raspberries
Lettuce	Peas	Celery	Rice
Onion	Radish	Corn	Ryegrass
Rapeseed	Soybean	Cotton	Sorghum
Spinach	Tomato	Flax	Sweet corn
	Turnip		Tomato
			Wheat

Molybdenum in Plants

The requirement of plants for molybdenum is lower than any other mineral nutrient except nickel (Ni). Plants differ in their ability to absorb molybdenum from the root medium, and the sufficiency range for molybdenum in plants varies widely. Mo is absorbed as the weak acid molybdate (MoO_4^{-2}) that can form complexes with other anions such as phosphomolybdate. Mo complexation may explain why Mo can be absorbed in relatively large amounts without any apparent toxicity. Mo content of plants is normally low (0.2-2 ppm), because of extremely low MoO_4^{-2} in solution. In some cases, Mo levels in crops may exceed 300 ppm. Mo toxicity symptoms in plants occurs at > 1, 000 ppm Mo.

Mo is an essential component of nitrate reductase, an enzyme concentrated in chloroplasts, which catalyzes the conversion of NO_3^- to NO_2^-. Molybdenum also is a structural component of nitrogenase, the enzyme essential to N_2 fixation by root-nodule bacteria of leguminous crops, by some algae and actinomycetes, by free-living and N_2 fixing organisms. Mo concentrations in legume nodules in 10 times more than the leaves. Mo requirement of plants decreases with increasing inorganic N availability. Mo is also reported to have an essential role in Fe absorption and translocation in plants, which explains why a common Mo-deficiency symptoms is similar to interveinal chlorosis in Fe deficiency. Excessive amounts of Mo are toxic, especially to grazing animals.

Role of molybdenum in plants

The transition element molybdenum is essential for most organisms and occurs in more than 60 enzymes catalyzing diverse oxidation-reduction reactions. Although the element is capable of existing in oxidation states from 0 to VI, only the higher oxidation states of IV, V, and VI are important in biological systems. The functions of molybdenum in plants and other organisms are related to the valence changes that it undergoes as a metallic component of enzymes. Several molybdoenzymes including nitrogenase, nitrate reductase, xanthine dehydrogenase, aldehyde oxidase, and possibly sulphite oxidase are of significance to plants. Because of its involvement in the processes of N_2 fixation, nitrate reduction, and the transport of nitrogen compounds in plants, molybdenum plays a crucial role in nitrogen metabolism of plants.

Nitrogenase

It is now well established that molybdenum is required for biological N_2 fixation, an activity that is facilitated by the molybdenum-containing enzyme nitrogenase. Several types of asymbiotic bacteria, such as *Azotobacter*, *Rhodospirillum*, and *Klebsiella*, are able to fix atmospheric N_2, but of particular importance to agriculture is the symbiotic relationship between *Rhizobium* and leguminous crops. Nitrogenases from different organisms are similar in nature, and they catalyze the reduction of molecular nitrogen (N_2) to ammonia (NH_3) in the following reaction:

$$N_2+8H^++8e^-+16ATP\ ' \rightarrow 2NH_3+H_2+16ADP+16Pi$$

One of the great wonders in nature is how the process of N_2 fixation takes place biologically at normal temperatures and atmospheric pressure, when in the Haber–Bosch process, the same reaction performed chemically requires temperatures of 300 to 500°C and pressures of >300 atm.

All nitrogenases contain the same two proteins, both of which are inactivated irreversibly in the presence of oxygen: an Mo–Fe protein (MW 200,000) and an Fe protein (MW 50,000 to 65,000). The Mo–Fe protein contains two atoms of molybdenum and has oxidation-reduction centers of two distinct types: two iron–molybdenum cofactors called FeMoco and four Fe-S (4Fe-4S) centers. The Fe–Mo cofactor (FeMoco) of nitrogenase constitutes the active site of the molybdenum-containing nitrogenase protein in N_2-fixing organisms.

Nitrate Reductase

Like other molybdenum enzymes in plants, nitrate reductase is a homodimeric protein. Each identical subunit can function independently in nitrate reduction and each consists of three functional domains: the N-terminal domain associated with a molybdenum cofactor (Moco), the central heme domain, and the C-terminal FAD domain. This enzyme occurs in the cytoplasm and catalyzes the reduction of nitrate to nitrite (NO_2^-) in plants:

$$NO_3^- + 2H^+ + 2e_2^- \rightarrow NO_2^- + 2H_2O$$

Nitrate and molybdenum are both required for the induction of nitrate reductase in plants, and the enzyme is either absent, or its activity is reduced, if either nutrient is deficient. In deficient plants, the induction of nitrate reductase activity by nitrate is a slow process; whereas the induction of enzyme activity by molybdenum is much faster. It has been demonstrated that the molybdenum requirement of plants is higher if they are supplied nitrate rather than ammonium (NH_4^+) nutrition an effect that can be almost completely accounted for by the molybdenum in nitrate reductase.

Xanthine Dehydrogenase

In addition to the enzymes nitrogenase and nitrate reductase, molybdenum is also a functional component of xanthine dehydrogenase, which is involved in ureide synthesis and purine catabolism in plants. This enzyme is a homodimeric protein of identical subunits, each of which contains one molecule of FAD, four Fe-S groups, and a molybdenum complex that cycles between its Mo(VI) and Mo(IV) oxidation states. Xanthine dehydrogenase catalyzes the catabolism of purines to uric acid: In some legumes, the transport of symbiotically fixed N_2 from root to shoot occurs in the form of ureides, allantoin, and allantoic acid, which are synthesized from uric acid. Although xanthine dehydrogenase is apparently not essential for plants, it can play a key role in nitrogen metabolism for certain legumes for which ureides are the most prevalent nitrogen compounds formed in root nodules. The poor growth of molybdenum-deficient legumes can be attributed in part to poor upward transport of nitrogen because of disturbed xanthine catabolism.

Deficiency Symptoms

Because molybdenum is highly mobile in the xylem and the phloem, its deficiency symptoms often appear on the entire plant. This appearance is unlike many of the other essential micronutrients where deficiency symptoms are manifest primarily in younger portions of the plant. Molybdenum deficiency is peculiar in that it often manifests itself as nitrogen deficiency, particularly in legumes. These symptoms are related to the function of molybdenum in nitrogen metabolism, such as its role in N_2 fixation and nitrate reduction. However, plants suffering from extreme deficiency often exhibit symptoms that are unique to molybdenum.

Legumes often require more molybdenum than other plants, particularly if they are dependent on N_2 as a source of nitrogen. Molybdenum deficient legumes commonly become chlorotic, have stunted growth, and have a restriction in the weight or quantity of root nodules. In dicotyledonous species, a drastic reduction in leaf size and irregularities in leaf blade formation (whiptail) are the most typical visible symptoms, caused by local necrosis in the tissue and insufficient differentiation of vascular bundles at an early stage of leaf development. Marginal and interveinal leaf necrosis is a symptom of extreme molybdenum deficiency and symptoms are often associated with high nitrate concentrations in the leaf, indicating that nitrate reductase activity is impaired.

The whiptail disorder is observed often in molybdenum deficient cauliflower, one of the most sensitive cruciferous crops to low molybdenum nutrition. In addition, molybdenum-deficient beans often develop scald, where the leaves are pale with interveinal and marginal chlorosis, followed by burning of the leaf margin. In molybdenum-deficient tomatoes, lower leaves appear mottled and eventually cup upward and develop marginal necrosis. Molybdenum deficiency also decreases tasseling and inhibits anthesis and pollen formation in maize. The inhibition of pollen formation with molybdenum deficiency may explain the lack of fruit formation in molybdenum deficient watermelon.

Molybdenum source

Some of the molybdenum containing commercial fertilizers are available (Table 1). Since it's requirement is very low, the quantities recommended vary from 75g to 1 kg/ha depending on the crop and molybdenum fertilizers.

Table 1. Fertilizer sources of molybdenum

Source	Molybdenum (Per cent)
Ammonium molybdate $(NH_4)_6Mo_7O_{24}.4H_2O$	54.0
Sodium molybdate $Na_2MoO_4.2H_2O$	39.0

❑❑❑

Chapter 15

Chloride

The earth's crust contains about 500 parts per million of chlorine, with average soil concentration estimated at 100 ppm. Chlorine is a "universal contaminant." It is present as chloride in ocean water and gets into the atmosphere as ocean spray. The names chlorine and chloride are frequently confused. The element chlorine exists in nature as chloride salts of calcium, magnesium, potassium, and sodium. Chlorine gas (Cl_2), which is used for water purification, does not occur naturally but is manufactured from chloride salts. It is extremely reactive and unstable, and it combines with many of the same elements as oxygen. The non-reactive chloride is the form present in soils and is therefore the form that concerns agriculture.

Chloride in Soil

Chloride concentration is 0.02% to 0.05% in the earth's crust, and occurs primarily in igneous and metamorphic rocks. Soil Cl^- commonly exists as soluble salts such as NaCl, $CaCl_2$ and $MgCl_2$. Chloride is often the principal anion in saline soil solutions. Solution Cl^- ranges from 0.5 ppm in acid soils to >6,000 ppm in saline/sodic soils. The majority of Cl^- in soils originates from salts trapped in parent material, from marine aerosols and from volcanic emissions. Nearly all of soil Cl^- has been in the oceans at least once, being returned to the land by uplift and subsequent leaching of marine sediments or by oceanic salt spray carried in rain or snow. Chloride is highly soluble in soils. Because of Cl^- mobility, it will accumulate where the internal drainage of soils is restricted and in shallow groundwater where Cl^- can be moved by capillarity into the root zone and deposited at or near the soil surface.

Factors affecting chloride availability

Most soil chloride is highly soluble and is found predominantly dissolved in the soil water. Chloride is found in the soil as the chloride ion. Being an anion, it is fully mobile except where held by soil anion exchange sites. In areas, where rainfall is relatively high and internal soil drainage is good, it may be leached from the soil profile. Also, where muriate of potash fertilizer is not regularly applied, chloride deficiencies can occur. Atmospheric chloride deposition tends to be rather high along coastal regions and decreases as towards inland.

Crop plant responses

Depression of Cl^- uptake by high concentrations of NO_3^- and SO_4^{-2} has been observed in a number of plants. The negative interaction between Cl^- and NO_3^- has been attributed to competition for carrier sites at root surfaces. The effect of Cl^- fertilization on root and leaf disease suppression has been observed on a number of crops. Crop plants exhibit differences in sensitivity to chloride deficiency (Table 1).

Table 1. Crop Sensitivity to Chloride Deficiency

High Sensitivity	Mild Sensitivity	Low Sensitivity
Avocado	Cotton	Barley
Bean	Oats	Maize
Citrus	Potato	Spinach
Legumes	Soybean	Sugar beet
Lettuce	Wheat	Tomato
Peach		
Potato		

Chloride in plants

Cl^- is absorbed by plants through roots and leaves. Active transport of Cl^- across plasma membranes occurs through other anion transporters (i.e., NO_3^-, SO_4^{-2}, $H_2PO_4^-$). Cl^- concentration in plants is about 0.2-2.0%, although levels as high as 10% are possible. All of these values usually exceed the strictly nutritional need of most plants. Although usually considered a micronutrient, its concentration in plants can be similar to sulphur. Tissue concentrations of 0.5% to 2.0% of sensitive crops can be toxic. Similar reductions in yield and quality can occur when $Cl^- > 4\%$ in tolerant crops.

In the soil solution, Cl occurs predominantly as the chloride anion (Cl''). The Cl'' anion does not form complexes readily, and shows little affinity (or specificity) in its adsorption to soil components. Thus, Cl^- movement within the soil is largely determined by water flows. Chloride is a major osmotically active solute in the vacuole and is involved in both turgor and osmoregulation. In the cytoplasm, it may regulate the activities of key enzymes.

Role of Chloride in Plants

Chlorine is readily taken up by plants in the electrically charged form as chloride ion (Cl^-). Although chlorine occurs in plants as chlorinated organic compounds, chloride is the major form within plants, where it is bound only loosely to exchange sites or is a highly mobile free anion in the plant water. As an essential element, chlorine has several biochemical and physiological functions within plants. Chloride appears to be required for optimal enzyme activity of asparagine synthethase, amylase, and ATPase. In photosynthesis, chloride is an essential cofactor for the activation of the oxygen-evolving enzyme associated with photosystem II. Chloride may bind to the polypeptides associated with the water-splitting complex of photosystem II, and it may stabilize the oxidized state of manganese by acting as a bridging ligand. Chloride concentrations required for biochemical functions are relatively low in comparison to concentrations required for osmoregulation.

In rapidly expanding tissues such as elongating cells of roots and shoots, chloride accumulates in the tonoplast, to function as an osmotically active solute. This transport of chloride into the tonoplast occurs in association with the proton-pumping ATPase activity at the tonoplast, being specifically stimulated by chloride. This osmoregulatory function in specific tissues requires concentrations of chloride that are not typical of a micronutrient. The accumulation of chloride in plant cells increases tissue hydration and turgor pressure. Chloride along with potassium participates in stomatal opening by moving from epidermal cells to guard cells to act as an osmotic solute that results in water uptake into the guard cells.

Relative differences in the uptake of cations (NH_4^+, Ca^{2+}, Mg^{2+}, K^+, Na^+) and anions (NO_3^-,Cl^-, SO_4^{2-}, $H_2PO_4^-$) by plants require the maintenance of electroneutrality in plant cells as well as in the external soil solution. As an anion, chloride serves to balance charges from cations. In plants well supplied with chloride, this inorganic anion may serve as an alternative to the formation of malate in its charge-balancing role. This role of chloride may be of greater importance when cation uptake exceeds anion uptake, as often occurs with plants provided with ammonium nutrition.

Deficiency Symptoms

Visible deficiency symptoms for chlorine have been well characterized in several crops by growth of plants in chlorine-free nutrient solutions. The most commonly described symptom of chlorine deficiency is wilting of leaves, especially at the margins. As the deficiency becomes more severe, the leaves may exhibit curling, shriveling and necrosis. Roots of chlorine deficient plants have been described as stubby with club tips. Deficiency symptoms of chlorine are not commonly exhibited visually in most crops growing in the field, but symptoms are sometimes observed in wheat and coconut palm. In chlorine-deficient wheat, the symptoms are expressed as chlorotic or necrotic lesions on leaf tissue. These symptoms that result from chlorine deficiency have been named 'Cl-deficient leaf spot syndrome'. In coconut palm, the symptoms are exhibited as wilting and premature senescence of leaves, frond fracture, and stem cracking and bleeding.

Symptoms of Excess

Chloride toxicity symptoms have been observed in many field, vegetable and fruit crops. Curling of the leaf margins, marginal leaf scorch, leaf necrosis, and leaf drop are typical symptoms. Older leaves are usually the first to exhibit symptoms that may progress upward, affecting the entire foliage. Dieback of the terminal axis and small branches may occur in cases of severe toxicity. These symptoms of chloride toxicity occur in the absence of sodium, but they are also similar to symptoms of salt toxicity that occur when chloride is accompanied by sodium.

Sources of chloride

Potassium chloride is the most widely applied chloride fertilizer. Although KCl is usually intended as a potassium fertilizer, it in effect supplies 0.9 kg of chloride for each kg of potassium. Other chloride fertilizers include NaCl, $CaCl_2$, $MgCl_2$, and NH_4Cl (Table 2). All these salts are soluble and readily available to supply chloride for plant uptake.

Table 2. Chloride containing fertilizers

Source	Chlorine (Per cent)
Potassium chloride (KCl)	47.0
Sodium chloride (NaCl)	60.0
Ammonium chloride (NH_4Cl)	66.0
Calcium chloride ($CaCl_2$)	64.0
Magnesium chloride ($MgCl_2$)	74.0

❑❑❑

Chapter 16

Nickel

Nickel comprises approximately 3% of the earth's crust composition and is the twenty-fourth most abundant element. Total Ni concentration commonly ranges from 5 to 500 mg kg^{-1}, with an average of 50 mg kg^{-1} in soils. However, Ni concentrations in dried biosolids (also referred to as treated sewage sludge) or soil near metal refineries range between 24,000 and 53,000 mg kg^{-1}. Soils for crop production contain 3–1,000 mg kg^{-1}. Ni^{2+} is the available form of nickel for plants. Excessive use of Zn and Cu may induce Ni deficiency in soil because these three elements share a common uptake system. Overliming, which raises pH excessively, also causes soil to be deficient in plant available Ni.

It is considered that the system of Ni^{2+} uptake by roots is similar to that of Cu^{2+} and Zn^{2+}, a conclusion obtained after confirmation of competitive inhibition in absorption of these three nutrients. When the available concentration of Ni^{2+} in the soil is low, the process of its absorption by roots is dependent on the expenditure of ATP, a characteristic that indicates the presence of an active transport of high affinity. Once Ni is absorbed by the root, its movement to the aboveground parts of plants is closely linked to the formation of organic complexes. Ni in the phloem may be retranslocated rapidly from the leaves to young tissues, especially during reproductive growth; this movement is associated with the formation of complexes with organic acids and amino acids. Thus, Ni is considered an element mobile in the phloem, whose mobility is higher than that of cobalt.

Factors affecting Ni availability

Like most micronutrients, nickel becomes less available for plant uptake as the pH of the growing medium increases. Higher levels of zinc, copper, iron, cobalt, cadmium or magnesium in the growing medium can induce nickel deficiency. Crops that are most sensitive to nickel deficiency include legumes (beans and alfalfa), pecans, plum, peach, citrus, barley, wheat and certain wetland crops. Nickel deficiency is induced by: (i) excessively high soil zinc, copper, manganese, iron, calcium, or magnesium; (ii) root damage by root-knot nematodes; or (iii) dry or cool soils at the time of bud break.

Ni in Plants

Nickel is taken up from soils as Ni^{2+}. It is readily mobile in plants and in some species is preferentially translocated to developing seeds. The Ni concentration of most plant material normally ranges from about 0.1 to 5 ppm dry weight, but can be highly variable depending on its availability in soils, plant species, plant part, and season. Concentration greater than 10 ppm is considered toxic in sensitive species, and when greater than 50 ppm, it becomes toxic in moderately tolerant species. Some plants can tolerate levels of Ni in tissue as high as 50,000 ppm dry weight. These are called "hyperaccumulators", and are defined as plants that can accumulate at least 1,000 ppm Ni without phytotoxicity.

Uptake of Nickel in Plants

The uptake of Ni in plants is carried out mainly by root systems via passive diffusion and active transport. The ratio of uptake between active and passive transport varies with species. Since Cu^{2+} and Zn^{2+} inhibit Ni^{2+} uptake competitively, these three soluble metal ions seem to be absorbed by the same transport system. In addition, soluble Ni compounds could also be absorbed via the Mg^{2+} ion transport system, because of the similar charge/size ratio of the two metal ions.

Ni is transported from roots to shoots and leaves through the transpiration stream via the xylem. This essential element is supplied to meristematic parts of the plants by retranslocation from old to young leaves, and to buds, fruits and seeds, via the phloem. This transport is tightly regulated by metal ligand complexes and proteins that specifically bind Ni. Metal ligands, such as nicotianamine (NA), histidine (His) and organic acids (citric acid and malate ions), can act as intracellular chelators, which bind Ni in the cytosol or in subcellular compartments for transport, translocation and accumulation within plants.

Over 50% of the Ni absorbed by plants is retained in the roots. This may be due to sequestration in the cation exchange sites of the walls of xylem parenchyma cells and immobilization in the vacuoles of roots. Furthermore, a high percentage (over 80%) of Ni in the roots is present in the vascular cylinder, while less than 20% is present in the cortex.

Role of Nickel in plants

Nickel is chemically related to iron (Fe) and cobalt (Co). Oxidation state of Ni in biological systems is Ni^{2+}, but it could also exist as Ni^{+} and Ni^{3+}. Nickel is a functional constituent of seven enzymes, six of which are present in bacteria and animals, while only one, urease (urea amidohydrolase, EC 3.5.1.5), occurs in plants. Of the seven Ni-dependent enzymes, two have non-redox functions (urease and glyoxylase) and the remaining five are involved in oxidation-reduction reactions (Ni-superoxide dismutase, methyl coenzyme M reductase, carbon monoxide dehydrogenase, acetyl coenzyme A synthase and hydrogenase). The role of urease is to catalyze the hydrolysis of urea $CO(NH_2)_2$ to ammonia (NH_3) and carbon dioxide (CO_2), a reaction that occurs mainly in leaves.

Nickle is also involved in symbiotic nitrogen fixation through its role as an active center of hydrogenase, a process documented in strains of nitrogen-fixing bacteria *Bradyrhizobium japonicum, Bradyrhizobium* sp., *Rhizobiumtropici, Rhizobium leguminosarum, and Azorhizobium caulinodans* . Hydrogenase is an enzyme responsible for oxidizing the hydrogen produced by nitrogenase during symbiotic nitrogen fixation resulting in the production of ATP and, therefore, this enzyme increases the efficiency of symbiotic process, and decreases the inhibitory activity of hydrogen in the bacteroids.

Deficiency Symptoms

In legumes and other dicotyledonous plants, nickel deficiency results in decreased activity of urease and subsequently in urea toxicity, exhibited as leaflet tip necrosis. With nitrogen-fixing plants or with plants grown on nitrate and ammonium, nickel deficiency results in a general suppression in plant growth with development of leaf tip necrosis on typically pale green leaves. These symptoms were attributed to the accumulation of toxic levels of urea in the leaf tissues.

In graminaceous species, deficiency symptoms include chlorosis similar to that of induced iron deficiency, including interveinal chlorosis and patchy necrosis in the youngest leaves. Nickel deficiency also results in a marked enhancement in plant senescence and a reduction in tissue-iron concentrations. In monocotyledons and in dicotyledons, the accumulation of urea in leaf tips is diagnostic of nickel deficiency. In early or incipient stages of nickel toxicity, no clearly visible symptoms develop, though shoot and root growth may be suppressed. Acute nickel toxicity results in symptoms that have variously been likened to iron deficiency (interveinal chlorosis in monocotyledons, mottling in dicotyledons) or zinc deficiency (chlorosis and restricted leaf expansion). Severe toxicity results in complete foliar chlorosis with necrosis advancing in from the leaf margins, followed by plant death.

Nickel Source

The most commonly used nickel fertilizers is Nickel sulphate (Table 1) which is required in very small quantities.

Table 1. Fertilizer sources of nickel

Source	Nickel (per cent)
Nickel sulphate ($NiSO_4 . 6H_2O$)	22.33

□□□

Part II :

NUTRIENT ABSORPTION AND ANALYSIS

Chapter 17

Rooting Pattern in Crops

A. Field Crops

Rice

The rice root system is basically composed of nodal roots. Each node usually develops about 5-25 roots. A leaf and thin roots emerge from the upper region of the internode; whereas a tiller and thick roots develop from the lower region of the internode. Roots developing directly from the culm's nodal regions are called primary roots. The diameter of the primary roots ranges from 0.5 to more than one mm. As growth advances, the primary roots develop branched secondary roots, which in turn develop teritary roots and so on. Under lowland conditions, the root diameter becomes successively smaller as the order of branching increases.

The formation of root hairs is greatly affected by the root environment. Aerobic conditions prevailing in upland soils favour root hair formation. The reducing conditions of flooded soils impair it. Rice root hairs are about 5-10 μ in diameter and about 50-200 μ long. Most rice varieties reach a rooting depth of one m or deeper in height in upland soils. In anaerobic environment, such as flooded soils with impaired water percolation, rice roots seldom exceed a depth of about 40 cm. Total root length increases as the shoot grows. At heading, it is about 2-3 km plant^{-1} when plants are grown singly in a root box. In an upland field, the total root length ranges from 15 to 35 km m^{-2} of soil surface.

A peculiar characteristic of mature rice roots is the presence of large air spaces in them. These air spaces are connected with those in the culms and leaves, providing an efficient air passage system from shoot to root. Rice varieties also vary in their rooting patterns in different soil zones. In all the varieties, more than 60 per cent of the roots are confined to within the surface 0-8 cm layer. The rice plant in general shows shallow rooting habit with more than 79 per cent of the roots residing in the 0-16 cm soil zone.

Wheat

The roots of the wheat plant are of two types; (i) the seminal roots, which belong initially to the embryo and later develop from the hypocotyl or near it and (ii) the adventitious roots, which spring from the nodes of the plant within the soil or just above it. The apex of each root is pointed, smooth and more or less mucilaginous and is protected by root cap. From each lateral root, arise a number of root hairs which are unicellular, thin-layered structures. As the root increases in length, penetrates into fresh regions of soil, new root hairs arise at a short distance from the tip, while those on the older portions of the roots lose their absorptive power and turn brown and die.

The efficiency of wheat roots to absorb water varies with the age and the region of the root. The water absorption through the older regions of the root was slower than through the young and rapidly extending regions. This may be attributed to the impregnation of the epidermis of the older regions with suberin.

Total dry weight of roots was observed to increase exponentially during the vegetative phase and then almost linearly to reach a maximum at anthesis. After anthesis, the total root dry weight decreased but roots continued to grow below 80 cm. From vegetative phase, approximately 65 per cent of the total root dry weight was in the 0-30 cm layer.

Maize

The primary root soon breaks through the coleorhiza. About 40-45 hours after germination, under favourable conditions, one to five and occasionally up to thirteen seminal roots develop and the coleoptile elongates upward. Crown (nodal) roots then develop successively on six to tenth nodes. The first root to emerge from the kernel is designated as primary root. It has also been referred to as the radicle.

The whorl of roots that emerges from the basal meristem of the first internode (first whorl adventitious roots) has been variously referred to as seminal roots. Another whorl of roots that develops from coleoptilar node of the stem (second whorl adventitious roots) has been variously referred to as adventitious roots. Subsequently, six to eight additional whorls may form as the plant develops and may be designated numerically. Usually, second to seventh nodes are very short and remain below the soil surface to form a cone shaped portion of the stem.

The main roots arising from the first three or four nodes are relatively small and of uniform diameter throughout their length. These roots develop an extensive growth of fine lateral rootlets. As the main roots from each higher node appear, there is a progressive increase in diameter, but this thickness is not maintained through their length.

Maize root growth period can be classified into five stages on the basis of the progressive development of the plant. The stages were characterized by i) down-ward diagonal root growth, ii) root filling in the upper soil layers and appearance of the brace roots, iii) pronounced extension of roots vertically into the deeper soil layers and the continued development of brace roots, iv) completion of brace root growth and v) no significant growth of the root system.

Stage 1 covered the period from the time the plants were seedlings until they were knee-high (up to 37 days). The maximum root mass occurred in the 7.5 to 15 cm soil layer. Stage 2 covered most of the vegetative growth and till the tassel was about to emerge (54 days). At the end of Stage 3, the plant attained the maximum height with the full development of the tassel (67 days). Profuse branching of brace roots was observed at this stage. Root growth below 30 cm occurred largely at this stage. Stage 4 covered the period from earhead development until early milk stage (80 days). Extension of brace root growth was observed at this stage which increased the root mass by nearly 50 per cent. Stage 5 covered the period from the 80th day until maturity.

The relative distribution of roots in the soil changed as plant growth progressed. The highest root density occurred at top 15 cm zone at 79 days. After this period. it decreased in all the depth zones.

Pearl Millet

In pearl millet, like other monocots, the radicle develops into the primary root which is soon replaced by other roots. The root system thus consists of a seminal axis (primary root) and a series of nodal axes which originate at the base of leaf nodes on the main stem and tillers. As the seminal axis develops, it produces several slender secondary roots. When the plant has produced about three to four leaves (6 to 14 days after emergence), the first adventitious roots begin to appear.

Rapid initiation of nodal roots is a prerequisite for successful establishment of the crop, especially in dryland situations where evaporation rate is very high and moisture depletion from the upper soil layers is rapid. The first pair of nodals appear at the first node. Each tiller produces whorls of nodal roots. When adventitious roots are initiated, their development reduces the growth of the seminal axis and its branches. The main roots mature to become coarse and woody while the laterals remain thin and fibrous.

Sorghum

With the emergence of two leaves on the plant (3-4 days after seedling emergence), the lower internode reaches its maximum length and nodal roots start elongating from the buds at this node. At each node, the first expansion of the leaf blade takes place, followed by expansion of the leaf sheath and then elongation of the lower internode. Soon the nodal roots differentiate and when the internode reaches its maximum length, nodal roots elongate. Subsequent production of nodal roots from the successive nodes follows the same pattern of development. In the beginning, roots are produced in concentric whorls rapidly (4-5 days). Later, the process slows down with a lag phase between successive whorl initiations.

Types of sorghum roots

i) **Seminal or Primary Roots:** A single seminal root is produced in sorghum, from the growth of the radicle. It supports the seedling for the first few weeks. Generally, it grows vertically downward and has a few slender branches. The seminal root growth in sorghum is similar to that of minor millets.

ii) **Mesocotyl or Subcoleoptile or Subcrown Internode Roots:** These roots are short and slender and are found in the surface soil layer. They are useful only temporarily during the early establishment phase. Their number varies considerably. depending on the conditions of seedling growth and sub-crown internodal length, the latter increasing with the depth of sowing and temperature

iii) **Crown or Nodal Roots:** These are the most important roots and form the bulk of the sorghum roots system. They are initiated in whorls.

iv) **Brace or Aerial Roots:** These are also nodal roots but are initiated in whorls at higher nodes up to (about) eighth node at 0.1-0.3 m above ground and rarely even at a height of 1.5 m. These are stiff, finger-like structures, derived from a ring of auxiliary buds at the base of the internode below the leaf sheath and emerge by piercing the sheath.

Root hairs of sorghum are short (0.2-0.4 mm) and appear close to (1.1 mm behind) the root tip. Root hairs of a seminal root penetrate the soil soon after seminal root enters the soil. At the tip of each root hair, a small droplet of thick mucilagenous substance is found, perhaps to help the root hair to stick to the soil noticed unique hydrophobic.

Pigeonpea

Pigeonpea has a well-developed tap root system, typical of dicotyledons, with considerable lateral branching. When hydration and base temperature requirements are satisfied. the root growth begins on the second day after sowing the seed. When its testa splits open near the micropyle and the radicle elongates and emerges from the seed coat. This eventually forms the tap root. The development of lateral roots occurs as soon as the primary leaves begin to unfold. The rate of elongation of seedling roots under stress-free conditions has been reported to be around 40 mm/day which is somewhat lower than for groundnut. Perhaps due to the perennial nature of the plant, the roots continue to accumulate dry matter and produce laterals throughout the growth of the plant until it is harvested or killed by diseases. The continuous production of roots enables the plant to exploit the receding soil moisture in the post-rainy season. Nevertheless, the rate of partitioning of the dry matter into the roots is somewhat slower than in some of the other annual legumes such as, cowpea and soybean. The total dry matter that pigeonpea plants can accumulate in their roots during a 5 month period was reported to be about 4.7 t ha^{-1}. In a two-year old crop, dry-matter accumulation in roots may go up to 6.2 t ha^{-1}. The root dry matter generally amounts to 15-30% of the shoot dry matter.

Pigeonpea roots may grow to a depth of 1. 9 m within 6 months and to a depth of 3 m in 2 years. A majority of the roots was, however, found in the upper 60-cm soil layer nearly 80% of the roots of a medium-duration pigeon pea and 60% of the roots of a short-duration pigeon pea were distributed in the upper 60-cm soil layer. The root system became extensive with time and maturity

Groundnut

The root system of groundnut is characterized by a deep tap root that grows up to one m and lateral roots that originate at different depths on the tap root but concentrate mostly on the top 30 cm length. The lateral roots may spread up to 40 to 60 cm on either side. The tap root and the primary roots provide strong anchorage to the plant; whereas the numerous secondary and tertiary roots and root hairs help the plant in water and nutrient absorption. In contrast to the aerial parts, groundnut roots are hard and lignified to a considerable degree.

At seedling stage, the root system is essentially a tap root, giving out primary and lateral roots. The tap root grows quickly reaching up to 60 cm in about three to four weeks after emergence and about a meter in five weeks. Much of the development of lateral root system occurs during 4 to 8 weeks time, though root growth may continue longer.

Sesamum

Sesamum has a pronounced branched tap root system with enormous number of secondary and tertiary roots close to the surface. The vertical growth of roots (root length) took place continuously from seedling phase to pod-forming phase. A similar trend was also observed for the growth of lateral roots. Root production was mostly observed during the stages up to flowering. Thereafter, the plant did not put forth many fresh roots. However, the increase in root dry matter even beyond this stage indicated thickening of the roots. The root-shoot ratio increased from seedling phase to vegetative phase (0.21-0.44) and decreased thereafter. These results indicated that the root production attained its peak just prior to flowering.

In seedlings, 95% of the roots were observed within 2.5 cm soil depth and up to 2.5 cm radial distance. During the vegetative phase, 94% of the roots were found within 5 cm soil depth and 2.5 cm lateral distance. In the flowering phase, 70% of the roots were located within 5 cm soil depth and 2.5 cm lateral distance. During pod formation phase, 55% of the roots were observed in this area. During the maturation stage, only 40% of the roots were observed inside 5 cm soil depth and 2.5 cm laterally.

Sunflower

The root system of sunflower is substantial but often shallow. The roots penetrate to a depth of 27-29 cm and the maximum growth of root has been noticed at 45 days after sowing i.e at the head formation and flower initiation stage. Maximum root weight has also been observed at this stage.

Cotton

Under favourable conditions, the cotton seed germinates and the sprout pushes through the pointed end of the seed within two or three days. The primary root develops from the radicle of the sprout and grows downward into the soil. The cotton plant has a tap root system which is also deep, its depth depending on the age and size of the plant and the moisture content and the physical nature of the soil. Cotton seedlings establish a deep root system very early. It was reported that plants of 20-25 cm height may have tap roots extending down to a depth of 120-150 cm from the soil surface. Because of the deep root system, cotton plants seldom need irrigation. Under conditions of moisture stress, the roots of cotton plant penetrate further deep into the soil and support the plant effectively.

The tap root has a proportionately thick cortex and the wood is found as a quadrangled structure with four well-defined ridges. The branches that arise from the tap root are irregularly grouped into four rows, only three of which are recognized because of the twisting and irregularities. Two rows of the branches stand at right angles to the other two. The tap root tapers rapidly, just 30 cm below the soil surface. The tap root is not larger than the secondary branches and at 60 cm beneath the surface. It has a diameter of about 1.5 mm. This slender root grows downward to a distance of more than a metre. In some upland cotton varieties, several laterals emerge, beginning at a depth of 10 cm below the soil surface within a space of about 25 cm.

Sugarcane

The plant is propagated vegetatively through stem cuttings, popularly termed as setts. Each sett contains usually two or three buds. The primary' stalk, which comes from a single bud, produces secondary and tertiary stalks each of which gets differentiated into nodes and internodes. While the internode is responsible for sugar storage, the node retains the root band with root primordia, a bud, a wax and a growth ring just above it. These root initials help in the rooting process, following planting of a sett in the soil.

A germinating bud puts forth sett roots rapidly from the well-developed root primordia occurring on the sett. These newly formed roots are essential during the early stages of cane development but, gradually, they die off with the appearance of plant roots over a period of 3 months. The sett roots start growing within 24 hours of planting and by the third day, the, rate of elongation reaches about 10 mm d^{-1}. These roots further grow at a faster rate of about 24 mm d^{-1} and develop into a network of secondary roots which are relatively thin and highly branched. They attain a maximum length of about 15 cm by the 60th day, cease growing thereafter and gradually disappear by 150 days. During the initial stages of growth, contribution of sett roots to water absorption would be significant.

In sugarcane, the bulk of the roots resided in the upper 20cm of the soil. At grand growth, more than 60% of the roots were present in the surface soil and the root mass decreased considerably with soil depth. Thus, more than 85% of the total root dry mass could be recovered from the 0-40 cm soil depth.

B. Horticultural Crops

Arecanut

Arecanut has adventitious root system, which is typical of monocots. Root has an epidermis, consisting of individual cells of almost the same size as the endo-dermal cells. The epidermal cells are devoid of root hairs. The cells of the endodermis, adjoining the epidermis. appear to become sclerenchymatous in nature, forming a hard surface subsequent to the death of the epidermal cells. The roots of arecanut in relation to their origin, size, shape. function, structure, etc and classified them into i) the adventitious roots emanating from the bole or the root-producing region, ii) the rootlets that branch from the main roots or their branches, iii) the pneumatophores or breathing roots and iv) the aerial roots which arise from the aerial parts of the stem. The main roots is developing with the production of the first root from the prostem. The thickness of these main roots in a tree varies considerably from 5 to 12 mm. The main roots penetrate very deep into the soil and live for long periods much longer than many of the rootlets.

The roots of areca palms radiate from all sides of the bole and generally grow in the direction in which they start. Most of the roots reside very dose to the palms within 30-60 cm of radius that a four year-old areca palm had 96 per cent of its roots spread. About 75 per cent of the roots were confined within a radius of 100 cm. The age factor in root distribution thus makes it necessary to open basins, at 50 and 100 cm radius for young and adult palms, respectively, for application of manure and fertilizers.

Oil palm

When the oil palm seed germinates, the first to emerge through the germ pore is the radicle, followed by the first set of adventitious roots from the circumference of the radicle-hypocotyl junction. The radicle continues to grow but is replaced, within six months by numerous primary roots which arise from the base of the swollen trunk. The primary roots grow vertically downwards at an angle of 45° and useful for both anchorage and absorption of nutrients for the seedlings. Consequent upon the development of the palm, the primary roots also spread horizontally at varying depths below the soil surface. Secondary roots are produced from the pericycle at right angles to the primary roots. Finer branches, originating at right angles to the secondary ones are called the tertiary roots. The finest roots are produced from the tertiaries.

A majority of the primary roots appears from the base of the stem close to the soil surface. They bend downwards from horizontal positions and are found below the soil surface down to one m. From the primary roots, the secondary roots of 5-10 mm diameter ascend and descend in equal numbers. These secondaries are 1-4 mm in diameter and give rise to horizontally growing tertiaries of 0.5-1.5 mm diameter and up to 15 cm in length. From these, develop the quaternaries which are up to 3 cm long and of only 0.2-0.5 mm diameter. The root density usually decreases with distance from the palm. In adult palms, the total quantity of absorbing roots in surrounding circles increases at least to a radius of 3.50-4.50 m. The greatest quantity of roots is in the top 15-30 cm of soil. Most of the absorption of nutrients has been found to be through quaternaries and absorbing tips of primaries, secondaries and tertiaries.

Cassava

Commercial propagation of cassava is through stem cuttings called setts. During the regeneration of the plants from the stakes, the first roots emerge from the nodes below the soil surface is about 5 days after planting. The buds also begin to enlarge simultaneously. By about 10 days, a callus is also formed at the cut edge of the stake below the soil surface. By this time, the first leaves also appear. The roots formed at the nodes may be two to three in number and are anatomically similar to the 20-30 roots formed at the base. The nodal roots continue to increase for about two weeks, while the production of the basal roots continues for at least two more weeks.

At three month stage, the maximum root density was observed in the 0-30 cm layer and within a lateral distance of 10 cm from the base of the plant. By six months, the root density significantly increased up to 60 cm depth and 60 cm lateral distance.

Cashew

Seedlings as well as vegetatively propagated planting materials generally exhibit difficulties in field establishment. Seedlings with good tap root and a number of laterals normally withstand transplantation shock and drought; whereas young seedlings, with poorly developed root system are prone to water stress, especially, during summer months. Poor establishment of vegetatively propagated planting materials in the field is also believed to be due to inadequate root development.

Over 67% of thick roots (tap root and secondary and tertiary branches) and 26% of the fine roots (fibrous roots developed from tap root and secondary and tertiary branches) were found within a radius of 50 cm from the base of the tree. In the 51-100 cm radial distance from the tree, about 16% of the thicker roots and 20% of the fine roots were found. The region beyond 2 m distance from the tree did not support many thick roots. In contrast, the fine roots were mostly concentrated within 25 to 200 cm area around the tree. Even beyond 2 m radial distance, the preponderance of fine roots was apparently more than that of the thick roots. The roots were found to penetrate vertically down to a depth of 9.5 m. Nevertheless, within the surface 50 cm soil layer, over 85% of the thick roots and 45 % of the fine roots could be extracted. Root penetration below 50 cm but within one m depth was accounted for by about 5 and 8% of the thick and the fine roots, respectively. Beyond 2 m depth, the root penetration decreased considerably. These results indicated that the thick roots were mainly confined to within 0-50 cm depth, while the fine roots extended down to a depth of 9.5 m. On volume basis, 512.73 g of the thick roots and 43.73 g of the fine roots were present per m^3 soil.

Coconut

The coconut palm has numerous fibrous adventitious roots produced from the bole of the tree. The total number of roots in a 25 year old palm could vary from 2000-4000. These roots are projected in all directions in the soil for anchorage and so absorb water and nutrients. The direction of root growth is often influenced by soil moisture availability, soil physical properties and gravity.

The first formed roots tend to grow vertically downwards into the soil. Others grow obliquely or horizontally. Their greatest concentration is in the top one meter of the soil and within a radius of about 2 meters from the base of the palm. The average length of a root ranges from 5m to 6m but some may extend laterally, especially in moist sandy soils, up to even 10m to 25m. Roots do not grow into dry soil or into stagnant water.

Generously, roots of the coconut palm are uniform in size and shape. Each root has at the extremity a growing point covered by a thimble shaped root-cap. Just behind this, a whitish thin walled region 2 mm to 5 mm in length. The thickness of the outer or epidermis of these roots varies depending on the age of the roots. The colour indicates the functional capacity of roots. Dark brown to reddish roots have lost the absorption capacity and they provide anchorage to the palm.

Coconut roots do not have root hairs. Absorption of water and nutrients take place in the cream coloured region just behind the root cap. It would therefore, be noted that the total absorption area of the coconut root system is very small compared to the total volume of the roots. Coconut palms grown in sandy soil have well distributed and deep root system; whereas palms grown in lateritic/gravelly soil with a hard pan have surface spreading root system with less vertical roots and confined to a limited space in the soil.

It is reported that the roots of a coconut palm up to a depth of about 25 cm and to a distance of 200 cm away from the bole are active in absorbing nutrients. Fertilizer application therefore, has to be concentrated within this area called the (manure circle), for efficient absorption. The ability of palms to withstand drought depends on the development of a deep, profusely branched root system that is capable of absorbing water and nutrients from the large volume of soil. Continuous renewal of roots in very important as roots cease to absorb water when they form a thick, impervious wall in the absorbing area. Therefore, roots depend on shoots for carbohydrates, growth regulators and amino acids while the shoots depend on roots for water, minerals and growth regulators.

Tea

Tea seed germinates epigeally by cracking seed coat and producing one white, thread-like rootlet from the radicle. The root descends vertically downwards (as a tap root) and reaches about 30 cm soil depth within a year. Two-three branch roots develop from the tap root and assume the size of the main tap root within five years. Consequently, the tap root loses its identity and cannot be distinguished from the branch roots. Thus, a mature plant possesses a 'multi main root' system, though the number of the main roots is much less than in clonal plants. Stem cuttings, right from the beginning, initiate more than five main roots which grow simultaneously from the callus.

Tea roots can be classified into three groups on the basis of their size and function and these are (a) main root, (b) subsidiary (lateral) roots and (c) feeder (feeding) roots.

a) **Main Root:** The main root system is about 7-8 cm in diameter and 90-150 cm in length. However, in exceptional cases vertical roots may descend to a depth of 6-9 m.

b) **Subsidiary Roots:** Subsidiary roots originate endogenously in the root pericycle. a uniseriate cylindrical layer of cells proximal to the root apical meristem and forming the outermost layer of the primary vascular cylinder. These roots help the main roots in providing anchorage. They branch and sub-branch, forming a reticular network mostly in the top 30 cm soil profile.

c) **Feeder Roots:** The white and creamy parts of fibrous roots are termed as 'feeding' or 'feeder' roots. They are present at the end of the branching system of subsidiary roots and concentrate (over

70%) at the top 30 cm soil profile. The distribution of feeder roots gradually decreases with depth only 5% of such roots were present below 60 cm soil profile. The feeder roots increase in thickness with age and on attaining 1 to 2 cm diameter, they act as principal storage organ for surplus carbohydrate.

Banana

Banana roots develop best in soils that are deep, porous, and have adequate aeration and good natural fertility. Any physical and chemical barrier limiting root growth significantly reduces yields.

Root colour, texture and size vary with age. Young roots are normally white and soft. As roots age, they become progressively harder and darker in colour. Those roots closer to the central axis of the plant are the first to become thicker, darker and harder. With increasing radial distance from the central axis, roots become softer, thinner and lighter in colour. The root system develops vigorously during the period between seeding and the beginning of floral differentiation, but subsequently diminishes once floral differentiation is initiated. Existing roots continue to develop after floral differentiation.

It is more appropriate to consider that the depth of the root system is controlled by soil conditions. Banana roots can reach depths of five to 10 meters when unrestricted by physical and chemical barriers. Root branching is also impaired by unfavourable physical and chemical soil conditions. Therefore, the banana root system can be extensive both laterally and vertically. Generally, 60 to 70 percent of the roots of banana are located in the top 30 cm of the soil profile, but studies have confirmed that normal growth requires at least 120 cm of non-restrictive soil. The feeding area around the corm varies with age. The zone of P uptake for two-month-old plants is located within a 150 cm radius around the plant; whereas the feeding area for five months old plants is located within a 230 cm radius.

Mango

In 8-year-old mango 'Alphonsa' tree budded on Bangalora rootstocks, the root activity is uniformly distributed up to 60 cm depth, indicating its deep-rooted nature. Bulk of the active roots are confined to 100 cm radial distance during late rainy and winter seasons which migrate towards drip circle during summer season.

Citrus

In 7-8-year-old sweet orange 'Mosambi' budded on Rangpur lime, majority of the active absorbing roots are located near the soil surface (10-15 cm depth). The drip circle (spread of the crown) extends to 2.0-2.5 m but the bulk of the root activity is confined to 1.0-1.2 m, indicating the need for placing the fertilizers near the soil surface midway between the trunk and the drip circle.

Grape

In grape 'Anab-e-shahi' and 'Thompson Seedless' (8-9 years old), the roots are located deeper in soil between 45 and 60 cm depths. Laterally, the bulk of active roots are located at 100 - 120 cm distance with slight change in the pattern of distribution during the non-cropping and cropping phases. In the former phase, the active roots clustered closer to the trunk while in the latter phase, they are located away from the trunk.

Guava

In 6-year-old guava, the root activity increases with increasing depth up to 45 cm. Laterally, majority of the active roots spread up to 100 cm during late rainy season while they move up to 200 cm distance during winter.

Vegetables

Potato

Roots originate at the base of the shoot and are mostly confined to the top 20-25 cm of the soil. Root system depends upon the variety and nature of the soil. Because of the superficial root system, the plant thrives best in soils where adequate nourishment is available in the top layers. Under good fertility conditions in the soil, the lateral roots usually spread out to nearly one meter but rooting is very profuse around the plant.

Tomato

Tomato has a strong tap root. It is often damaged at transplanting and instead a dense system of fibrous and adventitious roots is formed. Tomatoes have roots penetrating to a depth of 1 m with a lateral spread of 1.3 m.

Chillies

Chillies have a strong tap root which also usually gets broken or arrested in growth on transplanting. As a consequence, numerous, profusely branched laterals develop, extending to about one meter. The rooting pattern indicated that chillies had a limited root development. Only 80% of the actively absorbing roots were found in the uppermost 20 cm of soil, though measurable root activity was also recorded at 103 cm depth.

Brinjal

Brinjal has a strong, deeply penetrating tap root. Injury to the tap root during transplanting changes the type of root system to a rather fibrous one.

Cucurbits

Cucurbits develop a strong tap root system which penetrates the soil at a rate of 2.5 cm d^{-1} up to a depth of 90 to 120 cm in cucumbers and 180 cm in gourds. The root system of all the economic cucurbits was extensive but shallow.

❑❑❑

Chapter 18

Nutrient Absorption and Uptake

A. Nutrient absorption

Nutrient acquisition is the mechanism by which plants capture all those elements that are essential for their growth. Nutrient acquisition depends on a number of factors, including plant species, environmental conditions, nutrient supply and interrelationship among nutrients and between plant and soil, presence of microorganisms (e.g., fungi) in association with plant roots, etc.

Nutrient Bioavailability

The soil-plant system's capacity to supply/absorb nutrients is termed *soil nutrient bioavailability*, and is the ability of the soil-plant system to supply essential plant nutrients to a target plant, or plant association, during a specific period of time as a result of the processes controlling i) the release of nutrients from their solid phase in the soil to their solution phase; ii) the movement of nutrients through the soil solution to the plant root-mycorrhizae; and iii) the absorption of nutrients by the plant root-mycorrhizal system.

Release of nutrients from the solid phase to the soil solution is controlled by the physiochemical processes of desorption and dissolution. It is also a biochemical process by way of mineralization. The movement of nutrients through the soil solution to the root-mycorrhizal surface is determined by mass flow and diffusion, which in turn are controlled by the interaction of soil and plant characteristics.

Root Zone and Nutrient Acquisition

Mineral salts are absorbed from the soil solution in the form of ions. They are chiefly absorbed through the meristematic regions of the roots near the tips. Different zones in the roots are viz., Meristematic zone, Elongation zone and Maturation zone. (Figure 1).

Nutrient absorption zone

- Calcium - apical region
- Iron - apical region
- Potassium, nitrate, ammonium, and phosphate - all locations of root surface.
- In several species, root hairs are the most active phosphate absorbers.

Root tips are the primary site for the nutrient uptake because of following reasons:

- There are the tissues with greatest need for nutrients.
- The cell elongation requires potassium, nitrate, and chlorine to increase osmotic pressure within the wall.
- Ammonium is a good nitrogen source for cell division in meristem.
- Root apex grows into fresh soil and finds fresh supplies of nutrients.
- Nutrients are carried via bulk flow with water, and water enters near tips.
- It maintains concentration gradients for mineral nutrient transport and uptake.

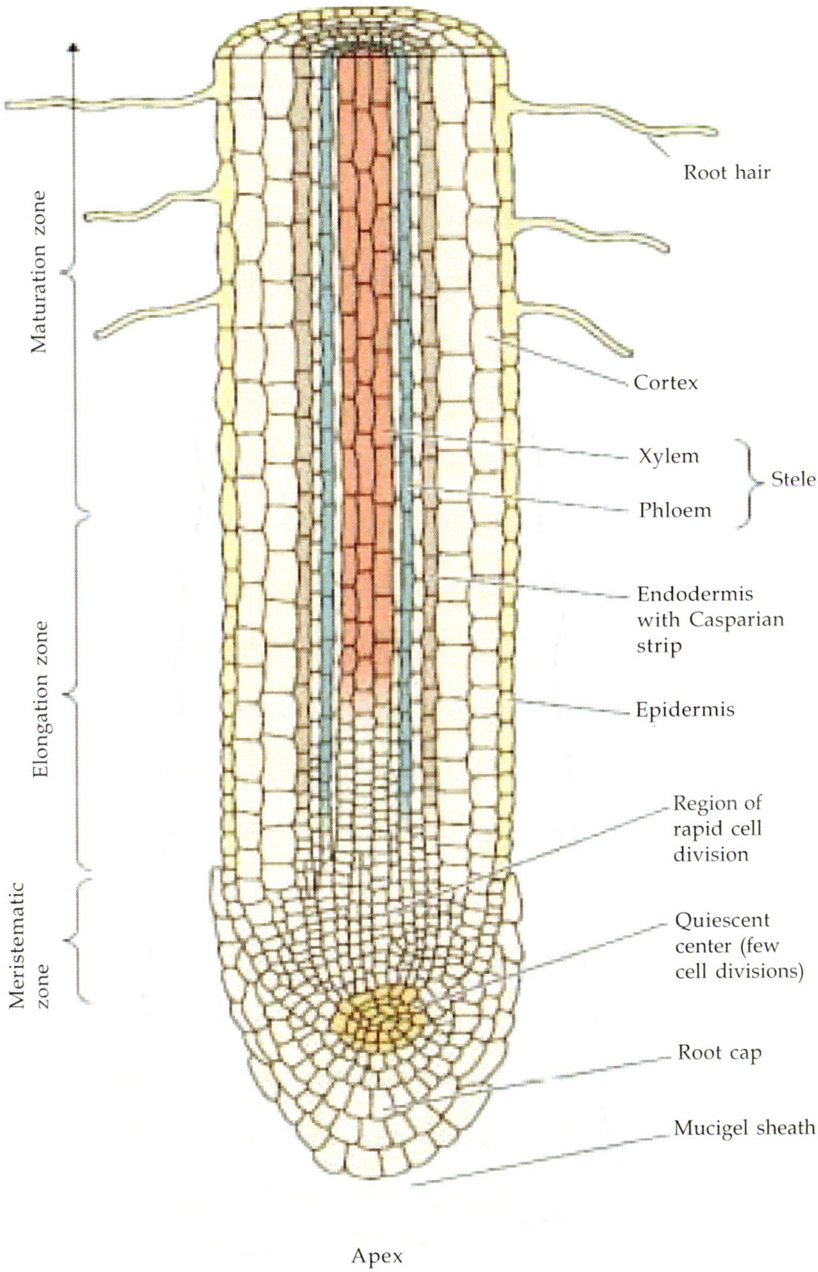

Figure 1. Different zones in plant root

i) **Nutrient Absorption:** Nutrient absorptions by plant depend on ion concentrations at the root surfaces, root absorption capacity and plant demand. It is a dynamic series of processes in which nutrients must be continuously replenished in the soil solution from the soil solid phase and transported to roots as uptake proceeds. After uptake by roots, nutrients are translocated to the various plant organs for utilization in different metabolic processes. In this way, nutrient uptake by plants involves several interconnected processes such as nutrient release from the soil solid phase to solution, transport to roots for absorption, and plant translocation and utilization. The consequence is that nutrient uptake by plants is simultaneously influenced by soil, climatic and plant factors. Nutrient transport to roots, absorption by roots, and translocation in the shoot, all occur simultaneously. This means that a rate change of one process will ultimately influence all other processes involved in uptake. In other words, if one process slows down, it may become the limiting factors in the uptake process. Only inorganic nutrients can be absorbed by plants.

ii) **Release of Nutrients from the Soil Solid Phase:** The key solid phase properties are i) the quantity of labile nutrients, and ii) the manner of nutrient release to the soil solution. The term labile, in this context, is defined as nutrients in a plant-accessible form during a given time period. Maintenance of the soil solution's nutrient concentration is a central concept of soil nutrient supply. The solid phase maintains the soil solution nutrient concentration. Release from the solid phase to the soil solution occurs biochemically (mineralization and immobilization) or physiochemically (adsorption and desorption, and precipitation and dissolution).

Mineralization and immobilization are transformations of nutrients between their organic to inorganic forms. These transformations are functions of the soil temperature regime, soil aeration regime, soil water regime, and the quality of the organic matter from which the nutrient is mineralizing because these factors control the population and activity of soil organisms.

Reaction of nitrogen with the mineral-organic surface complex of soil is chemically distinct from that of P and S. When nitrate is the primary N species produced by mineralization. Mineralization controls the soil solution's nitrate concentration. Since, there is little to no sorption of nitrate onto the solid phase, this is the only option. However, ammonium, phosphate and sulphate can absorb to the soil mineral surface subsequent to mineralization. The greater the sorptive capacity of the soil, the greater should be the potential for the mineralized ion to be sorbed on to the mineral surface. Therefore, the role of mineralization should be to i) control the soil solution concentration directly when the nutrient does not absorb on to the mineral soil, or ii) contribute to the soil solution concentration and to the nutrient concentration on the soil's mineral surface when the soil's sorption capacity is greater than zero. In the latter case, nutrient distribution between the solution and the mineral surface should be related to nutrient affinity for the mineral surface.

iii) **Nutrients movement to the root surface:** Transport of the nutrients to the vicinity of roots is the first step in the process of nutrient absorption by plants. Nutrient supply to the roots is governed by nutrient concentrations in the soil solution, nature of the nutrients, soil moisture status, and plant's absorption capacity (Table 1). The concentration of the nutrient in the solution immediately adjacent to a root appears to be the best measure of its availability for absorption, through many factors within the plant and the concentrations of other ions in the solution phase may influence the actual rate at which it is absorbed.

Table 1. Process and factors involved in nutrient transfer from soil to plant

Process	Factors
Root development	Root length, Root distribution Root morphology, diameter, and hairs
Nutrient uptake	Concentration at root surface Kinetics of uptake
Transport from soil to root mass flow diffusion	Transpiration Concentration of soil solution Concentration of gradient
Mobilization by roots	Depletion of soil solution Root exudates
Desorption, dissolution, hydrolysis or organic compounds	(H^+, HCO_3^-, reducing agents, chelates, organic anions) Chemical soil composition pH of soil solution Enzymes (e.g., phosphatases)
Mobilization by associated organisms	Mycorrhizal infection Bacteria

(Jung., 1991)

It is clearly evident that nutrient diffusion and mass flow, which account for some 70-98% of the nutrients other than copper and iron that are absorbed by roots, are more rapid in sandy soils than in the heavier loamy and clay soils, but that the nutrient reserves of the former soil are more rapidly exhausted (Table 2). This explains why nutrient deficiency symptoms appear much earlier in rapidly growing crops on sandy soils than on heavier ones, where nutrient replenishment is slower but more continuous. On light soils, however, only small amounts of plant nutrients are needed to replenish the nutrient reserves to a level permitting optimal growth. Owing to the restricted buffer effect of such soils, large amounts may even lead to excess and consequently to symptoms of toxicity. In contrast, large amounts of nutrients must be added to heavy soils to achieve the nutrient concentrations in the soil solution needed for optimal growth. Heavy soils also have a higher buffer capacity when nutrients are added in large amounts, and plant damage due to high nutrient concentrations need therefore rarely be feared. During unfavourable conditions, such as dry periods, nutrient intake is also more strongly inhibited in heavy than in light soils, however, and this may lead to deficiencies of boron, manganese, iron or magnesium and potassium.

Table 2. Percentages of nutrient uptake through roots accounted for by interceptive root growth, mass flow and diffusion

Plant nutrients	Approximately percentage nutrient uptake through		
	Interceptive root growth	**Mass flow**	**Diffusion**
N	2	98	-
P	3	6	91
K	2	20	78
Ca	28	72	-
Mg	13	87	-
S	5	95	-
B	3	65	32
Cu	70	20	10
Fe	50	10	40
Mn	15	5	80
Mo	5	95	-
Zn	30	30	40

(Bergmann, 1992)

a) *Mass flow:* Mass flow is the passive transport of nutrients to the root as soil water is absorbed by plants. The amounts of nutrients reaching roots by this process depend on the concentrations of nutrients in the soil solution and the rate of water transport into the roots. The amounts of nutrients supplied by mass flow are affected by soil properties, climatic conditions, solubility of the nutrient, and plant species. The level of a particular nutrient in the soil solution near the root may increase, stay the same, or decrease depending on the balance between the rate of supply to the root by mass flow and the rate of absorption into the root.

The contribution of mass flow in the process of nutrient supply to the roots can be calculated from the product of the soil solution concentration and the volume of water transpired by the plant. This, however, does not take into account mass flow of the soil solution, which is not due to the water uptake by the root, such as gravity-induced movement of water down the soil profile. Mass flow rate can be calculated with the help of the following equation:

$$MF = C \times WU$$

Where

MF is the contribution to ion uptake by mass flow

C is the solution concentration of any given ion

WU is the total water uptake, which is the water content in the plant plus the water transpired

Transpiration is not a constant process. It varies with plant species, climate, soil conditions, location of water source in soil, age of plants, and time. At night, mass flow is greatly restricted, which causes diurnal fluctuations at the root surface. Furthermore, nutrient concentrations in plants vary with development stages. Young plants usually have higher concentrations of nutrients. Therefore, mass flow may then contribute a smaller fraction of the demand.

b) *Diffusion:* Diffusion can be defined as the movement of molecules from a region of high concentration to a region of low concentration. When the supply of the nutrients to the root vicinity is not sufficient to satisfy the plant demand by mass flow and root interception, a concentration gradient develops and nutrients move by diffusion. Diffusion is described by Fick's first law (Barber, 1995).

$$F = -D\left(\frac{dc}{dx}\right)$$

Where

F is the flux

dc/dx - the concentration gradient

D is the diffusion coefficient that generally describes diffusivity in a homogenous medium

Since the soil is not homogeneous, the concept of a diffusive flux in the soil medium presents difficulties. Bye (1979) proposed the following formula to calculate the diffusion coefficient for soil medium:

In the process of diffusion, soil and plant factors are involved. The following equation illustrates the factors that are important in determining the rate at which a soluble nutrient diffuses to the root surface (Corey, 1973):

$$\frac{dq}{dt} = DAP\frac{C_1 - C_2}{L}$$

Where,

dq/dt represents the rate of diffusion to the root surface

D is the diffusion coefficient of the nutrient in water

A is the cross-sectional area considered, which can be assumed to represent the total absorbing surface of a plant root

P is the fraction of the soil volume occupied by water

$C\text{-}_1$ is the concentration of the soluble nutrient at a distance L from the root surface

$C\text{-}_2$ is the concentration of the soluble nutrient at the root surface

L is the distance from the root surface to where C_1 is measured

The distance for diffusive nutrient movement through the soil to the root is usually in the range of 0.1-15 mm. Hence, only soil nutrients within this soil zone contribute to diffusive nutrient supply to roots. Phosphorus has the slowest diffusion rate. Plant requirements for P and K are mostly met by the process of diffusion.

c) *Root Interception:* Nutrient interception by roots depends on the soil volume occupied by roots, root morphology and the concentration of nutrients in the root occupied soil volume. The amount of roots present per unit volume of soil can be measured in terms of root surface area, root length, or root volume. The root surface available for ion absorption is the function of surface area. Root density varies with soil properties, species, and management practices.

iv. **Ion absorption by plants:**

a) *Ion traffic in to root:* Nutrient uptake by plants can be described and modelled by soil supply (soil solution concentration, ability to replenish or maintain that concentration, ability to diffuse through the soil), root morphology (growth rate, diameter, and proximity to other roots), root nutrient-influx, and water influx (transpiration) parameters. The chemical composition of plant tissues does not reflect the availability of nutrients in the soil solution. This difference results from selective uptake and transport of nutrients by the root system. Along the path from the soil to the leaf, there are several control points and processes that allow plants to accumulate and exclude elements from their leaves.

At the root surface, ions enter the cell wall and follow one of two parallel pathways for movement of nutrients from the root surface into the plant (Figure 2). One is through the apoplasm, the cell wall and intercellular spaces, and the other is transport across the cell membrane into the symplasm and then cell to cell through the plasmodesmata. Ions can move from the root surface through the epidermis and cortex up to the endodermis, which is the inner most layer of cells of the cortex. The endodermis has a hydrophobic band of suberin in the walls known as the casparian band, which is a barrier to passive movement of ions into the stele. To move from the cortex to the xylem at the casparian band, ions must move through the plasma membrane of the endodermis, providing selectivity in ion transfer to the vascular system.

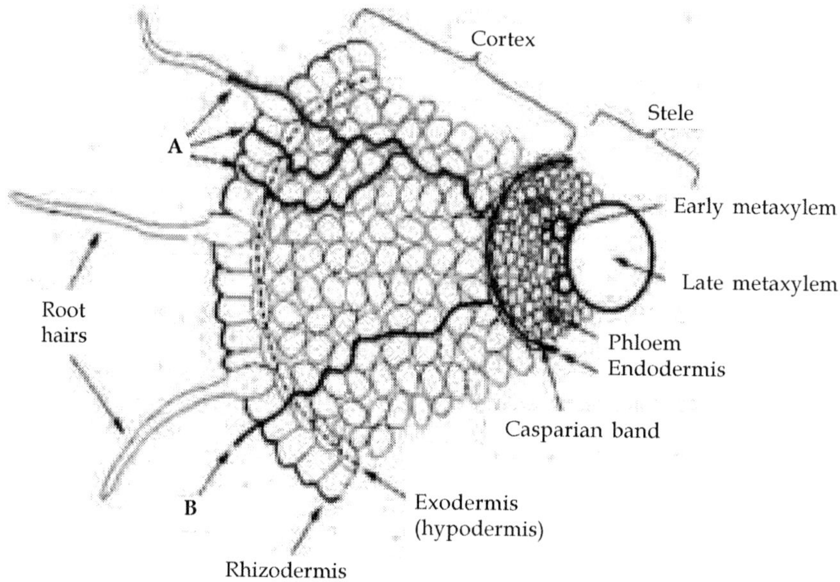

Figure 2. Part of transaction of a maize root showing the symplasmic (A) and apoplasmic (B) pathway of ion transport across the root.

The cell wall is like a filter with pores that restricts the uptake or movement of organic molecules based on size. Low-molecular weight ions, organic acids, amino acids, and sugars can readily move through the pores of the cell wall. However, high-molecular weight metal chelates, fulvic acids, and toxins or viruses and pathogens are too big to move through the cell wall pores. Since ions can readily pass through the cell wall, the wall does not function in the selectivity of ion uptake; Selectivity of ion uptake occurs at the plasma membrane of individual cells. The rate of nutrient uptake by roots depends on the concentration of the particular nutrient at the root surface, root properties or plant species, and requirements of the plant. At higher concentrations of nutrients, the rate of uptake is typically higher, but plants do exhibit selectivity in that they preferentially take up some elements and exclude others. Plants can take up nutrients in excess of their needs, termed luxury consumption, and store these in the vacuole. However, there are feedback mechanisms that reduce ion uptake as internal concentrations increase, maintaining a balance between demand and acquisition.

Plant roots can penetrate to significant depths and are branched so that they can explore large volumes of soil. Some nutrients (eg. nitrate) are only "weakly bound" in the soil, and can flow towards the root surface when a plant uses water. Other nutrients (eg. Phosphate) can be tightly bound in some soils, thus, continued uptake by plants relies on growth of roots into new areas of the soil. Root hairs, which greatly increase the volume of soil explored adjacent to the main root and also increase the root's surface area, may be particularly active in nutrient absorption. The degree of phosphate adsorption by soil particles depends on the soil type; sands adsorb phosphate only weakly; whereas clays rich in iron oxides absorb it very strongly.

b) *Nutrient flow in soil- plant system:* Nutrient absorption by plants is usually referred to as ion uptake or ion absorption because it is the ionic form in which nutrients are absorbed. After reaching plant roots, ions have to enter into root cells, be transported from cell to cell toward the xylem, be excreted into the xylem vessels, and finally be transported to the growing organs in the plant. The overall process of nutrient uptake in the soil-plant system is given below (Figure 3).

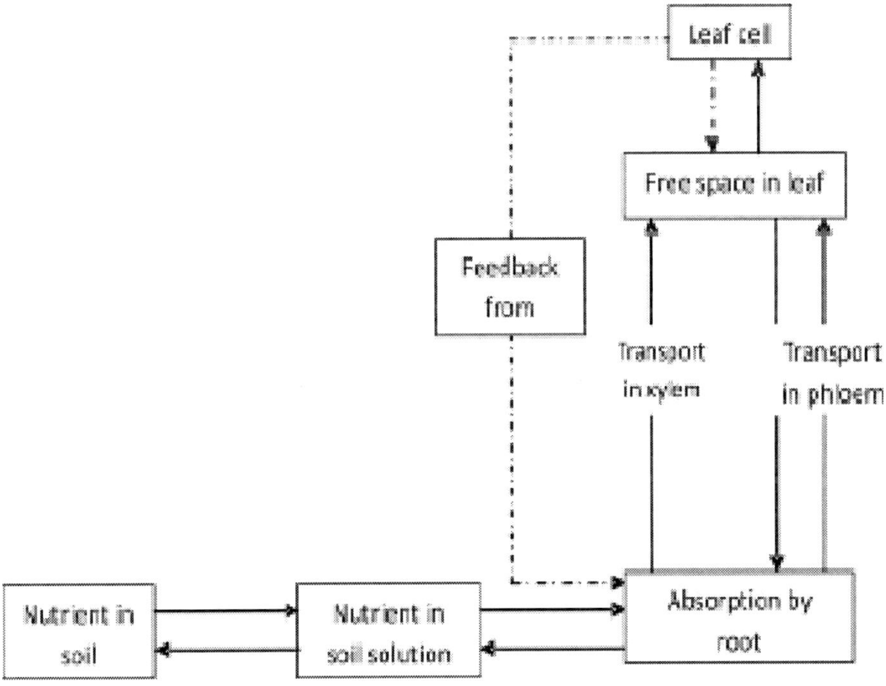

Figure 3. Flow chart of nutrient uptake in the soil - plant - system

B. Mechanism of nutrient absorption

Nutrients are absorbed from the soil solution in the form of ions. They are chiefly absorbed through the meristematic regions of the roots near the tips. However, some mineral ions may also be absorbed at other locations on the root surface or over the entire root surface including zone of elongation and root hairs that depends upon the high availability of such minerals around them and/or strong tissue demand at such locations. Plasma membrane of the root cells is not permeable to all the ions. It is selectively permeable. All the ions of the same salt are not absorbed at equal rate but there is unequal absorption of ions. Plants absorb minerals from the soil and translocate them to other parts of the body. Minerals are absorbed in the form of soil solution contained in the pore spaces between the soil particles and the root hair. The soil solution contains the mineral salts in the dissolved state.

Theories of nutrient absorption

Plants absorb minerals from the soil and translocate them to other parts of the body. Minerals are absorbed in the form of soil solution contained in the pore spaces between the soil particles and the root hair. The soil solution contains the mineral salts in the dissolved state. Several theories have been put forth to explain the mechanism of translocation of mineral salts. These theories can be placed under two headings i) Passive absorption and ii) Active Absorption.

i) **Passive absorption:** It is the absorption of minerals without direct expenditure of metabolic energy. In passive absorption.

- Mineral salt absorption is not affected by temperature and metabolic inhibitors.
- Rapid uptake of ions occurs when plant tissues are transferred from a medium of low concentration to high concentration.

The major hypothesis (theories) that explain the mechanism of passive transport of ions are:

a) Diffusion theory

b) Mass flow theory

c) Ion exchange theory

- Contact exchange theory
- Carbonic acid exchange theory

d) Donnan equilibrium theory

a) *Diffusion theory:* The concentration of ions inside the root cells is several times higher than in the soil solution. The mineral has to cross several barriers before entering the cells of the root. Therefore, mineral ions cannot directly move from the soil solution into the cells of the root by simple diffusion. They can pass through only from the free space by diffusion. The first barrier is the cell wall. Cell wall is permeable for the entry of ions. Therefore, they can freely pass through it by simple diffusion (Fig. 4). After crossing the cell wall, the ions move freely. They come across the cell membrane called ectoplast. It prevents further movement of ions by diffusion. It means ions can freely move from the soil solution up to this barrier. The term free space is used for the part of the plant cell and tissue. Where ions can freely move by diffusion. It is also called apparent free space (AFS).The inner boundary of free space is the ectoplast. Entry of ions into free space is very fast and it takes place by diffusion. Ions can also easily diffuse out. Therefore, ions can pass through free space area by diffusion. Ions cannot cross the barrier of ectoplast by this process and enter the protoplasm.

Figure 4. Diagrammatic representation of diffusion theory

b) *Mass flow theory:* The movement of ions along with bulk amount of water is called mass flow of ions. The increase of transpiration also increases the absorption of ions from the soil. Cohesion tension theory explains the role of transpiration in the mass movement of water. It is believed that mineral ions also move into the root along with this mass flow of water. Transpiration removes the minerals from the root xylem with the transpiration stream. So, it reduces its concentration in the root cells. As a result, root absorbs more minerals from the soil.

c) *Ion Exchange Theory:* The ions adsorbed on the surface of the walls or membranes of root cells may be exchanged with the ions of same sign from external solution. For example, the cation K^+ of the external soil solution may be exchanged with H^+ ion adsorbed on the surface of the root cells. Similarly, an anion may be exchanged with OH^- ion. There are two theories regarding the mechanism of ion exchange:

- *Contact Exchange Theory:*According to this theory, the ions adsorbed on the surface of root cells and clay particles (or clay micelles) are not held tightly but oscillate within small volume of space. If the roots and clay particles are in close contact with each other, the oscillation volume of ions adsorbed on root surface may overlap the oscillation volume of ions adsorbed on clay particles, and the ions adsorbed on clay particle may be exchanged with the ions adsorbed on root-surface directly without first being dissolved in soil solution (Figure 5).

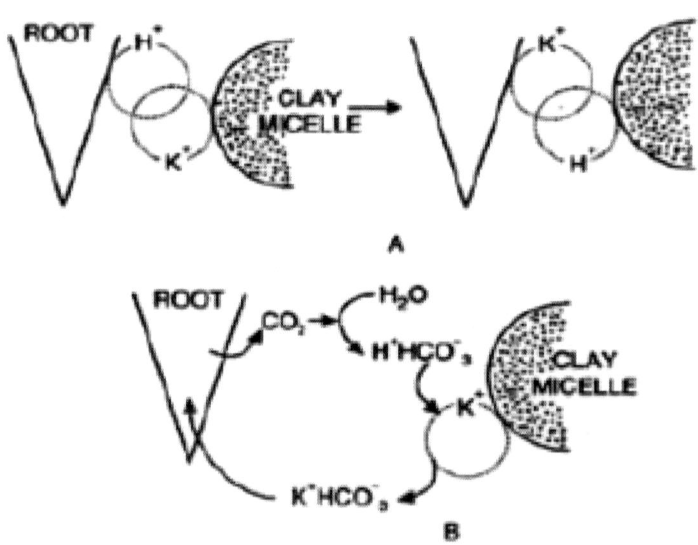

Figure 5. Diagrammatic representation of (A) the contact - exchange theory and (B) the carbonic acid exchange theory

- *Carbonic Acid Exchange Theory:* According to this theory, the CO_2 released during respiration of root cells combines with water to form carbonic acid (H_2CO_3). Carbonic acid dissociates into H^+ and an anion HCO_3^- in soil solution. These H^+ ions may be exchanged for cations adsorbed on clay particles. The cations thus released into the soil solution from the clay particles, may be adsorbed on root cells in exchange for H^+ ions or as ion pairs with bicarbonate (Fig. 5). Thus, soil solution plays an important role in carbonic acid exchange theory.

d) *Donnans' Equilibrium:* The accumulation of ions inside the cells without involving expenditure of the metabolic energy can be explained to some extent by Donnan's equilibrium theory (Figure 6). According to this theory, there are certain pre existing ions inside the cell which cannot diffuse outside through membrane. Such ions are called as indiffusible or fixed ions. However, the membrane is permeable to both anions and cations of the outer solutions. Suppose there are certain fixed anions in the cell which is in contact with outer solution containing anions and cations, normally equal number of anions and cations would have diffused into the cell through an electrical potential to balance each other, but to balance the fixed anions more cations will diffuse into the cell. This equilibrium is known as Donnan's equilibrium.

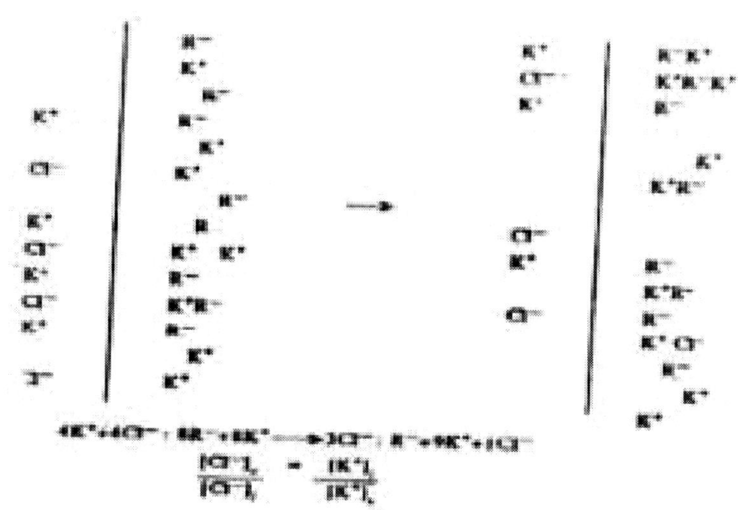

Figure 6. Diagrammatic representation of Donnan equilibrium.

ii) *Active absorption:* The concentration of ions in the cytoplasm is often much higher than in the soil solution, in extreme cases 10,000-fold higher. Therefore, the roots must be able to take up ions against a considerable concentration gradient. Uptake against a concentration gradient or, strictly speaking, against an electrochemical gradient requires metabolic energy, and the process is commonly termed active uptake. The energy required is obtained from respiration.

The carrier concept

According to this theory, the plasma membrane is impermeable to free ions. But some compounds present in it acts as carrier and combines with ions to form carrier- ion complex which can move across the membrane. On the inner side of the membrane, this complex leaves releasing ions into the cell while the carrier goes back to the outer surface to pick up fresh ions (Figure 7). They are two hypotheses based on the carrier concept to explain the mechanism of active salt absorption. Although they are not universally accepted.

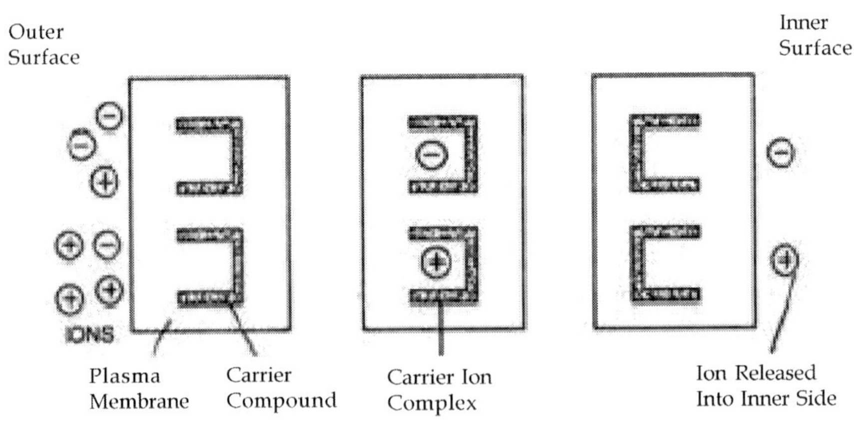

Figure 7. Diagrammatic representation of a model illustrating the carrier concept

Following observations strongly support the carrier concept of active absorption of min-eral salts:

- *Isotopic Exchange:* Several times, it has been found that actively absorbed radioactive ions cannot diffuse back or be exchanged with other ions in the outer solution indicating thereby that the plasma membrane is not permeable to free ions.

- *Saturation Effects:* Beyond a certain limit, increased concentration of salts in outer solution does not bring about an increase in the rate of mineral salt absorption. It is because the active sites on the carrier compound become saturated with ions.
- *Specificity:* Active sites on carrier compound may be specific which can bind only some specific ions. This also explains the selective and unequal absorption of ions by the plants.

a) *Lundegardhs cytochrome pump theory:* Lundegardh and Burstrom (1933) believed that there was a definite correlation between respiration and anion absorption. Thus, when a plant is transferred from water to a salt solution, the rate of respiration increases. This increase in rate of respiration over the normal respiration has been called as anion respiration or salt respiration. Lundegardh (1954) proposed cytochrome pump theory which is based on the following assumptions (Figure 8).

- The mechanism of anion and cation absorption is different
- Anions are absorbed through cytochrome chain by an active process.
- Cations are absorbed passively.

According to this theory :

- Dehydrogenase reactions on inner side of the membrane give rise to protons (H^+) and electrons (e^-).
- The electrons travels over the cytochrome chain towards outside the membrane, so that the Fe of the cytochrome becomes reduced (Fe^{++}) on the outer surface and oxidized (Fe^{+++}) on the inner surface.
- On the outer surface, the reduced cytochrome is oxidized by oxygen releasing the electron (e^-) and taking an anion (A^-).
- The electron thus released unites with H^+ and oxygen to form water.
- The anion (A^-) travels over the cytochrome chain towards inside.
- On the inner surface, the oxidized cytochrome becomes reduced by taking an electron produced through the dehydrogenase reactions and the anion (A^-) is released.

- As the result of anion absorption, a cation (M) moves passively from outside to inside to balance the anion.

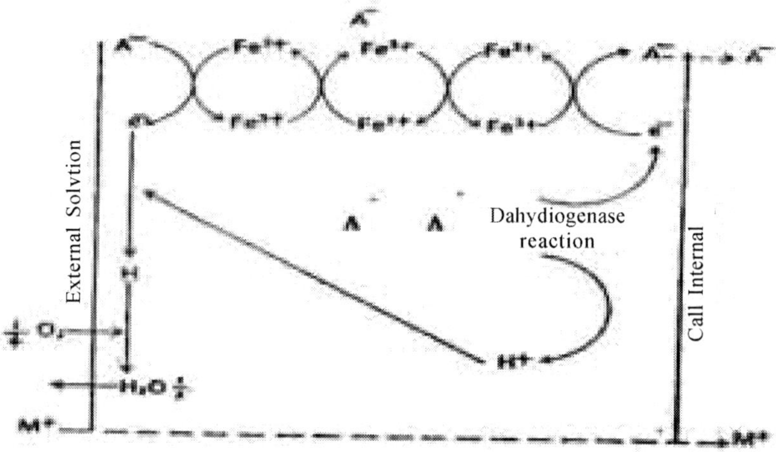

Figure 8. Lundegardhs cytochrome pump theory

b) *Bennert – Clark's protein Lecithin Theory:* In 1856, Bennet – Clark suggested that because the cell membranes chiefly consist of phospholipids and proteins and certain enzymes seem to be located on them, the carrier could be a protein associated with the phosphatide called as lecithin (Figure 9). He also assumed the presence of different phosphatides to correspond with the number of known competitive groups of cations and anions.

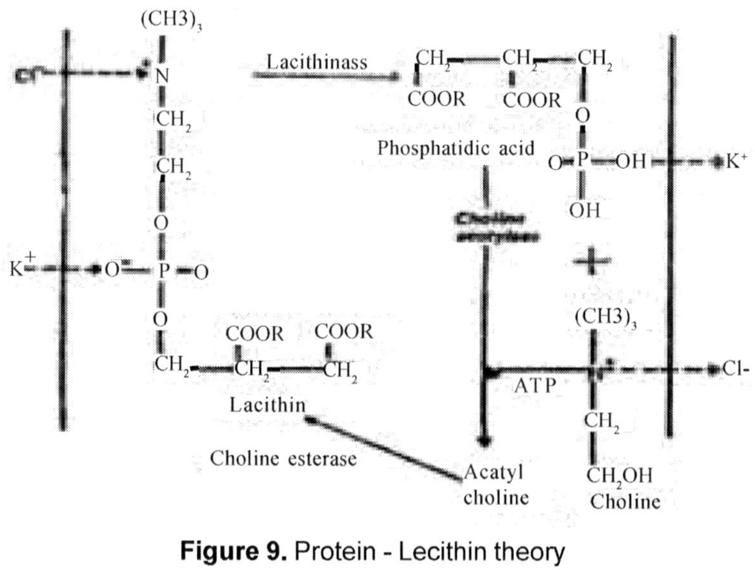

Figure 9. Protein - Lecithin theory

According to this theory :

- Phosphate group in the phosphatide is regarded as the active centre binding the cations and the basic choline group as the anion binding centre.
- The ions are liberated on the inner surface of the membrane by decomposition of lecithin by the enzyme lecithinase.
- The regeneration of the carrier lecithin form phosphatidic acid and choline takes place in the presence of the enzyme choline acetylase and choline esterase and ATP. The latter acts as a source of energy.

c) *Nutrient Uptake:* Nutrient uptake or absorption by plants is a dynamic and complex process. The amount of nutrient absorption by growing plants either from soil or other sources called nutrient uptake or nutrient absorption. The crops draw nutrients from soil. Uptake of the nutrients need not be in the same ratio as these occur in soil. The nutrient uptake can be calculated from the following formula and expressed in kg/ha.

$$= \frac{\text{Nutrient content (Per cent) x DMP (kg/ha)}}{100} = \text{kg/ha}$$

The nutrient uptake of primary and secondary nutrients by major crops are given below Table 3.

Table 3. Total uptake of major nutrients by crops

Group	Crop (main produce)	Total uptake (kg)/tonnes of main produce					
		N	P_2O_5	K_2O	S	Ca	Mg
Cereals	Rice (Paddy)	20.0	11.0	30.0	3.0	7.0	3.0
	Wheat (grain)	25.0	9.0	33.0	4.7	5.3	4.7
	Maize (grain)	29.9	13.5	32.8	-	-	-
	Sorghum (grain)	16.4	7.7	25.5	-	-	-
	Pearlmillet	31.8	17.4	61.3	-	-	-
	Fingermillet (grain)	24.2	9.5	30.6	-	-	-
Pulses	Chickpea (grain)	60.7	9.2	39.2	8.7	18.7	7.3
	Pigeonpea (grain)	70.8	15.3	16.0	7.5	19.2	12.5
	Lentil (grain)	57.0	14.9	21.6	3.0	7.5	2.0

	Greengram (grain)	106.0	48.1	73.2	12.0	71.0	43.0
	Blackgram (grain)	78.9	14.4	65.6	5.6	-	-
Oilseed	Groundnut (seed)	58.1	19.6	30.1	7.9	20.5	13.3
	Mustard (seed)	32.8	16.4	41.8	17.3	42.0	8.7
	Raya (seed)	64.5	20.6	53.4	16.0	56.5	9.5
	Rocket salad (seed)	70.0	26.0	61.1	20.7	19.3	9.3
	Soybean (seed)	70.7	30.9	57.7	6.7	14.0	7.6
	Safflower (seed)	38.8	8.4	22.0	12.6	-	-
	Sesame (seed)	51.7	22.9	64.0	11.7	37.5	15.8
	Sunflower (seed)	63.3	19.1	126.0	11.7	68.3	26.7
	Linseed (seed)	60.0	18.6	54.0	5.6	31.2	13.1
	Castor (seed)	40.0	9.0	16.0	-	-	-
Tubers	Potato (tuber)	3.3	0.9	6.2	0.4	1.0	1.8
	Cassava (tuber)	5.0	2.3	6.8	0.4	2.7	1.0
Sugar crops	Sugarcane (cane)	2.1	1.2	3.4	0.3	-	-
Fibre crops	Cotton (seed cotton)	43.2	29.3	53.3	-	-	-
	Jute (dry fiber)	35.2	20.3	63.2	-	39.7	8.0
Vegetables	Tomato (fruit)	2.8	1.3	3.8	-	-	-
	Cauliflower (curd)	4.0	2.0	4.0	-	-	-
	Cabbage (head)	3.5	1.3	4.2	-	-	-
	Beetroot (root)	4.4	2.0	6.7	-	-	-
	Carrot (root)	3.9	1.7	6.6	-	-	-
	Onion (bulb)	2.7	1.3	3.9	-	-	-

(Shivay and Kumar, 2008)

Micronutrient uptake and removal by crops has been an important aspect of overall micronutrients requirement. Some general estimates of micronutrients uptake by crops are given in Table 4.

Table 4. Average uptake of micronutrients by some crops

Crop	Economic yield (t ha^{-1})	Total uptake (grams)					
		Zn	Fe	Mn	Cu	B	Mo
Rice as paddy	1.0	40	153	675	18	15	2
Wheat	1.0	56	624	70	24	48	2
Maize	1.0	130	1,200	320	130	-	-
Sorghum	1.0	72	720	54	6	54	2
Pearl millet	1.0	40	170	20	8	-	-
Cassava	1.0	45	120	45	5	15	-
Potato	1.0	9	160	12	12	50	0.3
Chickpea	1.5	57	1,302	105	17	-	-
Pigeonpea	1.2	38	1,440	128	31	-	-
Soybean	2.5	192	866	208	74	-	-
Groundnut	1.9	208	4,340	176	68	-	-
Mustard	1.5	150	1684	143	25	-	-
Sunflower	0.6	28	645	109	23	-	-
Sesamum	1.2	202	952	138	140	-	-
Linseed	1.6	73	1062	283	48	-	-
Jute (oiltorius)	1.0	214	784	251	27	-	-
Jute (capsularis)	1.0	139	368	119	18	-	-
Coffee (Arabica)	1.0	35	83	62	82	-	-
Tea	1.0	276	2007	1933	632	101	-

(Tandon, 2009)

Factors affecting nutrient absorption

Absorption of mineral nutrients is affected by the number of external and internal factors. Some of them are listed below:

External factors

- Temperature
- Light
- Interaction.
- Hydrogen ion concentration (pH)
- Oxygen

Internal factors

- Growth and morpho-physiological status.
- Aging.

External factors

Temperature

Absorption of mineral nutrient is affected by change in temperature. In general, an increase in temperature results increase in the absorption of nutrients up to a certain optimum level. At very high temperature, the absorption is considerably inhibited. The inhibition might be due to denaturation of proteins which are directly or indirectly involved in mineral nutrient absorption. The change in temperature also affects the process of diffusion. The rate of diffusion depends upon the kinetic energy of diffusing molecules or ions which, in turn, dependent upon temperature.

Hydrogen ion concentration (pH)

The changes in the hydrogen ion concentration (pH) of the soil solution affects the availability of ions to the plants. In general, decrease in the pH of soil solution accelerates the absorption of anions.

Light

Light has no direct effect, but indirectly by transpiration and photosynthesis, influences salt absorption.

Oxygen

The active nutrient absorption is inhibited by the absence of oxygen.

Interaction

The absorption of one ion is influenced by the presence of other ions in the medium.

Internal factors

Growth

Active cell division, elongation and developmental processes promote the absorption of mineral nutrients.

Age of the plant

As the root matures, it increases the surface area which is favourable for nutrient absorption, but due to heavy suberization the mineral nutrient uptake is greatly reduced.

❑❑❑

Chapter 19

Plant and Leaf Analysis

Several tools are available to evaluate the nutritional status of plants. Plant analysis is an efficient one since it uses the plant itself as a nutrient extractor. Thus, it complements soil chemical analysis and makes it possible to predict nutritional disorders before the appearance of visual symptoms in the plant tissue. However, it is necessary to integrate both techniques, chemical analysis of plants and chemical and physical analysis of soil, besides visual diagnosis to maximize fertilization efficiency in terms of cost and prevention of environmental damage.

By plant analysis, it is possible, to determine nutrient needs and identify nutritional deficiencies that produce similar symptoms, evaluate nutritional status, help in the managing of fertilization programmes and diagnose about levels of nutrients in diverse plant organs. Several procedure, direct and indirect, are available to achieve these targets.

Nutrients Concentration

Usually, the plant exhibits a visual symptom indicating a deficiency in a specific nutrient which normally can be corrected or prevented by supplying that nutrient. Visual nutrient deficiency symptoms can be caused by many other plant stress factors. Therefore, caution should be exercised when diagnosing deficiency symptoms. The following terms are commonly used to describe levels of nutrients in plants (Figure 1, Table 1).

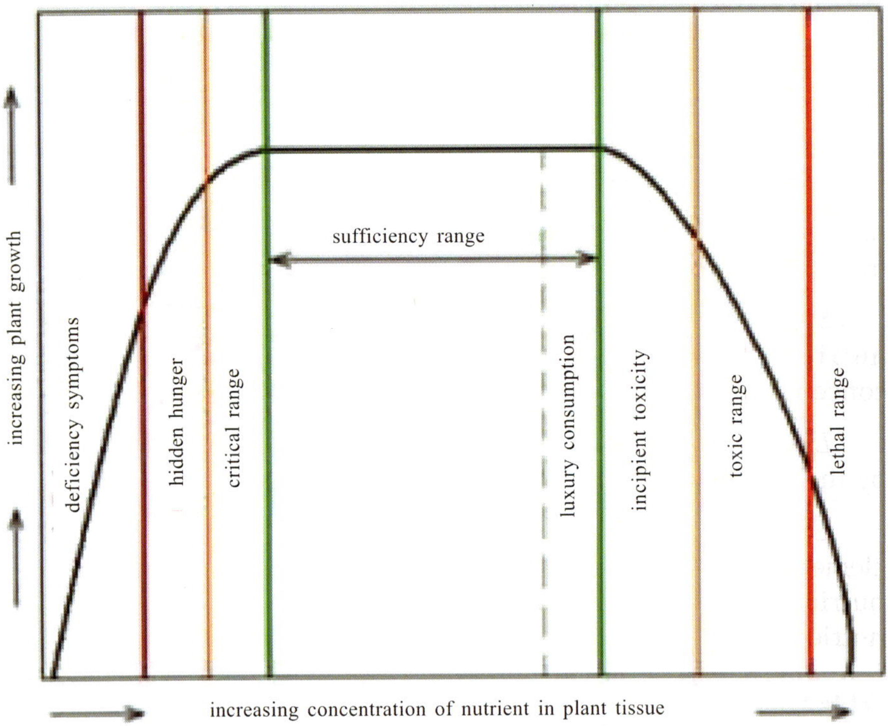

Figure 1. Relationship between field and nutrient concentration

Deficiency: When the concentration of an essential element is low enough to limit yield severely and distinct deficiency symptoms are visible. Extreme deficiencies can result in plant death. With moderate or slight deficiencies, symptoms may not be visible, but yields will still be reduced.

Hidden hunger

Deficiency symptoms always indicated severe starvation and therefore the crop may have suffered before the deficiency symptoms appear. Many crops start losing yields well before deficiency signs start showing. This yield limiting condition is called hidden hunger which refers to a situation in which a crop needs more of a given nutrient and yet has not shown any deficiency symptom. The nutrient content is above the deficiency symptom zone but still considerably below that needed for optimum crop production. In this case, significant responses can be obtained with application of nutrients even though no recognizable symptoms have appeared. To overcome such conditions, it is always recommended to confirm deficiency problem with other diagnostic techniques and the most common in soil testing.

Critical range: The nutrients concentration in the plant below which a yield response to added nutrient occurs. Critical levels or ranges vary among plants and nutrients, but occur somewhere in transition between nutrient deficiency and sufficiency.

Sufficient: The nutrient concentration range in which added nutrient will not increase yield but can increase nutrients concentration.

Luxury consumption: Often used to describe nutrient absorption by the plants that does not influence the yield.

Excessive or toxic: When the concentration of essential or other elements is high enough to reduce plant growth and yield. Excessive nutrients concentration can cause an imbalance in other essential nutrients which also can reduce yield.

Table 1. Concentration of the major and micronutrients in mature leaf tissue of crop plants

Essential elements	Deficient (per cent)	Sufficient or normal(per cent)	Excessive or toxic(per cent)
Major Elements			
Nitrogen (N)	<2.50	2.50-4.50	>6.00
Phosphorus (P)	<0.15	0.20-0.75	>1.00
Potassium (K)	<1.00	1.50-5.50	>6.00
Calcium (Ca)	<0.50	1.00-4.00	>5.00
Magnesium (Mg)	<0.20	0.25-1.00	>1.50
Sulphur (S)	<0.20	0.25-1.00	>3.00
Micronutrients	ppm	ppm	ppm
Boron (B)	5-30	10-200	50-200
Chlorine (Cl)	>100	100-500	500-1,000
Copper (Cu)	2-5	5-30	20-100
Iron (Fe)	<50	100-500	>500
Manganese (Mn)	15-20	20-300	300-500
Molybdenum (Mo)	0.03-0.15	0.1-2.0	>100

(Prakash *et al.*, 2013)

Plant Sampling procedure

Sampling techniques vary with the crop. In some cases, such as cereal crops, leaf samples are taken. In a few cases, for example suspected boron deficiency in sugar beet, the roots should be sampled. In any sampling of plant material, great care should be taken to avoid soil contamination. If a deficiency or toxicity condition is suspected in only a part of a field, 'good' and 'poor' areas should be sampled for comparison in the laboratory. Leaves or other parts of the plant of the same age or growth stage should be taken at the same time.

It is extremely important to use containers for transporting the samples which will not contaminate them. The other metal containers and even some paper bags can cause serious contamination of the sample. The best container is a clean polythene bag. After taking the sample, labels should be attached, the bag should sealed and transferred to the laboratory as quickly as possible. The basic principle behind this technique is that the nutrient concentration of plants is related to the amount of nutrient element available in soil. If the nutrient level in the tissues falls below the critical concentration, the soil may be deficient in that element for optimum plant growth.

Different plant parts of the same plant contain different concentrations of the same nutrient. Nutrient concentration varies with the stage of the crop. So, leaf samples for analyses should be selected on the basis of physiological age, i.e., developmental stage (Table.2). It is important that the sample must be free from diseases, insect damage and physical or chemistry injury. Leaf near the fruit or grain should not be sampled as the nutrients are often translocated to the fruits.

Table 2. Suggested plant tissue sampling procedures for some crops

Crop	Stage of growth	Plant part to sample	No. of plants to sample*
Maize	Prior to tasselling	The entire leaf fully developed below the whorl	15-25
Soybean or other bean	Prior to or during initial flowering	2 or 3 fully developed leaves at the top of the plant	20-30
Small grains (including rice)	prior to heading	The 4 uppermost leaves	50-100
Forage grasses	Prior to seed head emergence, or at the optimum stage for best quality forage.	4th uppermost leaf blade.	40-50
Sugar beet	Mid-season	Fully expanded and mature leaves midway between the younger center leaves and the oldest leaf whorl on the outside.	40-50
Tobacco	Before bloom	Upper most fully developed leaf	8-12
Sorghum	Prior to or at heading	2nd leaf from top of plant	15-25
Ground nut	Prior to or at bloom	Mature leaves from both the main stem and lateral branch.	40-50
Cotton	Prior to or at first bloom or when first squares appear	Youngest fully mature leaves on main stem	30-40

b. Vegetable crops

Crop	Stage of growth	Plant part to sample	No. of plants to sample
Potato	Prior to or during early bloom	3rd to 6th leaf from growing tip	20-30
Head crops (cabbage, etc.)	Prior to heading	1st mature leaves from center of whorl	10-20
Tomato (field)	Prior to or during early fruit set	3rd or 4th leaf from growing tip	20-25
Tomato (greenhouse)	Prior to or during fruit set	Young plants: leaves adjacent to 2nd to 3rd clusters Older plants: leaves from 4th to 6th cluster	20-25
Bean	Prior to or during initial flowering	2 or 3 fully developed leaves at the top of the plant	20-25
Root crops (carrots, onions beets, etc.)	prior to root or bulb enlargement	Center mature leaves	20-30
Celery	Mid-growth	Petiole of youngest mature leaf	15-30
Leaf crops (lettuce, spinach, etc.)	At heading	Wrapper leaf	10-12

Peas	Prior to or during initial flowering	Leaves from the 3rd node down	30-60
Melons (Water melon, cucumber melon, muskmelon)	Early growth prior to fruit set	Mature leaves near the base portion of plant on main stem	20-30
Sweet potato	Fourth to sixth leaf from the growing tip	Prior to root enlargement	20-30
Asparagus	Top 10 cm of the new fern branch	Mid-growth of the fern	20-30
c. Fruits and nuts			
Crop	**Stage of growth**	**Plant part to sample**	**No. of plants to sample**
Apple, apricot, almond, prune, peach, pear, cherry	Mid-season	Leaves near base of current year's growth or from spurs	50-100
Strawberry	Mid-season	Youngest fully expanded mature leaves	50-75
Walnut	6 to 8 weeks after bloom	Middle pair of leaflets from mature shoots	30-35
Fig	Mid-summer	Mature leaves from current	10-15
Lemon, lime	Mid-season	Mature leaves from last flush or growth on non-fruiting terminals	20-30
Orange	Mid-season	Spring cycle leaves, 4 to 7 months old from non-bearing terminals	20-30
Grapes	End of bloom period	Petioles from leaves adjacent to fruit clusters	60-100
d. Flowers and Ornamentals			
Crop	**Stage of growth**	**Plant part to sample**	**No. of plants to sample**
Ornamentals trees, shrubs	Current year's growth	Fully developed leaves	30-100
Turf	2-3 weeks after mowing	Leaf blades; clip by hand to avoid contamination with soil or other material	30-40
Roses	During flowering	Upper leaves on the flowering stem	20-30
Chrysanthemums	Prior to or at flowering	Upper leaves on the flowering stem	20-30
Carnations	Unpinched plants	4th or 5th leaf pairs from base of plant	20-30
	Pinched plants	5th and 6th leaf pairs from top of primary laterals	20-30
Poinsettias	Prior to or at flowering	Most recently mature fully expanded leaves	15-20

(Jones *et al.*, 1990)

*Number of plants to sample per acre. For smaller areas, the minimum number of plants to sample should not be less than five.

Sample handling and preparation for laboratory analysis

Each step in sample preparation may contribute errors which may have a cumulative effect on the final result. Hence, standard procedures should be employed to minimize or prevent such errors which are furnished below :

- Field sampling should be undertaken early in the week.
- The collected samples should be placed in labeled open paper bags and immediately placed in cooled containers (5°C) for transport from the field to the laboratory within 24 hours of sampling.
- If roots are to be included, uproot the whole plant carefully from wet soil, retaining even the fine/active roots. Gently dip the plant roots in water several times to remove adhering soil as far as possible.
- Wash the plant parts with water several times.
- Wash the samples with about 0.2% detergent solution thoroughly to remove the waxy/greasy coating on the leaf surface, which is often present.
- Wash with 0.1N HCl followed by thorough washing with plenty or water. Give final rinse with distilled water.
- Rinse with double distilled water, specifically if micronutrient analysis is to be carried out.
- Dry with good quality tissue paper.
- Air dry the samples on a perfectly clean surface at room temperature for at least 2-3 days in a dust-free atmosphere away from any kind of contaminant.
- Put the samples in an oven and dry at 60°C for 48hours.
- Grind the samples in an electric stainless steel mill using 0.5 mm sieve. Clean the cup and blades of the grinding mill before each sample.
- Put back the samples in oven and dry again for few hours more for constant weight. Store in well-stoppered plastic or glass bottles or paper bags for analysis.

The dried plant samples are analysed for N, P, K, Ca, Mg, S, Zn, Cu, Fe, Mn, B, and Mo. The N content of the plant samples was estimated by microkjeldahl method (Humphries 1956). For analysis of P, K secondary and micronutrients, a measured quantity of dried samples is digested with tri acid mixture of nitric acid, sulphuric acid and perchloric acid @ 9:3:1. This extract has been used for the estimation of Ca and Mg (Versanate Titration Method), phosphorus and potassium (Jackson, 1973), Zn, Cu, Fe and Mn (Atomic absorption spectro photometer) (Lindsay and Norvell, 1978). The sulphur content of the plant can be estimated by barium chromate colourimetric method (Palaskar et al 1981). Boron content is estimated by the method suggested by John *et al.* (1975).

Interpretation of nutrient analysis

The nutrient analysis data can be used as a guide for attaining improved crop quality and yield. For long-term orchard crops, plant tissue nutrient levels can be logged over time and used as a diagnostic tool to assist in developing a fertilizer programme. If a tissue level of a nutrient is below the lower end of the sufficiency range, the nutrient should be considered deficient, whereas if the level is above the upper end of the range, the nutrient can be considered as approaching a toxic level. The midpoint of the sufficiency range is the target to aim at. As the level approaches the lower limit, the nutrient should be added. As the level approaches the upper limit, additions of the nutrient should be withheld. It is important to be near the midpoint for most nutrients, because imbalances in the ratios of nutrient can affect crop growth. The nutrient concentration at sufficient range for the field and horticultural crops are given in Table 3 and 4.

Table 3. Sufficiency Range of Nutrient Concentrations in Crops at Critical Stage

Crops	N (%)	P (%)	K (%)	Ca (%)	Mg (%)	S (%)	Fe (mg/kg)	Mn (mg/kg)	Cu (mg/kg)	Zn (mg/kg)	B (mg/kg)	Mo (mg/kg)
Rice	2.0-2.5	0.2-0.4	1.5-2.5	0.3-0.5	0.2-0.3	0.2-0.3	100-150	50-100	7-15	25-50	6-15	NA
Maize	3.0-5.0	0.4-0.6	2.5-4.0	0.5-1.5	0.3-0.6	0.2-0.4	40-400	50-150	6-20	25-50	5-25	NA
Sorghum	2.5-3.0	0.4-0.6	2.5-3.5	0.5-1.5	0.3-0.6	0.2-0.4	50-500	50-200	5-20	25-50	5-25	NA
Millets	1.75-3.0	0.3-0.5	1.5-3.0	0.2-0.5	0.15-0.5	0.2-0.4	50-250	25-100	5-25	15-70	5-20	NA
Wheat	1.5-2.0	0.3-0.45	2.0-3.0	0.15-0.30	0.1-0.2	0.2-0.7	50-250	25-100	5-15	20-60	5-20	NA
Oats	2.5-5.0	0.2-0.4	1.2-2.5	0.5-1.0	0.1-0.3	0.2-0.7	50-200	60-150	7-15	20-50	15-50	0.5-4.0
Barley	1.5-2.0	0.2-0.4	1.5-2.5	0.2-0.3	0.1-0.2	0.2-0.6	50-250	20-100	5-10	20-50	15-80	0.05-0.2
Pulses	2.0-2.5	0.2-0.4	1.0-3.0	2.0-2.5	0.4-0.8	0.2-0.4	50-250	20-150	10-15	20-40	50-250	0.2-0.4
Soybean	2-4	0.2-0.4	1.0-2.5	0.5-1.0	0.4-0.6	0.2-0.4	100-250	15-150	10-20	20-60	20-150	NA
Cotton	2.0-3.0	0.2-0.4	1.0-1.5	2.0-2.5	0.4-0.8	0.3-0.7	50-250	20-100	10-25	20-50	150-350	0.4-4.0
Groundnut	2.5-4.0	0.2-0.5	1.5-2.5	0.5-2.0	0.2-0.6	0.2-0.4	50-250	30-150	10-25	10-50	25-100	0.15-2.5
Sunflower	2.02.5	0.2-0.4	1.5-2.5	1.5-2.5	1.0-2.0	0.25-0.5	50-250	25-50	10-30	20-50	20-120	0.2-0.5
Sugarcane	1.5-2.5	0.25-0.5	1.5-3.5	0.4-1.5	0.5-1.0	0.2-0.4	100-1000	30-100	10-25	10-20	15-20	NA
Sugar beet	3-4	0.2-0.3	2.5-5.0	0.5-2.5	0.5-2.5	0.5-1.5	50-500	100-1000	20-60	10-50	30-400	0.2-20
Sweet potato	1.5-2.5	0.2-0.4	2.0-3.0	1.5-2.0	0.5-1.5	0.3-0.6	200-500	50-250	20-50	10-40	50-125	NA
Crops	**N (%)**	**P (%)**	**K (%)**	**Ca (%)**	**Mg (%)**	**S (%)**	**Fe (mg/kg)**	**Mn (mg/kg)**	**Cu (mg/kg)**	**Zn (mg/kg)**	**B (mg/kg)**	**Mo (mg/kg)**
Apple	1.5-2.5	0.1-0.3	0.5-1.5	1.0-1.5	0.2-0.5	0.1-0.3	25-200	25-125	5-15	10-40	10-25	0.1-0.2
Avocado	1.5-2.0	0.15-0.3	1.0-1.2	1.0-1.5	0.4-0.8	0.2-0.4	50-80	50-100	5-10	25-50	10-20	NA
Banana	2.0-2.5	0.2-0.3	3.0-4.0	2.0-3.5	0.4-0.8	0.2-0.4	50-250	25-50	5-10	25-50	5-20	NA
Grapes	0.2-0.50	0.2-0.7	0.5-3.5	1.2-3.2	0.2-0.5	0.2-0.4	50-300	25-250	20-30	25-50	250-300	NA
Lemon	2.0-2.5	0.1-0.15	1.5-2.5	1.0-2.0	0.2-0.4	0.1-0.2	100-150	15-75	8-10	15-50	30-100	0.3-3.0
Lime	2.5-3.0	0.15-0.3	0.9-1.2	1.0-2.0	0.2-0.5	0.2-0.3	70-200	30-150	5-10	20-80	30-100	0.3-3.0
Mango	1.0-1.2	0.06-0.1	0.8-2.0	0.9-1.8	0.91-1.0	0.2-0.3	50-200	30-150	10-15	20-50	30-100	NA
Orange	2.0-2.5	0.15-0.3	1.2-1.7	3.0-5.0	0.2-0.4	0.2-0.3	80-300	25-250	5-25	20-100	20-80	0.1-0.3
Pineapple	1.3-2.5	0.2-0.3	2.5-4.0	0.75-1.5	0.25-0.8	0.25-1.0	50-250	25-100	8-12	25-150	25-100	NA
Tomato	2-2.5	0.2-0.8	1.5-3.0	0.8-1.6	0.5-1.0	0.2-0.4	100-250	80-400	15-50	25-50	35-150	0.5-1
Bean	2.5-5.0	0.3-0.6	0.5-1.5	1.0-1.5	0.4-0.8	0.2-0.6	50-300	200-1000	5-10	25-50	1.0-5.0	0.2-0.4
Cabbage	1.5-2.0	0.2-0.4	2.5-3.0	1.0-1.5	0.4-0.6	0.3-0.5	40-200	50-100	5-10	25-50	20-40	NA

Carrot	2.0-2.5	0.3-0.4	2.5-3.0	0.5-1.5	0.2-0.3	0.2-0.4	50-250	50-150	5-10	25-75	25-100	NA
Cauliflower	2.0-2.5	0.4-0.8	2.5-3.0	2.5-3.5	0.5-1.0	0.4-0.6	50-250	50-75	3-5	20-80	30-60	NA
Lettuce	2.0-4.0	0.2-0.6	2.5-3.5	0.2-1.5	1.5-2.0	0.2-0.4	50-250	50-150	5-10	20-60	25-50	NA
Onion	2.0-3.0	0.25-0.5	2.0-2.5	0.2-1.0	0.5-1.0	0.2-0.4	50-250	50-75	5-10	20-50	25-40	NA
Crops	**N (%)**	**P (%)**	**K (%)**	**Ca (%)**	**Mg (%)**	**S (%)**	**Fe (mg/kg)**	**Mn (mg/kg)**	**Cu (mg/kg)**	**Zn (mg/kg)**	**B (mg/kg)**	**Mo (mg/kg)**
Potato	3-4.5	0.1-0.4	4-6	2.5-3.5	0.3-0.9	0.4-0.6	50-200	25-75	2-6	30-90	20-60	NA
Coffee	3.0-4.0	0.2-0.4	1.5-2.5	1.5-2.0	0.2-0.4	0.2-0.4	100-300	50-150	10-25	20-50	40-70	3.0-4.0
Tea	3.0-4.0	0.2-0.4	1.0-1.5	1.0-1.5	3-5	0.2-0.5	100-500	100-250	10-20	15-30	20-50	NA
Tobacco	2.0-2.5	0.2-0.4	2-3.5	1.0-2.5	0.5-1.0	0.25-0.5	90-150	100-250	15-30	30-60	25-250	1-2
Cacao	2.0-2.5	0.2-0.4	2.5-3.5	1.0-1.5	0.2-0.3	0.2-0.4	50-250	50-100	5-10	25-75	20-50	NA
Coconut	1.5-2.5	0.2-0.4	0.5-1.5	0.5-1.5	0.25-1.0	0.2-0.4	50-400	50-100	5-15	20-60	10-30	NA

(Subramanian *et al.*, 2007)

Table 4. Sufficiency Range of Nutrient Concentrations in Crops at Critical Stage

Crops	N (%)	P (%)	K (%)	Ca (%)	Mg (%)	S (%)	Fe (mg/kg)	Mn (mg/kg)	Cu (mg/kg)	Zn (mg/kg)	B (mg/kg)	Mo (mg/kg)
Apple	1.5-2.5	0.1-0.3	0.5-1.5	1.0-1.5	0.2-0.5	0.1-0.3	25-200	25-125	5-15	10-40	10-25	0.1-0.2
Avocado	1.5-2.0	0.15-0.3	1.0-1.2	1.0-1.5	0.4-0.8	0.2-0.4	50-80	50-100	5-10	25-50	10-20	NA
Banana	2.0-2.5	0.2-0.3	3.0-4.0	2.0-3.5	0.4-0.8	0.2-0.4	50-250	25-50	5-10	25-50	5-20	NA
Grapes	0.2-0.50	0.2-0.7	0.5-3.5	1.2-3.2	0.2-0.5	0.2-0.4	50-300	25-250	20-30	25-50	250-300	NA
Lemon	2.0-2.5	0.1-0.15	1.5-2.5	1.0-2.0	0.2-0.4	0.1-0.2	100-150	25-75	8-10	15-50	30-100	0.3-3.0
Lime	2.5-3.0	0.15-0.3	0.9-1.2	1.0-2.0	0.2-0.5	0.2-0.3	70-200	30-150	5-10	20-80	30-100	0.3-3.0
Mango	1.0-1.2	0.06-0.1	0.8-2.0	0.9-1.8	0.91-1.0	0.2-0.3	50-200	30-150	10-15	20-50	30-100	NA
Orange	2.0-2.5	0.15-0.3	1.2-1.7	3.0-5.0	0.2-0.4	0.2-0.3	80-300	25-250	5-25	20-100	20-80	0.1-0.3
Pineapple	1.3-2.5	0.2-0.3	2.5-4.0	0.75-1.5	0.25-0.8	0.25-1.0	50-250	25-100	8-12	25-150	25-100	NA
Tomato	2-2.5	0.2-0.8	1.5-3.0	0.8-1.6	0.5-1.0	0.2-0.4	100-250	80-400	15-50	25-50	35-150	0.5-1
Bean	2.5-5.0	0.3-0.6	0.5-1.5	1.0-1.5	0.4-0.8	0.2-0.6	50-300	200-1000	5-10	25-50	1.0-5.0	0.2-0.4
Cabbage	1.5-2.0	0.2-0.4	2.5-3.0	1.0-1.5	0.4-0.6	0.3-0.5	40-200	50-100	5-10	25-50	20-40	NA
Carrot	2.0-2.5	0.3-0.4	2.5-3.0	0.5-1.5	0.2-0.3	0.2-0.4	50-250	50-150	5-10	25-75	25-100	NA
Cauliflower	2.0-2.5	0.4-0.8	2.5-3.0	2.5-3.5	0.5-1.0	0.4-0.6	50-250	50-75	3-5	20-80	30-60	NA
Lettuce	2.0-4.0	0.2-0.6	2.5-3.5	0.2-1.5	1.5-2.0	0.2-0.4	50-250	50-150	5-10	20-60	25-50	NA
Onion	2.0-3.0	0.25-0.5	2.0-2.5	0.2-1.0	0.5-1.0	0.2-0.4	50-250	50-75	5-10	20-50	25-40	NA
Potato	3-4.5	0.1-0.4	4-6	2.5-3.5	0.3-0.9	0.4-0.6	50-200	25-75	2-6	30-90	20-60	NA
Coffee	3.0-4.0	0.2-0.4	1.5-2.5	1.5-2.0	0.2-0.4	0.2-0.4	100-300	50-150	10-25	20-50	40-70	3.0-4.0
Tea	3.0-4.0	0.2-0.4	1.0-1.5	1.0-1.5	3-5	0.2-0.5	100-500	100-250	10-20	15-30	20-50	NA
Tobacco	2.0-2.5	0.2-0.4	2-3.5	1.0-2.5	0.5-1.0	0.25-0.5	90-150	100-250	15-30	30-60	25-250	1-2
Cacao	2.0-2.5	0.2-0.4	2.5-3.5	1.0-1.5	0.2-0.3	0.2-0.4	50-250	50-100	5-10	25-75	20-50	NA
Coconut	1.5-2.5	0.2-0.4	0.5-1.5	0.5-1.5	0.25-1.0	0.2-0.4	50-400	50-100	5-15	20-60	10-30	NA

(Subramanian *et al.*, 2007)

Practical Uses of Plant analysis

- Determining the element is essential for plant growth, development, and maturation or some process closely associated with nutrition and growth.
- Verifying the element associated with a phenotypic or apparent nutrient element deficiency or toxicity symptoms.
- Establishing optimum concentrations or critical values for elements associated with optimum or maximum economic yields.
- Determining availability of soil nutrients and/or availability and recovery of an applied element or elements in fertilizer in crop response experiments.
- Evaluating and comparing different areas or sites within a production field, orchard, or forest.
- Monitoring fields, vineyards, orchards, or forests to determine if deficiencies or nutrient imbalances exists so that corrective action may be taken if deemed necessary.
- Determining the total elemental uptake by a crop, particularly at high levels of production (by this process, an estimate of the nutrient element requirement per unit of yield).
- Determining the internal nutrient efficiencies (output per unit of nutrient uptake) of varieties or cultivars based on the recovery of soil and/or fertilizer nutrients by different methods of application, including foliar.
- Conducting field surveys of crops grown in production or soil areas to determine if an element or a combination of elements is or are deficient, sufficient, in excess, or toxic.

❑❑❑

Chapter 20

Rapid Tissue Test

The rapid tissue test would pave way for rectifying the nutritional problems for quick recovery. However, the quantitative estimation of both plant and soil for nutrients concentration will be more useful and economic for applying fertilizers either as basal or foliar and would be the long term strategy to cope up with nutritional problems. To assess the nutrient status by tissue test, different parts of plant should be taken as indicator tissue (Table 1).

Table 1. Leaf sampling for tissue testing.

Crops	Nutrients					
	N	P	K	Ca	Mg	S
Cereals	Stem/Midrib	Leaf blade	Leaf blade	Leaf lamina	Leaf lamina	Leaf blade
Pulses	Petiole	Leaf blade	Leaf blade	Leaf lamina	Leaf lamina	Leaf blade
Oil seeds	Petiole	Leaf blade	Leaf blade	Leaf lamina	Leaf lamina	Leaf blade
Cotton	Petiole	Petiole	Petiole	Petiole	Petiole	Petiole
Banana	Leaf lamina	Leaf lamina	Leaf lamina	Leaf lamina	Leaf lamina	Leaf lamina
Papaya	Petiole	Petiole	Petiole	Petiole	Petiole	Petiole
Vegetables	Petiole, Leaf blade	Petiole, Leaf blade	Petiole, Leaf blade	Petiole, Leaf blade	Petiole, Leaf blade	Petiole, Leaf blade

Fruit trees

Either leaf blade/mid rib/leaf lamina can be taken.

Ornamentals, Tea, Coffee, etc.

The leaf blade should be taken.

Procedure for tissue test

a) Nitrogen:

Reagent: 1% diphenylamine in concentrated sulphuric acid.

Small bits of leaf or petiole are taken in a petridish and a drop of 1% diphenylamine is added. The development of blue colour indicates the presence of nitrate-nitrogen. The degree of colouration indicates the amount of nitrogen present in the leaf.

Dark blue: Sufficient Nitrogen

Light blue: Slightly deficient Nitrogen

No colour: Highly deficient Nitrogen

b) Phosphorous:

Reagents: 1. Ammonium molybdate solution, 2. Stannous chloride.

Eight g of ammonium molybdate is dissolved in 100 ml of distilled water. To this, add 126ml of conc. Hydrochloric acid (HCl) and volume is made up to 300 ml with distilled water. This stock solution is kept in an amber coloured bottle and at the time of use it is taken and diluted in the ratio of 1:4 using distilled water.

A tea spoonful of freshly chapped leaf bits are taken in a test tube and 10 ml of ammonium molybdate reagent is added and kept for few minutes. After shaking a pinch of stannous chloride is added. Colour development is observed.

Dark blue: Sufficient Phosphorous

Bluish green : Slightly deficient Phosphorous

No colour: Highly deficient Phosphorous

c) Potassium:

Reagent: 1. Sodium Cobalt nitrate reagent, 2. Ethyl alcohol (95%).

Take 5 g of cobalt nitrate and mix with 30 g of sodium nitrate in 80 ml of distilled water. To this, 5ml of glacial acetic acid is added. The volume is made up to 100 ml distilled water. Dilute reagent is prepared (5 ml) with 15 mg sodium nitrate to 100 ml using distilled water.

Finally cut leaf bits are taken in a test tube and 10 ml diluted reagent is added and shaken vigorously for few a minutes and kept for 5 minutes. Then, add 5ml of ethyl alcohol reagent and allowed to stand for 3 minutes. The solution is observed for the formation of turbidity.

No turbidity : Deficiency of Potassium

Slightly turbidity: Moderate deficiency

High turbidity : Sufficient Potassium

d) Calcium:

Morgan's Reagent: 30 ml of glacial acetic acid and 100 g of sodium acetate are dissolved in a little of distilled water.

Procedure: 0.5 g of finely cut plant material is taken in a test tube and add 5 ml of Morgan's reagent. After allowing it to stand for 15 minutes, 2 ml of glycerin and 5 ml of 10% ammonium oxalate is added and the solution is shaken for 2 minutes. The turbidity resembling after 15 minutes indicate the amounts of calcium in normal plant tissue.

e) Magnesium:

Reagents

1. 5% pure sucrose solution
2. 2% Hydroxylamine hydrochloride
3. Titan yellow solution
4. 10% Sodium hydroxide

150 mg of Titan yellow is dissolved in 75 ml of 95% ethyl alcohol and add 25 ml distilled water. This solution is stored in darkness.

Procedure

To a teaspoonful of finely cut material, following reagents are added in sequence. One ml of 5% sucrose solution, 1 ml of 2% Hydroxylamine hydrochloride and 1 ml of Titan yellow. Finally, the solution is made alkaline with 2 ml of 10% NaOH. Red colour indicates the presence of magnesium and yellow colour indicates absence of magnesium.

f) Iron:

Finely cut leaf materials (0.5 g) are taken into a glass vial and one ml of con. HCl is added to it. After 15 minutes, Ten ml of distilled water and 2-3 drops of con HNO_3 are added. Ten ml of this solution is pipetted out into a specimen tube, after 2 minutes, 5 ml of 20% ammonium thiocyanate is added and stirred. Further, 2 ml of amyl alcohol is added, shaked well and allowed to stand for few minutes. The intensity of red colour in amyl alcohol layer indicates the presence of iron.

❑❑❑

Part III :
DIAGNOSIS OF NUTRITIONAL DISORDERS

Chapter 21

Nutrient Deficiency Symptoms of plants

A. Principles and Development of Deficiency Symptoms

Growing plants act as integrators of all growth factors and are the products in which the grower is interested. Therefore, careful inspection of the growing plant can help to identify a specific nutrient stress. If a plant is lacking in a particular nutrient, characteristic symptoms may appear. Deficiency of a nutrient does not directly produce symptoms. Rather, the normal plant processes are thrown out of balance, with an accumulation of certain intermediate organic compounds and a shortage of others. This leads to the abnormal conditions recognized as symptoms. Visual evaluation of nutrient stress should be used only as a supplement to other diagnostic techniques (i.e., soil and plant analysis). Nutrient deficiency symptoms may be classified as follows:

- Complete crop failure at the seedling stage.
- Severe stunting of plants.
- Specific leaf symptoms appearing at varying times during the season.
- Internal abnormalities such as clogged conductive tissues.
- Delayed or abnormal maturity.
- Obvious yield differences, with or without leaf symptoms.
- Poor quality of crops, including differences in protein, oil, or starch content, and storage quality.
- Yield differences detected only by careful experimental work.

Each symptom must be related to some function of the nutrient in the plant. A given nutrient may have several functions, which makes it difficult to explain the physiological reason for a particular deficiency symptom. For example, when N is deficient, the leaves of most plants become pale green or light yellow. When the quantity of N is limiting, chlorophyll production is reduced and the yellow pigments viz., carotene and xanthophylls are shown through a number of nutrient deficiencies produced such as pale green or yellow leaves and the deficiency must be further related to a particular leaf pattern or location on the plant.

Nutrient deficiency symptoms appear only after the nutrient supply is so low that the plants can no longer function properly. In such cases, it would have been profitable to have applied fertilizer long before the symptoms appeared. If the symptoms are observed early, it might be corrected during the growing season. Since the objective is to get the limiting nutrient into the plant as quickly as possible with some nutrients and under some conditions, this may be accomplished with foliar applications or side dressings.

Diagnostic Process

Many factors influence the environment in which crop grows. Visual stress symptoms on plants reflect these factors and may be related to specific identifiable causes. Some common plant symptoms and potential causes are reported in Table 1.

Table 1. Guide to plant diagnosis

Poor growth with no distinct leaf abnormal pattern	Soil water excess or shortage Soil compaction Excess salt or sodium Root disease organisms Nematodes Root insects Weed competition Herbicide or other pesticide injury
Distinct leaf chlorosis or necrosis	Nutrient deficiency Specific ion toxicity Viral or fungal disease Insect injury Herbicide or other pesticide injury Cold and wet soils

Discoloration or necrosis of internal tissue	Plant pathogens NematodesInsects Uptake of excess chemicals from soil
Discoloration, overgrowth, surface injury or roots with multiple growth and suckering	Plant pathogens Nematodes Nutrient deficiency Herbicide or other pesticide injury

(Swarup and Ganeshmurthy, 1998)

Development of deficiency

The expression of deficiency symptom is due to the altered physiology caused by deficiency of an element. Deficiency or absence of any one of the essential nutrient element leads to disruption in normal morphology and physiology of the plant. It is logical to believe that abnormalities do not appear suddenly, but there must be some period between the beginning of subtle change at the molecular level and manifestation of symptoms on the plant morphology. In other words, these changes affected through nutrition imbalance must begin early from the formation of cell components leading to the tissue and organ development. Therefore, considerable damage has already occurred to the plant by the time the abnormalities are macroscopically visible in the form of what is commonly called visual deficiency symptoms.

Some of the common symptoms for all the nutrients are given below

- **Chlorosis:**

  It is a physiological disease that occur due to deficiency of mineral elements. (eg. N, K, Mg, Fe, Zn, Mn, S). Leaves become abnormally yellow or white due to reduction of chlorophyll contents.

- **Mottling:**

  It is a condition of plant surface marked with coloured spots due to accumulation anthocyanin pigment (eg. due to deficiency of N, K, Mg, P, and S).

- **Necrosis:**

  It refers to the patch of dead tissues and occurs due to deficiency of K, Mg, Zn, Ca, Mo, Mn, Fe, or B.

- **Bronzing:**

  Development of bronze/copper colour on the leaves and various parts (eg. K).

- **Die back:**

  Collapse of growing tip, affecting the youngest leaves and buds (eg. K, Ca, B or Cu).

- **Scorching:**

  Burning of tissues accompanied by light brown colour. (eg. K, Ca, B or Cu)

- **Firing:**

  Burning of tissues accompanied with dark brown or reddish brown colour (eg. Mg)

- **Rosetting:**

  Clustering of leaves due to reduced leaf size and shorter internodal distance (eg. Cu, K,Zn)

- **Distortion of leaves:**

  Irregular shaping of leaves like cupping, twisting,, hooking or curling and also wavy margins.

  Cupping: B or Mo

  Twisting/Hooking: Zn/Ca/B

  Curling: Ca, B or Zn

  Head distortion: Cu (rice), Mn (sunflower)

  Wavy margins: Zn

- **Gummosis:**

  Oozing out of cell sap in the form of gummous nature (eg. Cu as in coconut etc.)

Distribution of visible deficiency symptoms

One of the criteria for mineral deficiency diagnosis is the mobility of nutrient in phloem vessels of plants. Based on the phloem mobility, essential elements are classified in to three groups as mobile, less mobile and immobile. The distribution of visible deficiency symptoms and remobilization of mineral nutrients vary among plant parts. Based on phloem mobility, remobilization of mineral nutrients and distribution of visual deficiency symptoms, quick diagnosis can be made. In order to achieve this objective, the following flow chart can be used. (Flow chart a and b).

Distribution of nutrient deficiency symptoms in crops

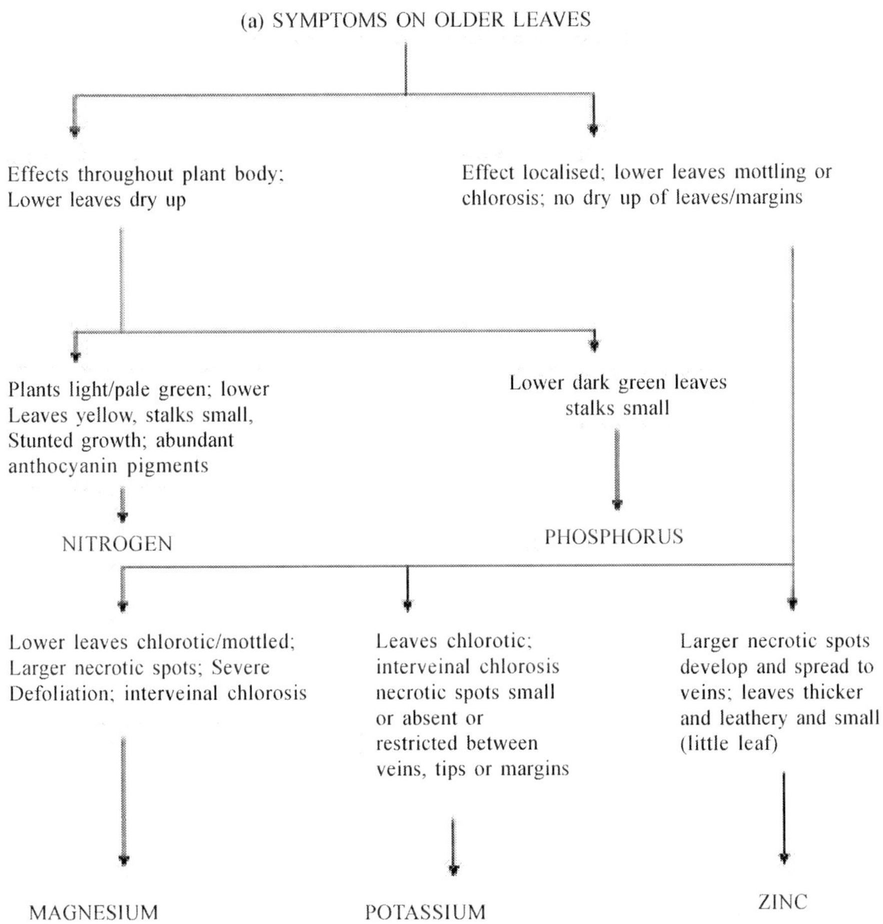

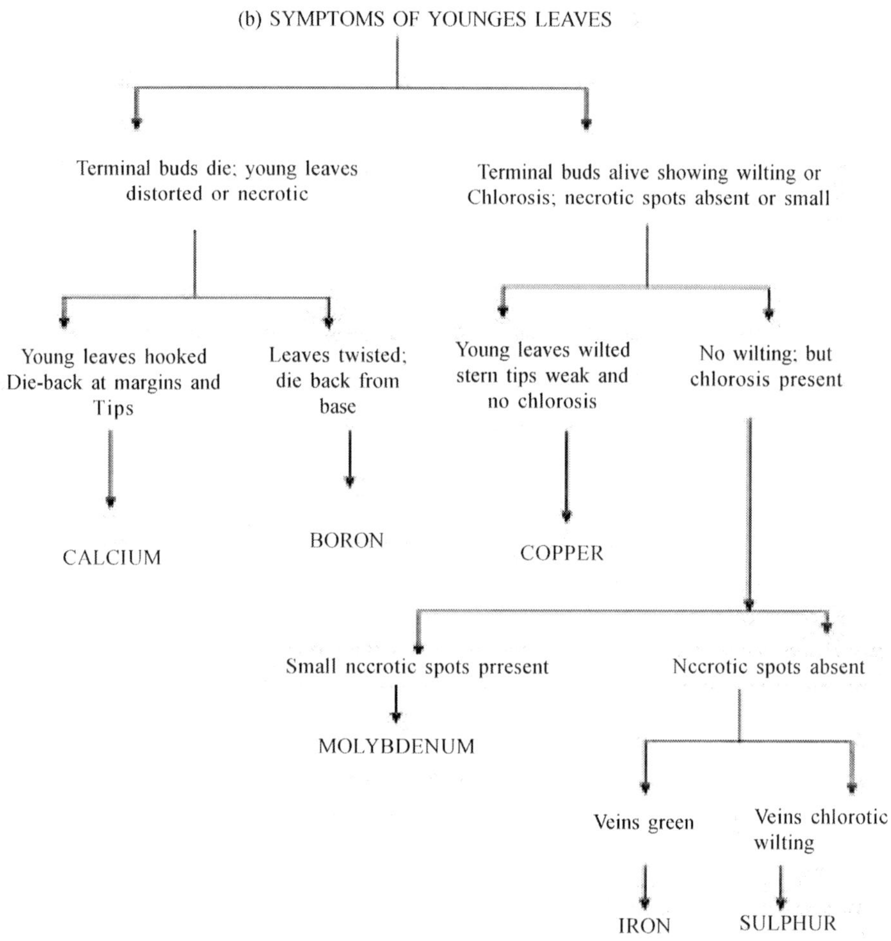

Similarity in deficiency symptoms

The deficiency symptoms in many cases are not always clearly defined and in cases, the symptoms can be common to other causes or may be masked by other nutrients. Some typical examples are given below:

- N deficiency can be confused with S deficiency.
- Calcium deficiency can be confused with B deficiency.
- Fe deficiency can be confused with Mn deficiency.
- Leaf stripe disease of oat can be confused with Mn deficiency.

- Effect of virus, little leaf etc. can be confused with Zn deficiency and / or B deficiency.
- Brown streak disease of rice can be confused with Zn deficiency.

Indicator plants

The genetically determined nutrient uptake, translocation and utilization of the different species and varieties play an important role in diagnoses of nutrient deficiencies. This property results in identification of indicator plants of particular deficiencies. These indicator plants help to diagnose the nutrient deficiency very easily. These plants are more sensitive or susceptible to particular deficiencies and the deficiency symptoms manifest themselves well ahead of other crop species. Indicator plants produces deficiency symptoms very clearly and characteristic to that particular deficient nutrient elements (Table 3).

Table 3. Indicator plants

On the basis of nutrients		**On the basis of plants**	
Nutrient	**Indicator plant**	**Plant**	**Nutrient deficiency/ toxicity**
N	Maize, Small grains, Cereals, Mustard, Apple, Citrus.	Oat	Mg, Mn and Cu deficiencies
P	Maize, Barley, Tomato	Wheat and barley	Mg, Cu and some time Mn deficiency
K	Maize, Potato, Tobacco, Cotton, Beans	Sugar beets	B and Mn deficiencies
Ca	Lucerne, Leguminous Crop	Maize	N, P, K, Mg, Fe, Mn and Zn deficiencies
Mg	Potato, Cauliflower	Potatoes	K, Mg and Mn deficiencies
S	Lucerne, Mustard	Rape	N, P and Mg deficiencies
Fe	Sorghum, Barley, Citrus, Peach	Brassica species	K and Mg deficiencies
Mn	Maize, Wheat, Oat, Citrus, Pea, Apple, Cherry, Radish	Sudan grass	Fe and Mn deficiencies

Zn	Maize, Paddy, Onion, Citrus, Peach	Sunflower	B deficiency
Cu	Maize, Barley, Oat, Onion, Citrus, Tomato, Tobacco	Cauliflower	B and Mo deficiencies
B	Lucerne, Turnip, Cauliflower, Apple, Peach	Flax	Zn deficiency
Mo	Legumes Cauliflower, Mustard, Citrus, Oat	Barley	B, Mn and Al toxicity
Cl	-	Cucumber	N and P excess
Ni	-	Sugar beets	Cu excess

(Subramanian *et al.*, 2004)

Drawbacks of using nutrient deficiency symptoms

- Similar symptom may be caused by several nutrient deficiencies.
- Deficiency of one nutrient may be related to excessive quantity of another. viz., Application of Fe to a marginally Mn deficient soil may induce Mn deficiency.
- Plant stresses (diseases, insects, herbicide damage etc.) can be difficult to distinguish from nutrient deficiencies. Example - Leaf hopper damage can be confused with B deficiency in lucerne.
- Visual symptoms may be caused by more than one factor. viz., Sugars in maize combine with flavones to form anthocyanins (purple, red, and yellow pigments), and their accumulation may be caused by low P availability, low soil temperature, insect damage to the roots, or N deficiency.
- Nutrient deficiency symptoms appear after the crop has been nutrient stressed therefore, it may be too late to correct the deficiency without yield loss.
- Toxicity of some nutrients and other elements can produce visual symptoms similar to nutrient deficiencies.

B. Field Crops

1. Rice

Nitrogen

Nitrogen deficiency is the most commonly observed nutrient disorder in rice. Old leaves and sometimes all leaves become light green and chlorotic at the tip. Leaves die under severe stress. Deficient leaves are narrow, short, erect, and lemon yellowish. The entire field may appear yellowish. Nitrogen deficiency often occurs at critical growth stages such as tillering and panicle initiation, when the demand for nitrogen is high.

Phosphorus

Stunted, dark green plants with erect leaves and reduced tillering may signal P deficiency. Leaves are narrow, short, very erect, and 'dirty' dark green. Stems are thin and spindly, and plant development is retarded. The number of leaves, panicles, and grains per panicle may also be reduced. Young leaves may appear to be healthy, but older leaves turn brown and die. Red and purple colours may develop in leaves if the variety has a tendency to produce anthocyanin. Leaves appear pale green when N and P deficiency occur simultaneously. Phosphorus is particularly important in early growth stages.

Potassium

Dark green plants with yellowish brown leaf margins or dark brown necrotic spots first appear on the tips of older leaves. Under severe K deficiency, leaf tips are yellowish brown. Symptoms appear first on older leaves, then along the leaf edge, and finally on the leaf base. Upper leaves are short, droopy, and "dirty" dark green. Older leaves change from yellow to brown and, if the deficiency is not corrected, discolouration gradually appears on younger leaves. Leaf symptoms of K deficiency are similar to those of tungro virus disease. Unlike K deficiency, however, tungro occurs as patches within a field, affecting single hills rather than the whole field.

Calcium

Calcium moves in the plants' through transpiration stream and is deposited mainly in the older leaves. Deficiencies are found in the youngest leaves and growing points, which have low rates of transpiration. Youngest emerging leaves show the following symptoms only under severe Ca deficiency: Interveinal chlorosis and leaf may bend downwards because the leaf margins have failed to expand fully. White or bleached, rolled, and curled tips of leaves. Necrosis along the lateral margins of leaves. Stunting and death of growing points. Mature and older leaves are generally unaffected, but in severe situations old leaves turn brown and die.

Magnesium

Magnesium deficiency include orange-yellow interveinal chlorosis on older leaves, plants pale-coloured with interveinal chlorosis first appearing on older leaves and later on younger leaves as deficiency becomes more severe and green colouring appears as a "string of beads" in which green and yellow stripes run parallel to the leaf. In severe cases, chlorosis progresses to yellowing and finally necrosis in older leaves. Magnesium deficient leaves are wavy and droopy due to an expansion in the angle between the leaf sheath and leaf blade.

Sulphur

Sulphur deficiency results in yellowing of the whole plant and chlorosis is more pronounced in young leaves, the tips of which may become necrotic. Other symptoms are: reduced plant height and stunted growth, reduced number of tillers, fewer and shorter panicles, reduced number of spikelets per panicle, Seedlings in nursery beds become yellowish and their growth is retarded. Seedling mortality after transplanting is high. Sulphur deficient rice plants have less resistance to adverse conditions.

Zinc

Deficiency symptoms appear between two to four weeks after transplanting. The symptoms are dusty brown spots on upper leaves of stunted plants, uneven plant growth and patches of poorly established hills in the field, but the crop may recover without intervention, increased spikelet sterility, chlorotic midribs, particularly near the leaf base of younger leaves, leaves lose turgor and turn brown as brown blotches and streaks enlarge and coalesce, white line sometimes appears along the leaf midrib and leaf blade size is reduced.

Under severe Zn deficiency, tillering decreases and can stop completely and time to crop maturity increases.

Copper

Chlorotic streaks on either side of midrib followed by necrotic regions on leaf tips. Deficient leaves appear bluish green. New leaves don't unroll. Leaf tips look like needle. High spikelet sterility. Susceptibility to lodging because of insufficient lignin synthesis.

Iron

The symptoms of iron deficiency are: yellowing or chlorosis of the interveinal areas of the emerging leaf. Later, the entire leaf turns yellow, and finally turns white. If the deficiency is severe, the entire plant becomes chlorotic and dies.

Manganese

Manganese deficiency affects photosynthesis and protein synthesis. Manganese deficient plants are often deficient in phosphorus. Manganese deficiency symptoms are pale grayish green interveinal chlorosis spreads from the tip to the leaf base, necrotic brown spots develop later and leaf becomes dark brown and newly emerging leaves are short, narrow, and light green. Manganese deficient plants are shorter, with fewer leaves, less weight, and have smaller root system at tillering. Plants are usually stunted but tillering is not affected. Affected plants are more susceptible to brown spot.

Boron

Boron deficiency symptoms in rice are: White and rolled leaf tips of young leaves, Reduction in plant height, Tips of emerging leaves are white and rolled. Death of growing points, but new tillers continue to emerge during severe deficiency. If affected by B deficiency at the panicle formation stage, plants are unable to produce panicles.

Molybdenum

Resembles N deficiency (older leaves turn chlorotic). In addition, necrotic spots are seen at leaf margins because of NO_3^- accumulation.

2. Millet Crops

a) Sorghum

Nitrogen

Deficient plants are pale green to pale yellow in colour, grow more slowly and often show delayed flowering and maturity. The stems tend to be thin and spindly and in some cultivars, red streaks develop on the lower leaf sheaths. Symptoms appear first on older leaves and then advance up to younger leaves. The leaf blades progressively change colour from dark green to pale green. A pale yellow chlorosis develops and this is followed by a brown necrosis. The chlorosis and necrosis begin near the tip of the leaf blade and advance towards the base until the whole blade turns yellow and brown. The chlorosis and necrosis may advance along the midvein in a V-shaped pattern, along the margins, or on a broad front. The boundary between the healthy green tissue and the chlorotic tissue is diffuse. In most cultivars, the chlorosis is a uniform pale yellow colour, but in others, small brown or dark grey necrotic spots or streaks appear within the chlorotic.

Phosphorus

Affected plants appear stunted with thin stems and dark green leaves. When the deficiency becomes more severe, growth is greatly reduced and turns of dark red or purple develop on the sheaths and blades of older leaves. Eventually, the red to purple pigmentation may extend the full length of the leaf blade. When the deficiency is very severe, a dark brown necrosis often develops near the tip of the leaf blade and advances inwards along the margins towards the base. Eventually, the affected leaves wither, die, turn dark brown, and hang down around the stem.

The degree of red or purple pigmentation varies from cultivar to cultivar. Some cultivars develop very little red or purple pigmentation. In these cases, the older leaves usually turn yellow and brown as a dark yellow chlorosis develops near the tip and advances inwards and along the margins towards the base of the leaf blade. The dark yellow chlorosis is usually followed by a dark brown necrosis. The boundary between the healthy green tissue and the chlorotic or necrotic tissue is usually sharp and distinct.

Potassium

Potassium deficient plants lack vigour, flowering and maturing more slowly than healthy plants. The stems are shortened and thin, and the older leaves display a marginal necrosis which may extend the full length of the leaf blade. However, the pattern of development and pigmentation of the necrosis differs considerably among cultivars. In some cultivars, a yellow to white chlorosis develops near the margins at the tip of the leaf blade. The chlorosis then advances along the margins towards the base of the leaf blade as the severity of the deficiency increases. The chlorosis is followed by pale brown necrotic lesions which often join together and may become continuous from the tip to the base of the leaf blade. A number of diseases of sorghum cause lesions which could be confused with the purple and brown lesions developed on some cultivars under potassium deficiency. These diseases include leaf blight, oval leaf spot, and anthracnose.

Calcium

Plants suffering from mild calcium deficiency or from a temporary shortage of calcium supply may show little depression in growth, but develop a characteristic 'torn leaf' symptom on their upper leaves. These younger leaves are short and are often held erect. The margins of the leaf blade usually turn pale green or pale yellow and tear easily to give a characteristic serrated or saw-tooth appearance. In some cultivars, the leaf tips become deformed and curl to form sword-like tips, or bend down and break off leaving blunt tips to the leaves. When the deficiency becomes severe, growth is greatly depressed. The upper internodes may be very short and the young leaves crowded together to give the appearance of a rosette. The young leaves often fail to unroll from the whorl and a number of leaves may become joined together at their tips. As the leaves try to expand, the blades become folded and sometimes assume a laddered appearance.

The apical meristem often dies when calcium is severely deficient. When this occurs, many tillers develop from the basal nodes and the newly formed young leaves on these tillers show symptoms of severe calcium deficiency. Flowering and maturity are delayed by mild calcium deficiency, but head size is often unaffected. However, when calcium deficiency is severe, the head may not form or it may be partly barren.

Magnesium

Plants suffering from magnesium deficiency lack vigour and are often stunted with thin stems. The leaves are usually pale green to yellow in colour with many brown lesions. The symptoms develop firstly on older leaves and advance upwards to younger leaves. When the deficiency is severe, the whole plant appears pale green or pale yellow in colour. The first symptom to develop is a yellow chlorosis which appears in interveinal tissues, usually about halfway along the leaf blade.

Sulphur

Sulphur deficient plants lack vigour, are stunted, pale green to yellow in colour and have thin stems. The blades of young leaves are shorter and held more erect than usual. Initially, these blades turn pale green in colour while the older leaves remain dark green. A pale yellow chlorosis develops in the basal tissues of these young leaves and advances in a broad front towards the tip until the whole blade turns a uniform pale yellow colour. In some cultivars, the chlorosis advances more rapidly down the interveinal tissues and a pale interveinal chlorosis appears before the blades turn a uniform yellow colour.

When the deficiency is prolonged or very severe, a second set of symptoms may develop on the older leaves of some cultivars. Initially, the older leaves turn from dark to pale green in colour. A dark yellow chlorosis then develops near the tip of the leaf blade and advances along the margins towards the base. Eventually, the chlorotic margins turn brown, and these older leaves wither, die, and hang down around the lower stem.

Zinc

Zinc deficiency causes the internodes to become very short, resulting in stunted plants. Flowering and maturity are also delayed by zinc deficiency. The first symptoms on the foliage appear on the youngest leaves even before they unroll from the whorl; a pale yellow chlorosis develops between the margins and midvein near the base of the leaf blade. As the deficiency becomes more severe, the chlorosis extends towards the leaf tip and often turns nearly white or sometimes pale brown. The midvein and margins usually remain green and the boundary between the healthy green tissue and the chlorosis is diffuse.

When a severe zinc deficiency continues for a relatively long time, symptoms develop also on some of the older leaves about halfway down the stem. A white or brown necrosis and purple pigmentation develop in interveinal tissues near the tip of affected leaves. The necrosis and pigmentation advance towards the leaf base.

Copper

A deficiency of copper depresses growth and affected plants become stunted with thin stems and pale green foliage. The symptoms appear only on leaves which are still within the whorl and on young mature leaves. In mild deficiencies, these leaves turn pale green in colour and an interveinal pale yellow chlorosis develops at the base of the blades. The whorl of expanding leaves may remain tightly rolled and become bent to one side.

When the deficiency is severe, a pale brown necrosis develops at the tips of the youngest leaves. The affected leaf tips wither and die, often twisting into a spiral shape. The basal tissue of the leaf blade usually remains green, and the twisted brown tip hangs down from the green basal tissue. In very severe deficiencies, this symptom develops on leaves still within the whorl and a number of young leaves often become joined together at their tips to give a laddered pattern resembling that of calcium deficiency.

Iron

The characteristic symptom of iron deficiency in sorghum is a pale yellow interveinal chlorosis which develops on the younger leaves. The chlorosis begins in interveinal tissues near the base of the leaf blade and eventually extends the full length of the blade. The distinctive striping of the leaf is caused by the sharp boundaries between the dark green veins and the yellow interveinal tissues. However, as the severity of the deficiency increases, the veins fade in colour and turn pale yellow. Eventually, white necrotic areas develop within the chlorotic tissues and the whole leaf blade turns pale yellow to white in colour. In some cultivars, pale red streaks may develop within the yellow chlorotic tissues, while in other cultivars, long brown lesions may develop within the white necrotic areas.

Manganese

Manganese deficient plants are pale green to yellow in colour and have thin spindly stems. If the deficiency is mild, growth may be depressed, but flowering and maturity will not be greatly delayed. However, severe deficiencies can cause death of young plants. The symptoms develop on young leaves which may be still emerging from the whorl. Initially, small discrete chlorotic lesions develop between the veins, usually about midway along or towards the tip of the blade. These chlorotic lesions join together to produce a faint interveinal chlorosis which often affects only a short length of the leaf blade but sometimes may extend almost the full length of the blade. Small, discrete dark reddish brown or black lesions develop within the chlorotic tissues. Eventually, these brown or black lesions join together producing brown interveinal stripes over the length affected by the interveinal chlorosis.

Boron

A deficiency of boron depresses growth and causes shortening of the internodes. It may lead to severe stunting. If heads are produced, they may be partly or completely barren. The leaves are short, dark green in colour and held erect. They are often brittle and easily torn or broken. When the deficiency is severe, the apical meristem often dies and tillers develop. If the deficiency occurs just before the boot stage of growth, the tillers may appear from nodes well above ground level.

Characteristic white or transparent lesions develop in interveinal tissues on the young leaves, often while they are still within the whorl. Initially, the lesions develop in an intermittent pattern and appear firstly in the midsection of the leaf blade. The boundary between the healthy green tissue and the white lesions is usually diffuse. As the deficiency increases in severity, these lesions become almost transparent and join together. In some cultivars, the transparent lesions may extend beyond the midsection of the blade. The veins appear to remain intact and green, even when the deficiency is very severe.

b) Pearlmillet

Nitrogen

Deficient plants appear stunted, thin and spindly, and grain head size and numbers of grains per head are severely reduced. In mild deficiencies, the entire plant appears uniformly light green in colour. If the deficiency becomes more severe, a pale yellow chlorosis begins at the tips of older leaves, then proceeds towards the base along the midrib in a V-shaped pattern - a specific symptom of nitrogen deficiency in pearlmillet. As the symptoms advance, the pale yellow chlorosis is followed by a pale brown necrosis. In the late stages, dead leaves hang down and form a thatch of dry leaves near the base of the plant.

Phosphorus

Deficient plants appear dark green, stunted, thin and spindly, and have delayed maturity. The size of grain heads and individual grains become small and both contribute to low grain yield. The deficiency symptoms appear first, and become more severe, on older leaves while young leaves usually remain unaffected. Dark green old leaves turn purple or purple-red in colour. Stem and leaf sheaths of lower leaves become purple-red in colour.

Potassium

Potassium deficiency causes shortening of the internodes, dwarfing of plants and a general loss of healthy and green growth. Affected plants produce small grain heads. Pearlmillet is very sensitive to K deficiency. Symptoms begin as a pale yellow chlorosis on the tips of older leaves covering marginal leaf tissue. In severe cases, the chlorosis is followed rapidly by pale brown necrosis and both (chlorosis and necrosis) advance down the margins towards the base.

Sulphur

Deficient plants appear pale green, stunted, thin and spindly with delayed maturity. The size of grain heads and individual grains are small and contribute to low grain yields. The deficiency symptoms appear first and become more severe on younger leaves. The younger leaves become pale green, then pale yellow and eventually turn yellow. The pattern of yellowing on the entire leaf appears uniform, covering both veins and interveinal tissues alike.

Zinc

Zinc deficiency prevents internode elongation and results in the crowding of the top-most leaves- producing a fan-shaped appearance. Like maize and sorghum, pearlmillet is sensitive to Zn deficiency. Since Zn is not readily transferred from older to younger plant parts, symptoms develop first, and are more severe on younger leaves- while older leaves can remain green and apparently healthy. White to yellow broad bands of bleached tissue appear on each side of the midrib, beginning at the base of the leaf. The midrib and the leaf margins remain green. Mild deficiency of Zn may show an interveinal striping on young leaves some what similar to Fe and Mn deficiency.

Iron

Deficiency symptoms appear first on younger leaves, while older leaves remain normal. Pearlmillet is relatively less sensitive to Fe deficiency than sorghum or maize. A pale yellow chlorosis develops in interveinal tissues, leaving the veins green and prominent. In mild deficiencies, the top most (youngest) leaves of plants develop temporary fading of interveinal tissues and maintain prominent green veins. Plants can recover and regain a normal appearance with time. When the deficiency persists, and becomes more severe, interveinal tissues of affected leaves turn a distinct pale yellow with prominent green veins. This interveinal chlorosis uniformly extends the full length of the leaf- interveinal chlorosis of the top (younger) leaves is a specific symptom of Fe deficiency in pearlmillet. In later stages of deficiency, the prominent green veins also fade and become light green to pale yellow.

c) Maize

Nitrogen

Under conditions of nitrogen deficiency, the proteins in the older leaves of the plant are hydrolysed and the soluble nitrogenous compounds move to the actively growing regions of the plant. The yellowing of the leaves is due to the breakdown of the chloroplast protein causing the loss of the chloroplasts and the disappearance of the chlorophyll. The older leaves are the first to show the symptoms which spread progressively to the young leaves of the plant and the degree to which these are affected is a measure of nitrogen deficiency. When the nitrogen deficiency becomes acute, the leaf turns brown and disintegrates.

Phosphorus

Phosphorus deficiencies are not always easy to diagnose. Slow growth of the seedlings and a dark green colour with a purplish tinge of stems and leaves are fairly typical. The purple colouration is due to a decrease in protein synthesis as a result of P deficiency leading to increase in sugar concentration in the plant tissues. Higher sugar concentrations promote the synthesis of anthocyanin, the purple pigment. However, the symptoms may be caused by factors other than P deficiency. In certain varieties, the purplish tinge of leaves and stems of the young plants (caused by the formation of anthocyanin) is a normal heritable character. Also, any factor that disturbs the metabolism of maize besides P deficiency such as, low soil temperature or even insect damage to the roots often effects the metabolic pathways in such a way that large quantities of anthocyanin are produced. In P-deficient plants, during pollination, the silks emerge slowly resulting in defective ears with poorly-developed grains.

Potassium

Characteristically, potassium deficient young maize plants develop slowly. The leaves are light green or streaked with yellow and the tips and margins of the leaves become necrotic. Older plants show similar symptoms but the marginal browning of the leaves is more conspicuous. Following prolonged deficiency, plants become stunted and weak. Though stunted, the plants wilt, tend to lodge because of the poorly developed root system and become more prone to the attack of stalk rots. Plants may be barren, ears when produced are small, pointed and barren at the tip and the grains are of poor quality.

Calcium

Calcium deficiency is rare in maize. If deficient, leaf tips stick to the next lower leaf, creating a ladder like appearance. Plants may be severely stunted because calcium is immobile in the plant. It is not translocated from old to growing plant tissue that needs calcium.

Magnesium

Magnesium deficiency is first observed as yellow to white interveinal striping of the lower maize leaves. Dead, round spots sometimes follow, which give the impression of beaded streaking. Older leaves become reddish-purple and the tips and edges may become necrotic if the deficiency is severe.

Sulphur

Sulphur deficiency shows as a general yellowing of the foliage, similar to nitrogen deficiency. Yellowing of the younger upper leaves is more pronounced with sulphur deficiency than with nitrogen deficiency because sulphur is not easily translocated in the plant. Stunting of plants and delayed maturity also are symptoms. Interveinal chlorosis of the youngest leaves may occur.

Zinc

Zinc deficiency symptoms may vary according to the degree of deficiency and the hybrid concerned. During the early vegetative development of the maize plant, a pale yellowish-green stripe appears on each side of the midrib on the older leaves of moderately deficient plants. These stripes extend from the base of the leaf blade to the middle or tip of the leaf and the midrib and leaf margin usually remain green. If the deficiency persists, these stripes may necrose or may change to a reddish-bronze colour. In case of extreme deficiency, the whole plant becomes light green and growth decreases. Plants exhibiting these symptoms may die before tassel emergence, may have barren ears or ears which are abnormally shaped and improperly filled. The new leaves may be nearly white in colour. Hence, the name of 'white bud' given to the Zn-deficiency.

Copper

Copper deficiency is rare in maize. The youngest leaves are yellow as they come out of the whorl, and the tips may die. Leaves become streaked, causing an appearance that is similar to iron deficiency. The stalk is soft and limp.

Iron

Iron deficiency turns the interveinal area along the length of the upper leaves pale green to nearly white.

Manganese

Moderate deficiency of manganese in maize is evidenced by a uniform streaking of all leaves, which are also lighter green than normal leaves. Severe deficiency results in long white streaks and the tissue in the middle of the chlorotic area dies and may fall out of the leaves. The stalks of affected plants are thinner than that of normal plants.

Boron

Maize is very sensitive to boron. Leaves have small dead spots and are brittle. Boron is not readily translocated in the plant and as a result, upper internodes do not elongate. Tassels and ear shoots are reduced and may not emerge.

Molybdenum

Maize is rarely ever deficient in molybdenum. However, if deficiency occurs, older tips die at the ends, along the margins, and then between the veins. Molybdenum deficiency also has a striking effect on pollen formation. Tasseling, anthesis, and development of anthers in maize are inhibited due to Mo deficiency. With Mo deficiency, the internodes are short, the leaf area is reduced, and leaves appear chlorotic, which is more severe in younger leaves than in older leaves. These symptoms of Mo deficiency in maize differed from those in other plants in the lack of typical mottling and whitish appearance and softness of the edges. Under severe Mo deficiency, young leaves failed to unroll, the older leaves shriveled, and the whole plant wilted and died.

d) Wheat

Nitrogen

Nitrogen deficiency first appear on the oldest leaves with the new leaves remaining relatively green. The older leaves become paler than newer leaves with chlorosis beginning at the tip and gradually merging into light green further down the leaf. As the chlorosis spreads to other leaves, the oldest leaves become totally chlorotic changing from yellow to almost white in colour.

Phosphorus

During the early stages of vegetative growth, the most noticeable feature of phosphorus deficiency in wheat is the reduced growth and vigour of the plant. The colour of all leaves of P-deficient plants becomes a dull dark green with slight mottling of the oldest leaf. Leaves appear coiled to a greater degree than normal; old leaves sometimes encase younger leaves. New growth can appear spindly and may remain folded for some time. Specific symptoms are, however, on the old leaves. Chlorosis begins at the tip of the old leaf and moves down the front of the leaf. Other common symptoms of P deficiency are delayed and irregular plant maturity and small heads.

Potassium

Potassium deficiency symptoms always appear in the oldest leaves of wheat, although growth of the whole plant can be affected prior to symptoms with all leaves having an unthrifty and spindly appearance. Under severe K deficiency, necrosis in the oldest leaves begins as a necrotic speckling along the length of the leaf, spreading quickly to the tip and the margins. Complete death of old leaves is common and plants in the field may appear to have dried prematurely due to drought stress.

Calcium

Symptoms of calcium deficiency always appear in new growth. As with B deficiency, the main roots become shortened with a proliferation of stunted laterals. Wheat leaves do not become chlorotic with Ca deficiency; old leaves, particularly, retain their dark green colour. The first definite symptom is a necrotic spotting about the middle of the leaf on the newest growth. This area quickly expands and the leaf collapses midway before unrolling.

Magnesium

The symptoms of magnesium deficiency resemble those of K and Fe deficiencies in a number of respects. Unlike K deficiency, the new leaves of Mg-deficient wheat plants are pale in contrast to the old leaves. This, however, is similar to Fe deficiency. The new leaves soon become chlorotic and remain unopened, resulting in a twisted appearance that gives the plant an unthrifty look reminiscent of drought stress. If magnesium deficiency is severe enough, the entire length of the leaf will remain folded or rolled. After some time, several leaves including the new shoot may be folded or rolled. Over time, the chlorosis of the new leaves referred to earlier becomes mottled and finally necrotic, although the leaves remain erect. Even the oldest leaves develop a mottled chlorosis and in some cases, a reddish colouration along the leaf margins.

Sulphur

Sulphur deficiency symptoms in wheat are similar to N chlorosis (i.e., a general chlorosis of the leaf). However, S deficiency does differ from N deficiency in that the whole plant is pale with a greater degree of chlorosis in the young leaves. The pattern of chlorosis of the new leaves may show gradation in intensity from tip to base, but they quickly become totally chlorotic and take on a light yellow colour. Leaf tip necrosis can occur, but this may be indicative of nitrate accumulation rather than a direct effect of S deficiency. Extreme S deficiency can result in a lack of head formation.

Zinc

Wheat is very susceptible to zinc deficiency. The first effect of zinc deficiency appears in the form of fading of the middle or the lower half of the lamina of the sub-terminal leaves. Generally, the effects appear first on the leaf next to the youngest. Soon afterwards, within 3 to 5 days from the commencement of the fading of lamina, the affected leaves develop minute reddish brown spots which tend to coalesce forming reddish-brown lesions. These lesions later turn light brown and limp. The leaf sheathes may also develop light brown or reddish-brown pigmentation. The leaves affected thus appear withered. As the symptoms intensify on the middle leaves, they spread to the younger ones. It has also been observed that these deficiency symptoms are more severe on culm than tillering. In zinc deficient plants, length of internodes, number of productive tillers, expansion of foliage, heading and grain formation are all markedly reduced. The number of spikelets in the inflorescence of zinc deficiency plans is also reduced. Grain yield may be reduced at marginally low levels of zinc supply even when foliage plant is apparently normal. Plants facing zinc deficiency may also exhibit poor setting and development of grain.

Copper

The first visual symptom of Cu deficiency in wheat is a general wilting of the whole plant occurring at early tillering, even though the soil may be at field capacity. Plants also appear lighter in colour where Cu is deficient. 'Withertip', the first characteristic symptom of Cu deficiency, appears on young leaves. It shows as a sudden dying and withering (curling) of the tip end of the leaf blade, sometimes up to half the length of the blade. The base end of the leaf, however, can remain green until normal senescence occurs. Under severe defi-ciency, as with Ca deficiency, the new growth can wither before unfolding and few heads will emerge. With less severe deficiency, heads may emerge but grains may not form in the spikelets at the tip of the head. The whole head takes on the appearance of a 'rat tail' with full grain in the base of the head, shrivelled grains in the middle of the head, and a withered necrotic tip. This rat tail appearance is sometimes confused with frost damage. Marginal Cu deficiency may decrease grain yield by producing shrivelled grain and a bending over of the stem and head. Cu is required for the lignification of cell walls and hence a deficiency decreases stem strength.

Iron

Iron deficiency is involved in the failure of plants to produce sufficient chlorophyll with the result that Fe and Mg symptoms can be similar for many plants. The nature of the chlorosis in the case of Fe deficiency is characteristic for cereals. Wheat leaves show a longitudinal interveinal chlorosis, resulting in a pattern of alternate green and yellow striping, and this pattern is more regular for Fe deficiency than for Mg or Mn deficiencies. Under severe Fe deficiency, the new growth may appear completely devoid of chlorophyll and turn white.

Manganese

In wheat, symptoms appear first in new leaves, which become pale. Light grey flecking and striping appear at the base of the youngest fully opened leaf. Under severe deficiency, the new growth may emerge with this flecking and striping over the entire length of the leaf.

Boron

As the deficiency becomes more severe, an increase in tillering occurs, new shoots have a water-soaked appearance and are paler than the older parts of the plant. Considerable distortion occurs along the margins of these leaves. Finally, new growth becomes necrotic and shoots wither.

Molybdenum

A characteristic symptom of Mo deficiency is a longitudinal yellow striping mostly on the middle leaves. Plants become very pale green and limp when compared to Mo-adequate plants. The old leaves remain greener than the remainder of the plant, although with time some plants develop necrosis of the tips and margins of the old and middle-aged leaves. Under severe deficiency, delayed maturity will occur and very deficient plants can produce empty heads. In the field, molybdenum deficiency can decrease yields by 30% before symptoms become apparent.

e) Barley

Nitrogen

Deficient plants are stunted with thin stems and pale green foliage. Affected plants lack vigour and produce small heads with few grains. In acutely deficient conditions, many young tillers fail to develop heads and die before maturity. The deficiency symptoms appear first and become more severe on older leaves. The younger leaves remain green and appear healthy.

Phosphorus

Deficient plants appear stunted with short, stout stems and dark green foliage. In acute cases, plants will often fail to produce grain heads. Deficiency symptoms appear first and become more severe on older leaves while younger leaves usually remain unaffected. Older leaves develop a dark purple colour on their tips which proceeds in a broad front towards the base. Stem and the leaf sheaths of lower leaves can also become a purple-red colour.

Potassium

Potassium deficiency causes short stem and stunted plants. In severely deficient conditions, many young tillers die before producing heads, while mature tillers produce small heads with few grains. Deficiency symptoms appear first on older leaves. Since younger actively growing leaves draw K from older parts of the plant, these young leaves often remain green and apparently healthy. On older leaves, a pale yellow chlorosis is usually followed by necrosis beginning at the leaf tip and this can advance along the margins towards the base of the leaf. Usually the mid-vein region remains green.

Sulphur

Sulphur deficiency causes poor plant growth, lack of vigour, and delayed maturity. Tiller production is depressed and often only a few tillers produce grain heads. In young crops, a general yellowing appears on the whole plant and symptoms are generally confused with N deficiency. Deficiency symptoms appear first, and become more severe on younger leaves. Younger leaves become pale yellow while older leaves usually remain green. The yellowing pattern appears uniformly across the entire leaf, affecting both the veins and interveinal tissues.

Zinc

Zinc deficiency prevents internode elongation resulting in a stunted and grass tufted appearance of plants. Crop yields are severely reduced even under mild Zn deficiency. Zinc is partly mobile in barley; therefore, the deficiency symptoms first appear on middle leaves that originate halfway up the shoot. Thus, youngest leaves at the top and oldest at the bottom, initially remain unaffected.

Iron

Iron deficiency in barley causes stunted growth, spindly stems, and pale green to yellow foliage. Iron deficiency sharply reduces tiller production and under severe cases of deficiency, many plant tillers die without producing a grain head. It is common for plants affected by severe Fe deficiency to have only a single main shoot. Because Fe is not transferred from older to younger leaves, deficiency symptoms appear first, and most severely, on younger leaves while older leaves usually remain dark green and healthy. A pale yellow chlorosis develops in interveinal tissues, leaving the veins green and prominent. Under severe Fe deficiency, new growth may appear completely devoid of chlorophyll and turn dark or pale yellow overall.

f) Oats

Nitrogen

Deficiency symptoms will be seen as yellowing first on older leaves and later on new growth. Nitrogen deficient stands will appear stunted with reduced vigour.

Phosphorus

Deficiency symptoms will appear first on tips of older leaves, base of the stem and will work its way up the plant. Purpling and browning will occur with eventual die back of the leaves.

Potassium

Symptoms first appear on both the older and actively growing leaves. Symptoms include burning or scorching of lower leaves beginning at leaf tip and continuing down the leaf margin. Plant stand will appear stunted, have delayed maturity and be more susceptible to lodging.

Sulphur

The symptoms are expressed on the newly emerging leaf tissue, because the youngest leaves cannot acquire the nutrient from the older leaves.

Copper

High nitrogen levels delay the movement of copper from old leaves to new growth, which magnifies the effects of the deficiency. There can be a 20% yield loss without showing any visual symptoms. Copper does not have movement within the plant. So, older leaves can remain relatively healthy, and the youngest leaves will show deficiency symptoms. Symptoms occur as limpness or wilting, leaf tips turn white and roll, commonly referred to as "pigtailing," poor seed set and uneven stand maturity.

Manganese

Oats are the most susceptible to manganese of the cereal crops. Manganese is partially mobile in oats, setting it apart from the other cereals. Symptoms will spread from the older to the youngest leaves. Grey to white specks will begin to form in the leaf surface, this is known as "Grey Speck."

Molybdenum

Mo deficiency in oats is known locally as 'blue chaff disease". Affected plants show a bluish colouration of the outer glumes, and the grain is pinched.

3. Pulses

a) Greengram and Blackgram

Nitrogen

The plants remain small and form only a small sized pods or seeds. The leaves remain pale green in colour. The older leaves also turn yellow and die prematurely.

Phosphorus

The symptoms of phosphorus deficiency are retarded growth with dark green foliage.

Potassium

The deficiency occurs in dry summer months because of the fixation of available potassium in the soil. Brown necrotic spots along the margin of the leaf are the typical symptoms of the deficiency. The leaves stand at an obtuse angle to the stem and the internodes remain short. At the advanced stage, the leaf margins whither.

Magnesium

The deficiency is manifested by a white-yellow discolouration of the lowest leaves in which the base of the leaf remains green longer. The interveinal region turns pale green or yellowish-green. Later, the browning can develop on the upper surface of the leaf between the veins. Superficial brown spots may also appear on the pods.

Sulphur

Sulphur deficiency can cause reduction in yield, even without having displayed the visual symptoms. Complete chlorosis of younger leaves including veins is the distinct feature of the deficiency.

Zinc

The symptoms are general chlorosis affecting most of the foliage. Defective development of pods in addition to the retarded onset of ripening leads to decline in yield.

Manganese

The deficiency first injures the younger parts of the plant with a display of sulphur-yellow colouration between the veins. In the centre of the injured leaf, often seen dark brown spots. The younger leaves often completely turn yellow and never attain their optimum growth.

b) Red gram

Nitrogen

Young leaves uniformly become pale-green, greenish-yellow or pale yellow. Old leaves show an interveinal chlorosis and only veins remain green, leaves soon abscise.

Phosphorus

Difficult to diagnose, it delays flowering. In general, plants are stunted due to failure of internode elongation. The foliage remains dark-green and old leaves shed.

Potassium

Leaf tips turn yellow or brown. Yellowing spreads from the tip outward along the margin and may coalesce with similar yellow areas at extremities of the lateral veins. The leaf tip becomes scorched as symptoms become severe. The scorching may spread around the leaf margin. There is a yellow band between the scorched area and healthy-green tissues in the early stages.

Calcium

Light-green patches, irregular in outline, appear randomly in interveinal areas or around the leaf margin. Patches may turn brown, particularly on the young and severely affected leaves; such leaves abscise. Stems are weak and prostrate.

Magnesium

Mild interveinal chlorosis, veins remain dark green. Interveinal areas become rusty brown or bronzed and may become necrotic so that narrow elongated streaks of dying tissue appear between the veins. The margins of the young leaves frequently in rolled.

Sulphur

Even yellowing across lamina of mainly young leaves. Eventually, old leaves become yellow.

Zinc

Stunted growth, narrowing of leaves with pale-green or yellow appearance. Interveinal chlorosis starting from tip of leaflets and spreading to the remaining area leaving only the midrib green. Curling and shedding of leaflet when it becomes two-thirds chlorotic.

Iron

Interveinal areas of young leaves become pale-green, veins and adjacent tissues remain dark-green. In severe cases, the entire area of the leaflet becomes chlorotic with small necrotic patches.

Manganese

Fading of the green lamina of the middle leaves without affecting the veins. The small white and brown spots, first appearing in the chlorotic areas, coalesce and form brown necrotic lesions. It reduces size, number and area of leaflets, and delays flowering by arresting the growth of apical shoots.

Boron

Boron deficiency results in dieback, rosetting, multiple branching and death of seedlings.

Molybdenum

Nitrogen fixation requires molybdenum for incorporation of the nitrogenase enzyme. Like nitrogen deficiency, it causes general yellowing of the plant.

c) Cowpea

Nitrogen

Little new growth, yellow leaves: this being more pronounced in older leaves. New shoots may be red to red-brown.

Phosphorus

Overall dark green with purple, blue or reddish cast to leaves particularly on underside, veins and stems. Foliage may be sparse, small and distorted becoming mottled and bronzy with maturity. Excess foliage with no flowers can also indicate lack of phosphorus.

Potassium

Sickly looking plants, undersized fruits, leaves showing marginal and interveinal yellowing. Yellowing starts on older leaves and progresses upwards. Leaves may crinkle, turn brown and roll upwards. Blossoms may be distorted and small. Plant has little resistance to heat, cold and disease problems.

Calcium

Young leaves are small and distorted with curled back leaf tips. Shoots may be stunted and show some dieback.

Magnesium

A lack of magnesium is characterized almost identically with iron deficiency but the older leaves, generally at the bottom of the plant, show marginal and interveinal reddening or yellowing with leaf base and midrib staying green. Later in the season, interveinal necrosis may occur. Leaves may be brittle and thin with leaf curling and stunted growth.

Sulphur

Leaves are pale yellow-green at any stage of development. Shoots are stunted.

Zinc

New and intermediate leaves are small, yellow, sometimes with a grayish cast. Narrow and older leaves may drop. Small shoots may show rosetting followed by dieback.

Copper

Small leaves with necrotic (dead) spots and brown areas near the leaf tips. Rosetting of the leaves and dieback of terminal shoots.

Iron

New leaves are the most symptomatic and when condition is most severe. Leaves can be all yellow or white but still have green veins. Yellow leaves with green veins leading to marginal scorching or browning of leaf tips. Tip leaves, especially basal areas of leaflets, intense chlorotic mottling; stem near tip also yellow. Fruits have poor colour. Shoot diameter is small.

Manganese

Leaves display marginal scorching, rolling and reduced width. Yellowing may also occur between leaf veins or total yellowing on youngest leaves.

Boron

Youngest leaves may be red, bronze or scorched also small, thick or brittle. New shoot tips may form what is called a 'witches broom'. Stems stiff; terminal buds die and growths die back; lateral shoots developed, giving plant flat top; leaves highly tinted purple, brown and yellow. Boron deficiencies are more pronounced during drought periods when root activity is restricted.

d) Chickpea

Nitrogen

Nitrogen deficient chickpeas grow slowly and leaves become yellow-green in colour. The older leaves near the base of the plant are more yellow than upper leaves. The colour is particularly intense in the leaf axils. It also forms on the rachis of the leaves.

Phosphorus

Plants have smaller roots and leaves are darker green in colour. Phosphorus deficiency appears first on the older leaves due to P being readily translocated from older to younger tissues. Leaflets lose chlorophyll and become yellow-green. This occurs initially with lower leaflets of lower leaves. Some lower leaves also have red-brown anthocyanin on their upper surfaces in pigmented plants. N-deficient plants are uniformly yellow with yellow veins in leaflets but P-deficient plants are irregularly yellow with green veins in leaflets.

Potassium

Potassium deficiency in chickpea develops first on the older leaves. The margins and tips of leaflets lose their green colour and become yellow-green. Reddish pigmentation is also seen on some leaflets. These leaflets often drop from the plant. These symptoms progress up the plant when deficiency persists. Necrotic patches develop on leaflets. Stems of some plants also develop reddish-brown anthocyanin pigmentation.

Calcium

Calcium deficiency in chickpea affects the younger tissues near the top of the plant. Roots become short and thick with dark brown or black tips. Calcium deficiency results in small leaves with fewer leaflets. The growing tip of the plants contains unexpanded leaflets and this young shoot often becomes rusty brown in colour and then dies. Leaflets on the tip of a leaf are much smaller than lower leaflets. These leaflets also have brown necrotic spots and then die. The older parts of the plants remain dark green in colour without evidence of anthocyanin pigmentation.

Magnesium

Magnesium deficiency in chickpea first appears on young leaves as a chlorosis. Chlorosis on some leaves later extends down the sides of the leaflets. The symptom on leaflets is yellow to yellow-green with a green area remaining around the midrib. Prevalence of Mg deficiency results in light brown necrotic areas developing on the leaflet tips or margins. The basal leaves usually remain green in severely affected plants.

Sulphur

Chlorosis on sulphur deficient plants firstly affects the leaves near the top of the plant, while leaves near the base remain green. With increasing severity of S deficiency, the chlorosis extends over the whole plant. Some leaflets become completely yellow, wither and drop from the plant. Reddish-brown pigmentation appears on the stems and leaves.

Zinc

Zinc deficiency appears as a reduction in inter-nodal growth and results in a rosette growth habit. The younger leaves of zinc deficient plants initially become pale green in colour. Pigmentation (reddish-brown) takes place on the margins of upper surfaces of leaflets and on the lower portions of the stems.

Copper

Copper deficiency symptoms appear on the younger leaves, while the plant remains a normal green colour. The leaflets on the top leaves of each stem are smaller than on normal plants. Cu-deficient plants produce small leaves with fewer leaflets on the young shoots. In such cases, these leaves usually wither and appear rust-brown in colour. Shoot elongation is also reduced due to insufficient development of the terminal growing point.

Iron

Iron deficiency appears as an interveinal chlorosis in younger leaves. Leaflets on Fe-deficient plants turn a yellow-green colour. White or light coloured necrotic patches develop on the leaflets of young leaves when Fe deficiency becomes more severe. Acute Fe deficiency, causes leaflets to wither and die. The young shoots also wither from the tip back towards the base. Fe chlorosis is transient, the deficiency symptoms largely disappearing during reproductive growth at a time coinciding with increases in soil temperature and day length conditions favourable for plant growth.

Manganese

Manganese deficient plants have a lighter green colour symptom on the younger expanded leaves. There are very small light brown spots on the surface of leaflets and on the stipules of these leaves, which develop later. These are generally found between the leaflet veins on the leaf surface. The leaves of Mn-deficient plants often lose many of their leaflets. Prevalence of Mn-deficiency results in symptoms similar to those of Ca deficiency. As a result of death of young shoots, there is enhanced axillary bud development on Mn-deficient plants.

Boron

The first symptom on boron deficient plants is observed on the tip and margins of leaflets of young entire expanded leaves with a yellowing and 'bronzing'. This 'bronzing' effect is partially due to red-brown pigmentation on the upper surfaces of leaflet margins. The tips and margins of affected leaflets start to die. The terminal buds turn rusty brown in colour. The young buds and leaflets die progressively from the tip. As a result of this, axillary bud development is enhanced. Plant roots are also stunted and thick with dark tips similar to those occurring on Ca-deficient plants.

Molybdenum

The symptoms start with chlorotic interveinal mottling of older leaves. In severe deficiency conditions, Mo causes leaf wilting and disorders. In addition to growth depression, flowers are less in number, smaller, and many failed to open/mature leading to lower seed yield. Molybdenum deficiency causes symptoms of N deficiency due to the role of Mo in N fixation. Leaves of N-deficient plants are lighter yellow than those of Mo-deficient plants. In many plants, there is an upward cupping of the leaves and mottled spots developing into large interveinal chlorotic areas under severe deficiency.

e) Peas

Nitrogen

In the rare occurrences of nitrogen deficiency, plants exhibit a uniform yellow appearance and reduced overall growth.

Phosphorus

Phosphorus deficiency in peas is difficult to diagnose; very slight yellowing of the lower leaves.

Potassium

Plants show retarded growth, yellowing of the lower leaves and poor pod filling.

Calcium

Deficient plants have reduced root and shoot growth and chlorosis of the leaves near the top of the plant.

Magnesium

Deficiency causes reduced vine growth, browning of the leaf tips, reduced pod formation and failure of seeds.

Sulphur

Deficient plants become chlorotic. New leaves are first affected, but gradually the entire plant becomes uniformly chlorotic.

Zinc

Deficiency is recognized by yellowing of the older leaves and interveinal chlorosis of the younger leaves. Flowering also is reduced.

Copper

Copper deficiency is recognized by wilting of the terminal stem tips, flower abortion and reduced pod formation.

Iron

Deficient plants become chlorotic when they reach the 5th-6th node stage of growth, affecting mostly the young leaves and growing tips. All subsequent leaves are chlorotic and some may appear bleached. Flowering and pod formation are reduced.

Manganese

Deficiency is recognized by interveinal yellowing of leaves and leads to formation of a brown spot in the center of the pea seed commonly termed "marsh spots" which makes the peas unsuitable for food or seed. Severely affected plants have very yellow leaves and die back from the growing tips, flowers abort, and pod formation is decreased.

Boron

Sprouts of deficient plants are some what pale, stunted and lack the recurved plumular bud typical of normal sprouts, and the buds fail to develop. Multiple branching and thickened stunted shoots may also be present.

Molybdenum

Symptoms of deficiency appear as reduced growth and general yellowing of the plants.

f) Faba beans

Nitrogen

Marked stunting and yellowing.

Phosphorus

Leaves are dark green, older leaves drop, the stems are thin and blossoms sparse.

Potassium

Older leaflets show a slight curling and then a distinct greying of leaf margins, eventually dying.

Calcium

Stunted growth; distortion of growing points and young leaves, possibly leading to the death of the plant. Pods wilt and blacken prematurely.

Magnesium

Leaf chlorosis, especially in the interveinal regions. Older leaves affected first.

Sulphur

Youngest leaves turn yellow, plants are slender and small.

Zinc

Plants are small, the areas between veins turn yellow, becoming yellower on the lowest leaves. Maturity can be delayed.

Copper

Loss of flower pigment.

Iron

Yellowing between leaf veins can progress to completely yellow plants.

Mangancse

Young leaves first show chlorosis, followed by small dead spots at each side of the mid-rib and lateral veins. The leaves can turn yellow and die.

Molybdenum

Molybdenum deficiency results in chlorosis, which is nearly always localized between the veins. The leaves are a paler, yellowish green and appear nitrogen deficient.

4. OIL SEEDS

a) Ground nut

Nitrogen

Plants are stunted and yellow in appearance. Stems are thin and elongated. The stem turns reddish. Yellowing or chlorosis appears first on the lower leaves while the upper leaves remain green.

Phosphorous

Mild deficiency of phosphorus are usually characterized by stunted growth of plants. Most severe deficiencies result in dark green leaves with a leathery texture and reddish purple leaf margins.

Potassium

Lack of potassium causes stunted plants with small branches and reduced vigour. Potassium deficient plants exhibit chlorosis or yellowing along the leaf margins or tip starting with the bottom leaves and progressing up the plant. In severe cases, the leaves appear dry and scorched at the edges and the surfaces are irregular chlorotic.

Calcium

Calcium deficiency results in aborted and shrivelled pods. The kernels have dark plumule. The germination and seedling survival is poor. Symptoms like pitted areas on lower surface of older leaves. Later on, large necrotic spots are found on both the surfaces, which give leaves a bronze colour. Youngest foliage appears distorted.

Magnesium

Older leaves become chlorotic, beginning at margins and spreading towards midribs, followed by orange colouration of margins under acute deficiency conditions. Young leaves are also affected.

Sulphur

The leaves turn pale. Young as well as middle leaves show chlorosis. Under severe conditions, the leaves become papery. The deficiency symptoms of iron and sulphur appear together in groundnut.

Zinc

In zinc deficient leaves, wider chlorotic strips are found, which may run the entire length of the leaflets. The bands will usually be on the leaf portion nearest to the petiole. Zn deficiency symptoms may occur concurrently with Fe deficiency symptoms and the two can easily be confused. The entire leaflet will become chlorotic, if the deficiency is severe and prolonged.

Copper

The copper deficient plants are stunted and rosette. The stunted plants are green and wilted in early growth stage. The entire leaflet becomes cupped and the leaflet margins turn upward. Some bronzing and necrosis of outer edge of the leaflet may occur.

Iron

Iron deficiency is characterized initially by interveinal chlorosis on the younger leaves. Under severe deficiency, the veins also become chlorotic and leaves become white. Iron deficient plant has limited root system which may cause other nutrient deficiencies.

Manganese

The typical manganese deficiency is interveinal chlorosis, but there may also be a distinguishing characteristic of brown spots on the leaf margin.

Boron

Boron deficient groundnut cotyledons are depressed and discoloured. This is classified as a form of internal damage and has been termed **"hollow heart"**. Necrotic areas are localized near the leaf margin.

Molybdenum

The role of molybdenum in enzyme system of symbiotic nitrogen fixation suggests that the Mo-deficient groundnut become deficient in nitrogen. The deficiency symptoms are paling, wilting, marginal rolling and scorching of leaves.

b) Sesamum

Nitrogen

Plants are light green, stalks slender, branching absent, and leaves erect. Lower leaves become lemon yellow turning to orange. Discoloured leaves are shed.

Phosphorus

Branching suppressed, stalks slender, lower leaves dull dark, grayish green. Necrosis of lower of majority of leaves is followed by defoliation.

Potassium

Plant dwarfed and margins of lower leaves become wavy and cupped upward. Light lemon yellow chlorotic mottling appears, which later turns bright orange and finally copper-coloured.

Calcium

Terminal bud dies out following distortion of the tips and bases of young leaves. Hooking downward of the young leaf tips followed by twisting and puckering.

Magnesium

Lower leaves develop interveinal chlorosis, light yellow in colour becoming orange later. Green colour persists in midrib and veins giving a characteristics pattern.

Sulphur

Sulphur deficiency produce smaller new leaflets with yellow and erect petioles than the normal. Plants are smaller in size and nodulation is poor.

Iron

Iron deficiency decrease the dry weight of leaves and stem. Decrease in taproot length and its dry weight. The leaves show deficiency symptoms of mild chlorosis.

Manganese

Leaves show interveinal chlorosis which become chlorosis and changes into light brown, a husk coloured necrotic lesion.

Boron

Yellowing of plant tops and of the youngest leaves. Upper leaves became dark green, coriaceous, with edges curved down, inhibits root elongation leading to the death of root tips.

c) Sunflower

Nitrogen

Chlorosis of young leaves and old leaves, marginal chlorosis, whole leaf become chlorosis, chlorosis followed by death, thin stems, spindly stems pale green stems are seen with fewer or smaller mature heads.

Phosphorus

Interveinal chlorosis yellow - pale green leaves, Interveinal necrosis on stems: Short and thin stems. Bent petioles (resulting in leaves pointing downwards), Reproductive stage symptoms expressed: delayed maturity, fewer of smaller mature heads which set few grains/ seed. Lower grain yield, dark brown necrotic lesions, grey-dark brown necrotic lesions.

Potassium

Deficiency symptoms appear on old leaves, whilst young leaves remain green and healthy, chlorosis of old leaves. Marginal chlorosis with yellow-pale green leaves and puckering of the leaves, necrosis on older leaves and interveinal necrosis. The stem show stout and stunted stem growth. Leaf tips and or margins curl up or down giving a wavy appearance. Fewer or smaller mature heads which set few grains/seed.

Sulphur

Plants showing paling or yellowing of leaves. Yellowing spreads from the base to the apex. Growth of plants is reduced. The size of capitulum is severely restricted. Inflorescence may remain covered within the bracts. Maturity of flowers is delayed.

Copper

Deficiency symptoms appear as interveinal chlorosis of the upper half of the old leaves starting from the tip of the leaf. Chlorosis is generally restricted to the apical half of the lamina. Chlorotic areas later develop dark brown necrotic scorching, which spreads along the margins towards the base of leaf. Marginal scorching of the leaves becomes more pronounced and brown necrotic spots develop near the midrib.

Manganese

Deficiency symptoms appear in the form of interveinal chlorotic mottling of apical part of the second set of leaves. The entire laminae become severely mottled and the interveinal chlorotic areas develop light brown irregular necrotic patches. These patches, which are initially more conspicuous near the leaf apices, later spread to the entire laminae, which eventually turn severely necrotic. Symptoms gradually spread from the middle to the young and old leaves.

Boron

Boron deficiency will cause dieback of terminals (buds) and of the connective tissue between stems and stalks. Leaves will curl downward, show discolourations and will develop necroses on the surface of the leaf blades and between veins (insides of veins drying up). These symptoms will first appear on the plant's lower parts (older leaves) and will slowly move up to its tip. Flowering will be impaired.

Molybdenum

Pale green foliage, interveinal and marginal chlorosis of old leaves and inward curling of leaf margins of the young leaves giving them a "cupped" appearance.

d) Mustard

Nitrogen

Light green colouration to the leaves and stem. Leaves become chlorotic, often showing a purple colouration. Older leaves may wither, poor growth with short, thin main stems and few branches.

Phosphorus

Dark bluish-green colouration of the leaves with purplish tinges. In severe cases, marginal necrosis of the leaves occur. Leaves narrow, stem thin, branching is restricted.

Potassium

Prostrated appearance of the young plants, upward convexity of the interveinal areas arching of the leaf blades. Darkening of the leaves, marginal or interveinal chlorosis patches of necrosis.

Calcium

Stunted growth, small leaves, rounded appearance of the top (leaves). Yellowing of veins with browning of leaves, arrested growth of younger leaves with purple pigmentation on petiole, fibrous and weak root system.

Magnesium

Stunted growth, prominent interveinal chlorosis, yellowing of leaves and vein at advance stage followed by whitish spots on terminal leaves.

Sulphur

Deficiency symptoms are shown on younger leaves. Chlorosis formation from leaf margins, spreading inward and/or develop purple pigmentation. Inward curling of young leaf lamina giving these leaves a cupped appearance followed by scorching and withering. Flowering is delayed, lacking normal pigmentation. Pods are borne on short peduncle and their development is restricted. Seed setting is poor and their maturity is delayed. Sulphur deficient plants show marked retardation in height, thickness of stem and size of leaves.

Zinc

Leaves are small in size with pinkish margin. Interveinal tissues turn yellowish white to papery white, with veins remaining green. There is upward or downward cupping of leaves. Severely affected leaves die. Flowering and fruiting is delayed.

Copper

Young leaves develop interveinal chlorosis. Chlorotic areas later turn papery and necrotic. Growth of plants is also reduced which is more pronounced at the time of flowering and thereafter. Inflorescence is poorly developed and most of the floral buds of the plants wither before opening. Pod formation and seed setting are also reduced.

Iron

Chlorosis starts from the base and spreads towards the apical part of the lamina. In case of severe deficiency, leaves become bleached and puckered. Newly emerging leaves are completely bleached, those subtending Leaves also develop pigmentation. Flowering, pod number and size of pods is reduced and pods appear chlorotic.

Manganese

Chlorosis in apical part of the middle leaves, followed by developing small greyish-brown spots which coalesce to form large necrotic lesions, Leaf margins also turn necrotic and the lamina curl downwards. Later, these symptoms spread to old and young leaves. Flowering is also reduced and large percentage of flowers shed prematurely resulting in poor fruit set and pod formation.

Boron

Dwarfing, wrinkling or rolling of the leaves. Leaves become chlorotic and crackling of the petioles. Leaves are small and dark green. Seed set is poor.

Molybdenum

Mustard is very susceptible to molybdenum deficiency. Growth is markedly reduced and plants develop foliar symptoms like cupping, marginal scorching and loss of lamina.

e) Soybeans

Nitrogen

Nitrogen deficient plants become pale green. Later, the leaves turn distinctly and uniformly yellow. Symptoms appear first on the basal leaves and quickly spread to upper parts. The plants eventually defoliate and often are spindly and stunted.

Phosphorus

The main symptom is retarded plant growth; affected plants are spindly and have small leaflets. Leaves turn dark green or bluish green. The leaf blade may curl up and appear pointed. Blooming and maturity are delayed. The dark green of the leaves gives the impression that the plants are quite healthy.

Potassium

Potassium deficiency symptoms are well defined. Symptoms appear first on the older leaves. In early stages of growth, an irregular yellow mottling appears around leaflet margins. The yellow areas coalesce to form a more or less continuous irregular yellow border. This symptom often is followed by necrosis of chlorotic areas and downward cupping of leaf margins. The dead tissues then drop away, so that the leaves appear ragged. Chlorosis and necrosis may spread inward to include half or more of the leaflet. The basal portions remain green. Severe deficiency tends to produce misshapen and wrinkled seeds. Injury from leaf-hoppers feeding on cultivars with glabrous leaves often mimics potassium deficiency.

Calcium

The emergence of primary leaves is delayed, and they are cup-shaped when emerged. Terminal buds of primary leaves become necrotic, and other parts are chlorotic. Tissues between veins tend to be ridged. Terminal buds wither and petioles fall. Later, the primary leaves become flaccid and drop prematurely.

Magnesium

In early stages of crop growth, magnesium deficiency is recognized by pale green to yellow interveinal tissues on the leaf blade. Leaves later become deep yellow, except at the base. The lower leaves show symptoms first later, rusty specks and necrotic blotches may appear between the veins and around the edges of middle and upper leaves. Leaf margins curve down with a gradual yellowing from the margin inward, followed by bronzing of the entire leaf.

Sulphur

The main symptom of sulphur deficiency is small, yellow-green leaves at the top of the plant. Stems are thin, hard, and elongated. Leaf symptoms resemble those produced by deficiencies of other elements, such as N and P, but stem elongation is characteristic of sulphur deficiency.

Zinc

Zinc deficient plants have stunted stems and chlorotic interveinal areas on younger leaves. Later, entire leaves may turn brown or gray and drop early. Few flowers are formed, and the pods that set are abnormal and slow in maturing. Yield reduction is primarily due to the result of the reduced number of seeds.

Copper

Research on Cu deficiency in soybeans is limited. The deficiency results in reduced seed yields, but plants may not show obvious symptoms.

Iron

Early leaf symptom of Fe deficiency is yellowing of interveinal areas on younger leaves. Later, even the veins may become chlorotic; the whole leaf finally turns ivory to white. Brown, necrotic spots may occur near leaf margins. Plant growth is suppressed.

Manganese

Soybeans are sensitive to Mn deficiency. Interveinal areas of leaves become light green to white; the veins remain green. The symptoms resemble the early symptoms of Fe deficiency, except the veins of Mn-deficient leaves remain green and stand out prominently. Necrotic, brown spots develop as the deficiency becomes more severe. Leaves may drop prematurely. The symptoms are pronounced in cool weather. When temperatures rise, affected plants generally outgrow mild deficiency symptoms. Manganese deficiency symptoms frequently appear if the Fe concentration is increased.

Boron

Symptoms of the deficiency boron include shortened internodes and yellowing or reddening of upper leaves. Flowers fail to develop, particularly in dry weather. Boron deficiency may be confused with leafhopper injury and K or Ca deficiency.

Molybdenum

The symptoms of molybdenum deficiency resemble those of nitrogen deficiency and likely are caused by reduced nitrogen utilization. Deficiency reduces plant growth, number of pods, number of seeds per pod, seed size, nodulation, and total nitrogen and protein content of seeds. Leaves are pale green or yellow, necrotic and twisted, and the necrosis is confined largely to the margins, midribs and interveinal areas.

5. Fibre Crops

a) Cotton

Nitrogen

Nitrogen deficiency symptoms in cotton are characterized by relatively meager growth and yellowish green colour of foliage. The older or lower leaves are most affected. In severe cases of deficiency, the leaves dry up and shed prematurely.

Phosphorus

The symptoms of lack of phosphorus in the cotton plant are not so marked as those caused by nitrogen deficiency. The clearest indications of phosphorus deficiency are dark green leaves of reduced size and a stunted stalk, the affected plant being in a decidedly dwarfed condition. An extreme deficiency of phosphorus not only reduces the number of cotton bolls but also delays the maturity of the bolls.

Potassium

Potassium deficiency is known as "cotton rust". First, the leaves turn yellowish green. Yellow spots appear between the veins and eventually turn black. The edges of leaves die and become black. Finally, the whole leaf dies and falls off. In acute cases, the whole field may shed all its leaves. Without leaves, the manufacture of plant food is stopped and bolls open before they are mature. This reduces down the yield and produces weaker and shorter fibre of poor quality.

Calcium

Calcium deficiency is generally manifested in early stage of the crop. The leaf petioles start bending and collapse later. These symptoms are seen in cotyledons as well as in true leaf stage of the crop. It also cause plant to have weak stems and leaves. In the case of severe deficiency, bending of the leaf and subsequently leads to defoliation.

Magnesium

The characteristic symptom of magnesium deficiency in the cotton is appearance of purplish red leaves with green veins. The discolouration occurs on the lower leaves first causing premature leaf shedding and then progress upwards. Through the loss of leaves and the resultant reduction in photosynthetic activity, the yields of both fibre and seed are greatly reduced. The magnesium deficiency is also associated with red leaf disorder of cotton. The common symptoms of leaf reddening include: Leaf margins turn yellow and later red pigmentation is formed on the whole leaf area. Reddening progresses to stems and leaves. Shedding of leaf and boll and the poor plants growth. Fewer bolls and in extreme cases no bolls and wilting of the plant.

Sulphur

Cotton plants suffering from severe sulphur deficiency show fairly characteristic symptoms. However, these symptoms are not always readily distinguishable and are frequently confused with the rather similar symptoms of nitrogen deficiency. Uniformly chlorotic plants characterize sulphur deficiency, which has a pronounced retarding effect on plant growth which are stunted, thin stemmed and spindly.

Zinc

Zinc deficient cotton plant shows a general bronzing of the first true leaves, and often a pronounced interveinal chlorosis. The leaves become thick and brittle, with their margins turned upward. Stalk elongation practically ceases and the shortened internodes tend to give the plant a small bushy appearance. Growth and fruiting are delayed and yields are drastically reduced.

Iron

Chlorosis (yellowing) of leaves is the symptom of iron deficiency.

Manganese

Chlorosis is the most common symptom of manganese deficiency. A chlorotic pattern and stunting of growth are usually the first visual symptoms. As the deficiency becomes more severe, necrotic spots appear in the chlorotic areas. The upper or young leaves are the first to be affected. They become yellowish grey or reddish grey with prominent green vein in the vegetative stage and fail to set fruit.

Boron

The first symptom is generally an excessive shedding together with the deformed condition of the relatively few buds and young bolls. This shedding, together with the deformed condition of the relatively few buds and bolls that may persist, results in extremely low yields. In the more severe cases of boron deficiency, the internodes are shorter and the terminal bud often dies, producing a short, busy plant with excessive branching. These rosette types of plants are characteristic deficiency symptom of boron deficiency. Another characteristic symptom associated with boron deficiency is excessive flow of plant juice from nectaries of leaves, bracts and bolls. Internal damage to the cell wall and cracking of tissues are responsible for this. This assumes serious dimensions very often and fruiting forms dry up and may stick on or fall to the ground. Even the remaining bolls are somewhat deformed showing a flat (or) till-hook appearance. A black discolouration is seen at the base of the bolls and such bolls are smaller in size and tend to open partially.

6. Sugar Crops

a) Sugar cane

Nitrogen

Nitrogen deficiency in sugarcane exhibits the following symptoms: Premature leaf senescence. Pale yellow leaves rigid with low moisture later turning to reddish purple, Premature drying of older leaves, Retardation of growth resulting in reduced upper internodal length and thinner stalks.

Phosphorus

Deficiency symptoms of phosphorus in sugarcane are: colour of older leaves become blueish green to reddish in contrast with wide dark green leaves of healthy cane plant. Older leaves indicate a definite drying at the tips and margins. The leaves are reduced in length and breadth. The length and diameter of the stalks are reduced, resulting in shorter, slender stalks that taper rapidly at the growing point. P deficiency stools consist only of a few primary stalks with small internodes indicating a very slow rate of growth and often no secondaries or suckers. Poor stooling is always observed with P deficiencies, resulting in low number of millable canes and poor cane yield. Root systems of P deficient cane are very poorly developed. The development of secondary rootlets is very poor, resulting in a minimum of root surface for soil contact.

Potassium

Deficiency symptoms are yellowing and spotting of older leaves followed by reddish discolouration which is confined to the epidermal cells. Latter, the leaves begin to die back from the margin and tips. Accumulation of free amino acids in the plants is reducing nitrogen utilization in the plant. Potassium deficiency results in depressed growth and development of slender stalks, Stalk elongation is slow and young leaves appear to develop bunchy top appearance. Plants deficient in K are more susceptibility to pests and diseases and moisture stress conditions.

Calcium

Under calcium deficient conditions, the young leaves become chlorotic. Minute chlorotic spots develop on the older leaves with dead centres and later turn dark reddish brown. These spots coalesce to give rusty appearance. Pre-mature death of old leaves also occur. Stalks are pithy and thin but tough and woody and characterized by soft rind. In acute conditions, the spindle dies and simply hangs down like a piece of straw coloured twine resulting in the death of the meristem. Root system is severely affected. Discolouration of roots is also observed.

Magnesium

Magnesium deficiency appear first in the basal leaves of the plants. When the deficiency is acute, the symptom will be seen in young leaves also. During initial stage of the deficiency, the young leaves are light green in colour; whereas the older leaves exhibit yellowish green colour. As the stage of deficiency is prolonged further, small chlorotic spots will be developed in the older leaves which later turn to dark brown. The lesions are uniformly distributed over the leaf surface and vary in size depending on the variety. When the spots coalesce, they give a rusty appearance to the leaves.

Sulphur

Sulphur deficiency initially resembles that of nitrogen deficiency but later on differentiated by light purplish tinge in the light yellow green colour due to the accumulation of carbohydrates and anthocyanin. The leaves become narrow and shorter in length. Stalks become stunted due to shorter and thinner internodes accompanied by light yellowing of newly emerged parts. About 30 kg of sulphur is needed to produce 100 tonnes of millable cane.

Zinc

Young sheath as well as young leaf blades serve as useful index tissues for zinc. The stalks of zinc deficient plants exhibit distinct pipe like structure in the centre with spongy and light brown tissues around the pipe. Zn deficient plants show a marked loss of auxin and therefore, results in diminished growth and internode elongation. Leaf blades are shortened with typical light yellow stripes along the vein. In severe cases, interveinal necrosis is observed. Young tillers exhibit chlorotic symptoms.

Copper

Copper deficiency in sugarcane results in poor development of stool. The spindle leaves fail to unroll. Leaves will have the tendency to droop and the condition is described as "droopy top" Leaves become chlorotic and broad with rectangular blotches. The stalk and spindle become soft and rubbery.

Iron

Pale chlorotic leaves alternating with green stripes along the full length of the leaves giving chlorotic streak between parallel veins of leaf blade called interveinal chlorosis. The leaves become sun burned and the affected plants remain stunted and die in severe cases. The symptom appears first in young leaves because of its immobility in the plant system. Therefore, iron deficiency occurs more frequently in young and ratoon crops.

Manganese

The deficiency of manganese is called "Pahala blight". At early stage, the deficiency symptom resembles to that of iron deficiency. However, it can be differentiated by the interveinal chlorosis confined to middle and tip of the leaves in contrast to iron deficiency symptom where the interveinal stripes extend to full length of the leaf. Under acute conditions, reddish brown necrotic spots appear causing longitudinal splitting of leaves.

Boron

Distortion of growing top. Young leaves become chlorotic and small. Curling of leaves, decrease in cell turgidity, exudation of water from leaves, development of water spots on young leaves are some of the deficiency symptoms associated with boron. Leaves become chlorotic and apical growth ceases. Internodes turn brownish red with water soaked spots on the stalk. At advanced stages, pith formation takes place just below the growth point in the stalk.

Molybdenum

Deficiency symptom will be noticed when molybdenum is less than one ppm in leaf. Deficiency is observed as yellow streaks in fully matured leaves. In acute conditions, the streaks show reddish tinge at the centre which becomes necrotic. In severe cases, leaves show whip-tail symptoms with mottling and necrosis.

b) Sugar beet

Nitrogen

Nitrogen deficiency in sugar beet shows as a general yellowing of the foliage. Usually, leaves and petioles of all ages are fairly uniformly yellow when the plant is nitrogen-deficient, although the youngest leaves may be somewhat greener than the old ones.

Phosphorus

Irrespective of the age of the plant, phosphorus deficiency is characterised by dark green leaves and stunting of the whole plant. In the seedling stage, the cotyledons and primary leaves of phosphorus-deficient plants are dark green and then become pitted and may die. The petioles curl upwards and a blackish-brown necrosis develops from the tips and edges of the leaves without previous chlorosis. In severely-deficient plants, a brown netted veining forms in the tissues of the fully expanded leaves when they dry up and die. Root growth is also stunted by shortage of phosphorus and the tap root often forms a mass of fibrous secondary roots.

Potassium

The symptom which appears first in apical and marginal chlorosis with a dull olive green appearance of the leaves. The surface of the leaf becomes dull and bronze in colour and later small clusters of diffuse buff-coloured spots appear, generally arranged in the form of rough triangles, the bases of which are towards the margins of the leaf with the apices projecting inwards between the veins. On some plants, there is a continuous band of shrunken chlorotic tissue round the margin of the leaf and the plants look stunted. Brown striped lesions commonly appear on the petioles. The scorch symptoms develop after the chlorosis or may develop independently of it. Necrosis generally follows, both forming an unbroken border round the leaf and lobes between the veins. The necrotic tissue is dull or reddish brown, it is tough but soft to handle, and it does not crumble or disintegrate like the necrosis associated with magnesium deficiency. All the leaves on affected plants are thin and flaccid. The plant is normal in habit above ground but the tap root development is very poor.

Calcium

Sugar beet show characteristic symptoms (often called 'tip-burn') when the supply of calcium is short. The first signs of calcium deficiency are the crinkling, downward 'cupping' or 'hooking' of young leaves. The laminae may be normal in size, with little crinkling; when the deficiency is more severe they can be reduced in size to a small area of necrotic tissue at the end of the petiole. Later, the growing point is damaged and abnormal lateral shoots appear.

Magnesium

The first symptoms of deficiency is the appearance of small, pale yellow areas at the distal margins of the middle-aged leaves. Tissue in these areas enlarges and the edge of the leaf becomes fluted. Lemon yellow areas then spread down between the veins and appear as well-defined lobes orientated towards the mid-rib, with distinct demarcation between yellow and green areas due to thickening and enlargement of affected tissue. The chlorotic areas on each leaf are usually continuous, but isolated patches may develop between the veins. Typically, the colour of these areas is pale yellow, distinguishing the symptoms from those of virus yellow which are orange-yellow, but individual plants sometimes have leaves which are dark yellow or orange. Usually within a few weeks, the yellow areas become necrotic, beginning at the edges of the leaf. The necrotic tissue is dark brown and brittle so that it breaks away easily and the leaves have a ragged appearance.

Sulphur

Young leaves become a light green or yellow. Brown blotches of an irregular pattern on older blades.

Zinc

Light greening or yellowing first appears in the younger leaves near the center of the plant, followed by pitting, collapse and drying of interveinal tissue, leaving the veins green and prominently displayed. Leaves become erect.

Copper

The veins remained dark green, contrasting with the light yellow areas. Older leaves became blue-green colour. Root development is retarded by copper deficiency and yields are much decreased. Another characteristic of copper-deficient plants is that the necrotic areas on old leaves are greyish-brown or greyish-white and the dead leaves looked bleached.

Iron

The blades of young leaves change from green to a light green and finally to a uniform bleached yellow. The veins remain green at first, but finally they too, become bleached. Eventually, the bleached blades become necrotic.

Manganese

A chlorosis begins in blades of young leaves that can be seen through transmitted light as a netted veining not readily distinguishable from that of Fe and Cu deficiency. The chlorosis gradually increases to a uniform yellow. As the deficiency continues to increase, a gray metallic, sometimes purplish luster develops on the upper surface, followed by a gray-to-black freckling along the veins. Upward curving of blade margins occurs frequently.

Boron

Deficiency of boron causes typical symptoms to appear not only on the leaves but also on the petioles, crowns and roots of sugar beet. Boron deficiency is the cause of 'heart rot' and 'dry rot'. Heart rot is the termed applied when the growing point becomes blackened and dies; dry rot describes the symptoms on the tap root which usually appear subsequently. Boron deficiency not only decreases yield but damages the roots, decreasing their value and keeping quality. When the supply of boron to sugar beet is insufficient for normal growth, the first sign of deficiency is the death of the growing point and the youngest leaves. As the deficiency becomes more serious, the older leaves have a silvery appearance and are often covered with a network of fine cracks and sometimes show interveinal chlorosis. Occasionally, the grooves in the petioles of affected plants have transverse cracks (laddering) and the tissue is brown and cork-like. These leaves are usually more or less prostrate. Boron deficiency seems to cause unequal growth of the two faces of the leaves, resulting in torsion of the petioles and veins and, finally, in abnormal leaves. As affected plants age, many of the leaves die and fall off.

Molybdenum

Symptoms of molybdenum deficiency first appear as a general yellowing, similar to that caused by sulphur deficiency. As the symptoms of Mo deficiency increase in severity, pitting develops along leaf veins. In Mo-deficient plants, the leaves turn golden yellow, with inward curling of margins, and leaves are arrow-shaped because of the curling of lamina.

7. Fodder Crops

a) Forage Grasses

Nitrogen

Nitrogen deficient plants become stunted and spindly and the foliage turns light green or yellow. The lower leaves become yellow and drop off. Premature maturity of the crops may also occur. Deficient supply of nitrogen makes the leaves harsh and fibrous due to small cells and thick walls. The older leaves develop chlorosis first and turn pale green to yellow, while the young upper leaves remain green. In case of severe deficiency, the whole plant become chlorotic and eventually dies.

Phosphorus

Phosphorus deficiency is more common in seedling forage grasses than in established stands. Symptoms of phosphorus deficiency in young plants are variable and not always easily identified. One pattern which develops in some species, particularly in cool weather, is that of slow growth, accompanied by the appearance of an abnormally dark green colour, while stems and lower leaves acquire a reddish or purplish cast. On more mature plants, all leaves will have a uniformly dark green colour, which tends to turn purple for some distance back from the tips of the older leaves. When viewed in full sunlight, the undersides of all leaves may exhibit a rather uniform purplish tinge. In some phosphorus deficient grasses the blade margins tend to curl upward, regardless of whether the front or the back of the leaf is turned up.

Potassium

Potassium starved plants, are likely to stunted with short narrow leaves. Susceptibility to drought is another symptom.

Magnesium

Magnesium deficiency in the larger leaved grasses starts with a stunting and yellowing of the older leaves. As this symptom progresses, a definite striping effect occurs. Green or yellowish green stripes alternating with yellow to whitish stripes run the full length of the leaves. In the smaller leaved grasses, this striped effect is less pronounced or entirely missing. Instead, whole leaves bleach to pale yellow or almost white.

Sulphur

Initially, the leaves become pale green, a symptom usually not distinguishable from the early stages of nitrogen deficiency. As the condition becomes more acute, however, affected plants develop a pattern quite different from those suffering from nitrogen deficiency. The result young leaves are often more severely affected and may become chlorotic before older leaves.

Zinc

Symptoms in the larger leaved grasses develop with the appearance of light yellow streaks between the veins of the older leaves. Sometimes accompanied by reddening of leaf sheaths and blades. White spots may appear on the leaves and some small areas may Lacks chlorophyll. Young bud leaves may become practically white in extreme cases.

Copper

The younger leaf tips and growing points are the first affected. In grasses, entire plants become stunted and yellow in colour, followed by necrosis (dying) of growing points and the tips of the younger leaves.

Iron

Iron deficiency of this element usually is reflected in a decrease in chlorophyll content. This results in a lighter colour of the inter-veinal areas of the younger leaves, which change to yellow as the deficiency advances.

Manganese

Symptoms include chlorosis of the younger leaves, with yellow-green spots on the older leaves and withering of leaf tips. Gray specks of dead tissue appear between leaf veins, merging lengthwise to form bands parallel to the veins. Later, these bands may join across the leaf resulting in withering and downward bending of the tip portion.

Boron

Deficiency symptoms to develop first in the younger leaves and also at growing points. Boron deficiency usually results in plants with shortened internodes in their upper parts and some form of chlorosis in their younger leaves.

Molybdenum

Symptoms described are development of pale yellow foliage similar to nitrogen deficiency, followed by extensive bleaching and withering of leaves.

b) Lucerne

Nitrogen

Stunted or spindly growth; light green to yellow colour.

Phosphorus

Blue-green or dark coloured leaves, stunted, stiff, or erect growth. Leaflets often fold together, and the undersides and stems may be red or purplish.

Potassium

White spots around edge of leaves. Spots begin on lower leaves but are most obvious on upper leaves. In advanced cases, leaves turn completely yellow and drop off.

Calcium

Impaired root growth or rotting. Petioles collapse on youngest mature leaves.

Magnesium

Interveinal chlorosis of lower leaves, margins initially remain green.

Sulphur

Overall light green colour; somewhat similar to nitrogen deficiency; spindly stems and weak growth.

Zinc

Small leaves, youngest leaves curl upward.

Copper

Severe curvature of petioles, grayish spots in midleaf.

Iron

Interveinal chlorosis of youngest leaves, bleached appearance.

Manganese

Interveinal chlorosis of younger leaves.

Boron

Yellow or reddish-yellow upper leaves; plant top may appear bunched with short internodes; can be confused with leafhopper damage. In advanced cases, leaves may turn bronze colour and the growing point may die.

Molybdenum

Molybdenum deficiency is manifested as general yellowing of the whole plant.

c) Stylosanthes

Nitrogen

Uniform chlorosis in the whole plant from the beginning. Under severe conditions, lower leaves with red pigmentation and border necrosis.

Phosphorus

Reduced plant growth, dark bluish green leaf colour, thick leaves, and drooping of lower leaves.

Potassium

Reduced plant growth, small and chlorotic upper leaves, marginal yellowing and border necrosis of lower leaves.

Calcium

Apical chlorosis and marginal yellowing of younger leaves, malformation of newly expanded leaflets, abundance of new regrowth which turns brown quickly, dark green colour of lower leaves.

Magnesium

Marked interveinal chlorosis on the lower and mid-positioned leaves. Under severe conditions, yellowing of young leaves and border necrosis on lower leaves.

Sulphur

General yellowing of upper leaves, which soon spreads throughout the plant, under severe conditions, lower leaves develop marginal necrosis, become yellow in colour and fall off.

Zinc

Interveinal chlorosis in top leaves associated with a slight degree of bronzing of the young emerging leaves. Under severe conditions, leaflets curved epinastically.

Iron

Uniform chlorosis of upper leaves which become white under severe deficiency; at this stage, marginal necrosis also occurs in young leaves.

Boron

Fast necrosis of young leaves, short petioles, irregular growth of young leaves, lateral leaf-lets show unequal size and are malformed; root growth restricted, dark in colour, and suppressed lateral root development.

8. Mulberry

Nitrogen

Retarded development of roots, reduction in size of branches and stunted appearance of the plant indicate deficiency of nitrogen. The leaves turn yellow, become small and thin, mature faster and are shed early. They become deficient in water and protein content and cannot support the growth of silkworms.

Phosphorus

Phosphorus deficiency can be recognized from stunted growth of the plant in which the stem is slender with very small, dark green young leaves and older leaves showing intravenous chlorosis, marginal necrosis and discolouration to a reddish-purple colour.

Potassium

Weaker stem and root systems, small leaves, young leaves with scorched margin and coarse, non-juicy and necrotic mature leaves indicate potassium deficiency. Further, deficiency induces intravenous chlorosis, retards photosynthesis and production of new tissues.

Calcium

The first symptom of calcium deficiency is necrosis along the veins which in later stages leads to abscission of the leaves. The plant shows stunted growth with a woody stem and yellowish tip. Roots become stubby and dry.

Magnesium

Appearance of yellow intervenous areas on mature leaves, spreading across the entire leaf and leaving the veins green. These symptoms spread upwards in the branch until premature defoliation occurs.

Sulphur

Yellowing of tender leaves with subsequent abscission. Abnormally long and woody stem and roots.

Zinc

Appearance of light yellow streak between the veins in older leaves, irregularly shaped leaves and short internodal distance giving the plant a bushy appearance.

Copper

Wilting of tender leaves and desiccation of leaf tips with no change in colour of the leaves.

Iron

Deficiency of iron in mulberry is well marked by the appearance of chlorosis on the younger leaves but their main veins remain green.

Manganese

Appearance of chlorosis and rough-margined leaves.

Boron

Stunted growth, abnormal shape and disintegration of young tissue, which leads to cracks in the plant, its blackening, and yellowing and death of the terminal buds.

Molybdenum

Necrosis, chlorosis and suppression of leaves which become thin and dry like paper. The activity of symbiotic nitrogen fixing organisms is reduced. Development of leaf lamina is restricted.

9. Masticatories

a) Tobacco

Nitrogen

Deficient plant is small, pale leaves which tend to grow upright at an acute angle to the stem. General growth is on a small scale with a thin stem and maturity is delayed. Leaf harvested from such crops cures out pale, thin and lacking in texture.

Phosphorus

Plants are very much stunted in growth. Deficient phosphorus supply tended to produce dark colours of dull luster in the cured leaf. Extreme deficiency gives dark-green, stunted growth of short, rather pointed leaves which tend to lie nearly horizontal, giving an open, flat, staring appearance and very delayed maturity. The lower leaves may have numerous, small, circular light brown spots.

Potassium

Severe field deficiency is shown by stunted growth and a characteristic chlorosis which spreads from the tip and later the sides of the leaves. Immediately in the wake of this chlorosis, necrotic spots start to coalesce and the tissue disintegrates. The browning of the leaf tip edge may be called "rim-firing".

Calcium

Deficiency symptoms occur in the upper portion of the plant indicates that during stress, there is little calcium transportation from the older tissues to the growing points. Leaves of calcium-deficient plants are dark green colour and in the advanced stage, the terminal bud dies. When this condition occurs early in the development of the plant, the top leaves may become much thickened. Hooking down at the tips of the bud leaves and ultimately a breakdown of the tissue at the tips and margins of the leaves are some prominent symptoms in the early stages of calcium deficiency. Complete breakdown and death does not occur and growth takes place later. Older leaves on calcium-deficient plants generally retain their normal shape.

Magnesium

Magnesium deficiency (called sand drown) is shown by a characteristic chlorosis which starts at the tip of the leaf and spreads to cover the whole area. The chlorotic areas often become white, rather than yellow. The older leaves are first affected and in extreme cases, the chlorosis spreads up the plant.

Sulphur

Deficiencies are most likely to occur in the early stages of growth or during dry weather. When they do occur, the first evident symptom is a lighter colour of the young leaves compared with the older ones.

Flower emergence is delayed when sulphur deficiency occurs early in plant development.

Zinc

Deficiency symptoms are mild chlorosis, leaf spotting and then necrotic destruction of the leaf. These symptoms start on the older leaves but the tops show a decrease in water content.

Copper

Deficiency resulted in a breakdown of the tissue in tobacco leaves. The affected leaves are high in total and protein nitrogen and-low in sugars.

Iron

Iron deficiency is shown by chlorosis of young leaves with main veins remaining green and contrasting with interveinal tissues. Subsequently, the entire lamina may discolour.

Manganese

The first apparent symptom is chlorosis of the young leaves. The tissue between the veins becomes light green to almost white, giving the plant a checkered appearance. In the advanced stages, chlorosis may be followed by necrotic spotting and the affected tissue may drop out, producing a ragged appearance.

Boron

Young leaves are distorted. The lamina at the leaf base turns light green and can break down later. The leaves show large deformations and the terminal bud may die or stop growing.

Molybdenum

The first indication of molybdenum deficiency is mottling of the lamina in leaves at the mid stalk position. Bending and twisting of the leaf lamina are usually present and are followed by the appearance of small necrotic areas which gradually enlarged until the entire leaf was withered. These symptoms gradually spread to other leaves. Deficiency causes a delay in blooming and in extreme cases a loss of flowers.

C. Horticultural Crops

1. Fruit Crops

a) Mango

Nitrogen

Nitrogen deficiency leads to pale green coloured undersized leaves which mature without attaining full size and the leaf stalks form narrow angle with the stem. The symptoms first appear near the base of the current growth and progress towards the tip. The mature leaves become yellow. Twigs become stiff and woody, limited in length and diameter. Under acute deficiency, the fruit size is greatly reduced which drop prematurely.

Phosphorus

The phosphorus deficiency produces leaf tip necrosis and premature abscission. The younger leaves turn excessively green and slightly smaller. The wood of the main trunk become thin and results in die-back of some of the branches. In some varieties, P deficiency causes development of reddish purple colour on the underside of the leaf, starting as spot, later spreading over the entire leaf with veins finally becoming reddish purple. It also results in stunted growth and may cause premature defoliation followed by die-back.

Potassium

The leaves located in the middle of the current season growth are first affected and the symptoms spread towards the distal and basal part of the affected twig. The scorching appears along the leaf margin and is preceded by dark purplish discolouration.

Calcium

The calcium deficiency rarely causes any definite leaf symptom in mango, however, the trees supplied with low calcium remain small.

Magnesium

The magnesium deficiency leads to stunted plant growth. Plants turn pale in colour and are smaller in height and girth. On the maturing leaf, yellow white layers appear on both sides of the midrib which runs parallel towards the margin. Sometimes, there are yellowish specks.

Sulphur

Sulphur deficiency leads to retarded growth and severe defoliation. The leaves turn dark green but upon reaching maturity, they develop necrotic areas along the margins.

Zinc

Zinc deficiency results in reduction in the leaf size. The young and middle leaves turn olive green in colour. Later, brown spots are observed along the margins of leaf lamina, such spots coalesce and large area of the leaf turn brown and necrotic. With the exception of the young leaves, all other leaves are usually misshapen, thick, brittle and puckered. In some cases, the young leaves curl inward from the margins and form a cup shaped structure. Large number of lateral branches appear having small misshapen leaves giving a bushy appearance.

Manganese

Manganese deficiency results in mild chlorosis of the middle and young leaves. Minute light buff-coloured spots appear along the major veins near the leaf margins. Such spots coalesce and form buff-coloured areas, which finally turn brown. The affected young and middle leaves shed prematurely. In severe Mn deficiency, the tips of the affected leaves also become necrotic. Ultimately, whole leaf becomes necrotic.

Boron

Besides mild chlorosis, boron deficiency causes a marked reduction in leaf size. Later, the chlorosis becomes intense. The main stem shoot becomes black and necrotic and growth is completely checked. The leaves on small lateral branches are highly chlorotic and such branches do not grow. The chlorotic middle leaves turn brown and finally necrotic. The remaining leaves become rough, leathery to touch and sickle shaped.

b) Citrus

Nitrogen

Nitrogen deficiency first appears on older leaves and proceeds towards younger leaves. Deficiency symptoms are characterized by light yellowish green leaves, veins are only slightly lighter in colour than the tissues between. New leaves are small in size, are thin, fragile and light green in colour. Deficiency can also be distinguished by general nature of yellowing of foliage over the entire tree with absence of any distinctive leaf pattern.

Phosphorus

Symptoms appear first on older leaves which loose their deep green colour. Leaves are small and narrow with purplish or bronze, lusterless discolouration, some leaves may later develop necrotic areas and young leaves would show reduced growth rate. Fruits would be coarse and rough in texture with a coarse, thick rind and hollow core.

Potassium

Early potash deficiency symptoms in the form of stunted growth, sparse foliage and somewhat bronzed and lusterless appearance of leaves. Under acute deficiency, the leaves wrinkle and twist and only weak new lateral shoots emerge because of lack of mechanical strength. These shoots have a tendency to be S-shaped. General leaf pattern begins with yellowing of tips and margins, yellow area then gets broader, necrotic areas and spotting can develop on leaves, with first appearance on older leaves. Small leaves, fine branches, compact tree appearance, an increase in susceptibility to drought, and cold, reduced fruit size, very thin peel of smooth texture and pre-mature shedding of fruits with lower acidity.

Calcium

Fading of chlorophyll appears along the leaf margins and between the main veins. Small necrotic spots develop in the faded areas. Small thickened leaves cause loss of vigour and thinning of foliage under severe conditions. Develops twig dieback and multiple growth of new leaves; undersized and misshapen fruit with shrivelled juice vesicles.

Magnesium

Magnesium deficiency is commonly referred to as "bronzing" or "copper leaf". Disconnected yellow areas or irregular yellow blotches start near the base along the midribs of mature leaves that are close to fruits. They become gradually larger and eventually coalesce to form a large area of yellow tissue to each side of the mid rib. This yellow area enlarges until only the tip and the base of the leaf are green, showing an inverted 'V' shaped area pointed on the mid-rib.

Zinc

The appearance of rosetting, premature shedding, whitish chlorotic streaks between veins in older leaves and whitening of upper leaves are most commonly observed due to zinc deficiency in citrus. Irregular green bands along the mid-rib and main vein on a background of light yellow to almost white appear on leaves. Relative amount of green and yellow tissues vary from a condition of mild Zn-deficiency in which only small yellow splotches between the larger lateral veins to a condition in which only a basal portion of the mid-rib is green and the remainder of the leaf is yellow to white. In very acute stage, the leaves are pointed abnormally narrow with the tendency to stand upright, and extremely reduced in size.

Copper

Wilting of terminal shoots, frequently followed by death. Leaf colour often fades, carotene and other pigments contents are reduced. First symptom is the formation of unusually vigorous large dark green foliage with a bearing mid-rib. Brown stained areas of hardened gum on the rind of the fruit may precede the appearance of leaf and twig symptoms. Fruit splitting with a part of splitting starting at the blossom end in the usual way or around the gummy stained areas. In severe case, twigs remain covered with reddish brown droplets of gum. The brown stained areas on the fruits darken and later turn black by the time fruit matures.

Boron

Terminal leaves turn necrotic and shed prematurely. Internodes of terminal shoots shorten, usually form resetting in addition to apical meristems turn blackened and later die. Flower development and seed production is usually impaired or lacking, death of terminal growing point of the main stem; thickened leaves and tendency of leaves to curl downward. Young leaves show small water soaked spots or flecks become translucent as the leaves mature. Fruit symptoms are more reliable as hard fruit and dry due to lumps in the rind caused by gum impregnations.

Molybdenum

The symptoms appear on the leaves as large interveinal chlorotic spots. Although, the characteristic yellow spots may be found any time during the year on deficient trees, they usually appear first as water-soaked areas in the leaves. Following the summer growth flush, bright yellow spots develop on older leaves with deposits of brown gum on the lower leaf surfaces. The affected leaves usually drop during the winter and in some instances, the trees become completely defoliated. When heavy leaf drop is caused by the deficiency, the fruit usually develops large spots similar to those caused by sunburn.

The leaves of the new cycle attained full size, or nearly so, without developing and unusual appearance other than a rough texture, probably due to enlarged oil glands, but, instead of maturing the dark green healthy leaves, they rapidly become mottled. The leaves of this mottled cycle may or may not develop necrotic spots in the mottled areas. Apparently, if the Mo deficiency is severe enough, these necrotic areas appear, enlarge, and extend to the margins.

c) Guava

Nitrogen

When nitrogen is deficient, the upper leaves turn pale yellow and the lower leaves becomes completely yellow. The leaves are reduced in size and are more upright. The purple patches on the leaf are concentrated on both sides of the midrib and principal veins. As the deficiency advances, such patches changes to dark brown and are blown away leaving holes in the leaf lamina.

Phosphorus

The characteristic symptoms of phosphorus deficiency is purpling of lower leaves between the veins starting from the margins and gradually extending towards the center. With advanced starvation, the purple patches enlarge, coalesce and the whole leaf attains deep brown colour.

Potassium

The upper leaves become light green and the lower leaves partially yellow along with brown edges. Necrotic spots are scattered all over the terminal end.

Calcium

Calcium deficiency appeared in recently mature, normal-sized leaves. It was also characterized by an interveinal chlorosis starting from the main vein towards the margins and the development of necrotic spots. The leaves later acquired a freckle like appearance. The spots later coalesced to form a large brown necrotic area and even the veins turned yellow and brown.

Magnesium

Magnesium deficiency is found in mature normal-sized leaves. It is characterized at the beginning by a mild interveinal chlorosis from the margin towards the midrib. At the intense stage, the chlorosis covered the entire leaf, except an area of green tissue in the main vein having an inverted wedge shape.

Sulphur

Sulphur deficiency of in guava is characterized by the occurrence of necrotic spots that vary in shape, size, number and colour, localized principally in the middle underside of the leaf. These spots are more clearly seen when the leaf is examined against the light. The colour is purplish along almost the full extension of the main vein and the secondary veins. The under surface of the leaf, other than being a little lighter than the upper side has chlorotic spots, although these are no easy to see. Only the secondary veins have a lighter purple on the upper surface.

Zinc

Zinc deficiency in guava is characterized by reduction in leaf size, interveinal chlorosis, suppression of growth and die-back of leaders. The 'Fatio' disease which has been attributed to zinc deficiency along with nitrogen and boron is characterized by red spots on leaves which later dry up and cracks in branches develop leading to die-back.

Copper

Copper causes absence of chlorophyll in young leaves and result in interveinal yellowing. Copper is relatively immobile in plants and tends to remain in older leaves, while newly formed leaves starve and become yellow. The top leaves appear deep yellow and short. Later, there is shedding of leaves from top to downwards resulting in defoliation of plants.

Iron

Iron deficiency is characteristic in the young terminals. It started with interveinal chlorosis in young leaves, acquiring a netlike appearance. After the pruning, all new shoots developed an albino appearance of the leaves because of a lack of chlorophyll. The leaves of young shoots are cupping down with marginal necrosis.

Boron

It's deficiency in guava fruits results in internal necrosis and affected portion becomes hard. Fruit size is reduced and leaves start falling. Cracking of fruits are seen in extreme cases. In boron deficient plants, the transport of sugar is poor and also caused degradation of cells.

d) Banana

Nitrogen

The most evident symptom of nitrogen deficiency is yellow leaves resulting from a reduction in chlorophyll content. The normal progression of the deficiency is yellowing to begin in the older, lower leaves and subsequently to affect younger leaves as the deficiency become acute. Petioles of the most affected leaves show a pinkish discolouration. Nitrogen deficiency significantly delays plant growth and development by reducing leaf production rates as well as the relative distance between leaves. A rosette appearance often results from each leaf emerging and developing on top of each other. Plant height and leaf lengths are also considerably reduced.

Phosphorus

Phosphorus deficiency symptoms show on the old leaves as chlorosis of their margins. Purplish brown flecks develop there as well. With severe deficiency, the affected leaves curl, the petioles break, and the younger leaves have a deep green colour. P deficiency causes complete cessation of elongation, at a height of about two feet rosetting of leaves with older leaves becoming increasingly irregularly necrotic, leaf production is reduced and marginal chlorosis and in severe cases, premature death follows.

Potassium

In potassium deficient plants, leaves are profusely small. There is premature yellowing with reduced growth rate. Once the K content is reduced, deficiency symptoms appear suddenly starting with yellowing of the tips and distal margin of the older leaves, closely followed by necrosis. The yellowing and necrosis spread rapidly in the proximal direction until the whole leaf has withered standing in a normal position. Split develop parallel to the secondary veins and the lamina folds downwards, while the midrib bends and fractures, leaving the distal half of the leaf hanging. Purplish brown patches appear at the base of the petiole and in severe cases, the center of the corm may show brown water-soaked areas. Fruits are badly shaped and poorly filled.

Calcium

The lack of Ca causes secondary veins to thicken, particularly in the leaf section adjacent to the central vein. This thickening is accompanied by marginal interveinal chlorosis as the plant grows. If the deficiency is severe, the leaves become completely deformed with margins becoming serrated or saw-like. Prolonged and severe Ca deficiency can prevent the emergence of new leaves and ultimately the death of the plant. A lack of Ca reduces leaf emergence and stunts plant growth, which gives the plant a flimsy appearance.

Magnesium

The most typical visual magnesium deficiency is yellowing of the outside section of the lamina in older leaves, which is due to its mobility in the plant. As the deficient leaf matures, chlorosis of its affected parts becomes more intense and dark spots form that later becomes necrotic. Affected leaves become an intense golden yellow colour. Magnesium deficiency changes the arrangement of leaves in the pseudostem, producing a rosette appearance. Another Mg deficiency symptom is a bluish-purple colouration of the petioles. In some countries this is known as "blue sickness".

Sulphur

Sulphur deficiency symptoms occur as a pale yellowing of younger leaves. If the deficiency is severe, the margins of the leaves become necrotic and develop a slight thickening of the veins. At times, leaf morphology changes so that leaves form without lamina. As the plant ages, these symptoms may disappear because the roots may remove SO_4^{2-} located deeper in the soil.

Zinc

Zinc deficiency typically occurs as white stripes, approximately one cm wide and parallel to the veins of new leaves. Symptoms develop in young tissue because Zn does not translocate from older to younger emerging tissue. These white strips are alternated with parallel dark green strips. As the deficiency progresses, the new leaves become narrower and a reddish colouration appears on the back of the leaves due to accumulation of anthocyanins. Zinc deficiency can be easily confused with those caused by viral infection or damage caused by herbicides.

A lack of Zn delays plant growth and development. Affected leaves align in the same plane to form rosettes. Another Zn deficiency symptom is the presence of small and deformed bunches. Zn deficiency shortened the separation between hands, giving the fruit bunch a compact appearance. Affected fruit remain in a horizontal position, possibly due to their low weight and short fingers that are elongated at the terminal end, all of which reduce quality.

Copper

Copper deficiency is characterized by the leaf tip chlorosis, rusetting, increased lamina, withering and drying of leaves, poor stem growth, detopped appearance, presence of strong water shoots and ribbon like roots. Die-back and defoliation with or without eruptions has also been recorded. Copper deficiency also causes drying of flowers and inhibits seed setting. The fruits produced on such plants exhibit irregular brown spots with splitting of skin.

Iron

Iron deficiency in banana manifests itself as interveinal chlorosis of young leaves, turning into a general severe chlorosis. Roots show decay symptoms. Chlorosis is visible on the first leaves and spreads quickly to both younger and older leaves. The chlorosis starts from leaf margins and spreads towards the midrib but more quickly to the main veins in comparison to the interveinal areas which remain green for a long time giving a "comb teeth" effect.

Manganese

Manganese deficiency causes marginal interveinal chlorosis on younger leaves and chlorosis advance towards the midrib. Pin head sized dark brown and black spots appear on the upper surface of the leaf mainly within the chlorotic area. All such spots coalesce and form a brown margin often around the entire leaf. The margin appear crinkled, drawn together and gives the leaf a boat shaped appearance. Affected plants produce bunches of fruits with widely distributed dark brown to black raised spots. The yellow pulp or green ripe disorder is also due to Mn deficiency.

Boron

Boron deficiency leads to marked deformation of young leaves which are reduced in size. Sometimes they consist only of midribs. Leaf shape becomes irregular with a curved and undulating margin. Proliferation of the shoots is also evident. Acute deficiency results in cessation of all growth. Leaf tip break, leaf ends are torn to pieces and turn brown. In severely affected plants, suckers do not root well. A severe deficiency of B can result in serious deformation of fruit bunches, but even mild B deficiencies affect the quality of the fruit bunch. Boron deficiency results in poor root system development with very few absorbing root hairs. Continued B deficiency ultimately leads to severe necrosis of the root system.

e) Pomegranate

Nitrogen

Nitrogen deficiency symptoms in pomegranate are first noticed in old leaves which become uniformly chlorotic yellow with light green midrib and veins. In extreme cases, necrotic spots develop on the chlorotic margin of the leaves. The overall growth of plants is checked and few thin branches develop. The leaf size is reduced and defoliation occurs. The yield is reduced considerably. Uniform yellowing of whole leaves appeared initially on lower and mature leaves. Afterwards, these leaves became stiffer and broke into pieces on folding. At advanced stage, young leaves became uniformly chlorotic and light brown followed by drying up from the tip.

Phosphorus

Deficiency symptoms prominently appeared on young leaves. Yellowing of leaf started from tip and other part remained green. Leaves became slender, elongated and are smaller in size. Leaf margins turned upward and got tunnel like shape. At an advanced stage, total leaves became yellow followed by appearance of chlorotic spots, which later on turned dark brown.

Potassium

Deficiency symptoms appeared initially on older leaves. Many brown spots appeared on dorsal side of leaves along the leaf margin starting from tip. At an advanced stage, leaf margin became yellow followed by scorching.

Calcium

Interveinal yellowing started from leaf tip and advanced towards midrib from margin. Veins remained green during initial stages and became yellow at later stages. The yellow portion of the leaf tip acquired inverted 'V' shape. Also, pinkish tinge appeared on lower yellow portion of the leaf. At an advanced stage, yellow portion of leaves turned dark brown in colour and half of the leaves from the tip dried up.

Magnesium

Leaf margin and veins turned light green in colour. Initially, grey patches appeared on side margin of the leaves and subsequently spread on whole leaf. Leaves dried up and the colour of dry leaves turned.

Sulphur

Sulphur deficiency results in pronounced effect on chlorophyll content of the leaves with suppression of the growth. The plants are stunted having few branches and leaves. Heavy defoliation starts at early stages. The younger leaves turn chlorotic yellow and chlorosis spreads to older leaves. Necrotic spots develop on the margin of the chlorotic leaves.

Zinc

Deficiency is first pronounced as interveinal chlorosis in young and mid-shoot leaves. Interveinal yellowish areas started appearing at leaf tip margins, which eventually affected all growing points of the plant.

Copper

Copper deficiency caused irregular growth and pale green leaves that withered at leaf margins. Leaf margins become yellow and leaves at top of the plant wilted first followed by chlorotic and necrotic areas on leaves and necrosis of the apical meristem. Leaves on top of the plant showed unusual puckering with veinal chlorosis.

Iron

The symptomology of iron deficiency is usually manifested as interveinal chlorosis of young leaves, while the veins remained green.

Boron

Fruits small, harder, abnormal and crack. Scattered yellow spots on the leaf surface and the leaf tip shown burn symptom. Leaves thick and brittle.

Molybdenum

Pale leaves with marginal leaf chlorosis are typical molybdenum deficiency symptoms. Interveinal yellow spotting and mottling of older leaves are also due to molybdenum deficiency.

f) Grapes

Nitrogen

The classic symptom is a uniform light green colour of leaves as compared to the dark green of vines that receive adequate nitrogen. Other clues pointing to nitrogen deficiencies are slow shoot growth, short intermodal length, and small leaves. Insufficient nitrogen can also reduce crop yield through a reduction in cluster, berries, or berry set.

Phosphorus

Phosphorus deficiency results in internal pink colouration of mature/old leaves, though it is a rare symptom as grapes normally do not exhibit P deficiency. When P is deficient to sub-optimum, the buds turn vegetative rather than fruitful. When P is sub-optimum, the canes take longer time to mature.

Potassium

Foliar symptoms of potassium deficiency become apparent during as a chlorosis or fading of the leaf's green colour. This yellowing commences at the leaf margin and advances toward the centre of the leaf. Leaf tissue adjacent to the main veins remains darker green, at least when the potassium deficiency is mild. Mid shoot leaves are the first to express these symptoms. With advanced or more severe potassium deficiency, the affected leaves will have a scorched appearance where the chlorotic zones progress to brown necrotic tissue. Leaf margins will curl either upward or downward. Severe potassium deficiency also reduces shoot growth, vine vigour, berry set and crop yield. Fruit quality suffers from reduced accumulation of soluble solids and poor colouration. The symptoms described can also appear under conditions of extreme drought or extreme moisture. Furthermore, leaf scorching can also occur under some conditions from pesticide phytotoxicity.

Calcium

Calcium is immobile in the phloem. Hence, in case of deficiency, the terminal buds of the shoots and apical tips of the roots fail to develop. A healthy cation exchange promoted by Ca and water in soil contributes to grape health and disease resistance. Some times, deficiency has also been noticed in clusters/bunches even in calcareous soils, particularly during cold season. The growing clusters dry up starting from the tip, which is similar to severe stem necrosis.

Magnesium

The foliar symptoms first appear as interveinal yellowing on the more rapidly expanding young leaves. This results into a pale yellow leaf having green network of veins, including the very small ones. As the deficiency becomes increasingly severe, more of the leaf area becomes yellow and ultimately ivory or white. Severely chlorotic areas on leaves often turn brown and necrosis occurs. Growth is reduced on severely affected shoots and the flowers and the stem structure on the clusters may become pale yellow. Fruit set can be poor on such shoots.

Zinc

Symptoms may vary depending on the degree of deficiency and the variety. Foliar symptoms such as mottling, usually appear in early summer at about the time when lateral shoot growth is taking place. The new growth on both the primary and secondary shoots has smaller, somewhat distorted leaves, having a chlorotic pattern exposing the veins as a marker green colour. Even the small veinlets retain a uniform width border of green unless the deficiency is quite severe. The term "little leaf" explains the gross appearance of stunted shoots, which have closely spaced, small distorted leaves. Zinc deficiency can seriously affect the set and development of the berries. Vines deficient in Zn tend to produce clusters with fewer and smaller berries than the normal vines.

Copper

Copper deficiency symptoms consist of poor root development, small pale green leaves, rough bark, short canes and short internodes. The yield and fruit quality is quite poor.

Iron

In case of iron deficiency, initially a pale yellow with green network of veins is observed on the leaves. As the deficiency becomes severe, the leaves become yellow or even white. Severely chlorotic areas on leaves often turn brown and necrotic. Shoot growth is reduced only in severe deficiency.

Manganese

The symptoms can appear in 2-3 weeks after bloom on severely deficient vines. A mild to moderate Mn deficiency will not appear until late summer. The symptoms start on the basal leaves which become chlorotic between the veins. Increasing chlorosis develop between the primary and secondary veins; the veinlets tend to retain a green border.

Boron

The presence of small sized fruits and large sized fruits in the same bunch is known as hen and chicken disorder. The fruits are sour in taste. The shoots are dwarfed because of the abnormally short internodes, which may grow in a zig-zag manner. Numerous lateral shoots grow from the stunted shoot giving the vines a bushy appearance. On severely affected shoots, the growing tip may die and such shoots are unfruitful. Severely affected vines, appear to burn off or dry around bloom time and may have occasional berries.

Numerous small, seedless berries may sometimes persist and ripen. Such clusters may have a few normal sized berries. Unlike the oval or elongated normal berries, these clusters have short berries which are round to somewhat flattened.

g) Pineapple

Nitrogen

Plants grown in nitrogen deficiency soils are slow in growth with pale pinkish green leaves of smaller size. The flowering and fruiting is delayed considerably. Reduced leaf size, leaf number and crown mass. Crowns absent. Leaves turn green to greenish yellow and yellow. Reduced fruit quality.

Phosphorus

The symptoms of phosphorus deficiency include dark, bluish-green foliage, which is more pronounced in the presence of excess N, and leaves that completely dry out at the tips, starting from the older ones. The old leaves have dry, brownish-red tips and brown transverse striations and the borders of these leaves turn yellow from the tips. Plants suffering from P deficiency also have more erect, long and narrow leaves and roots with longer, more coloured and less ramified hairs. The fruit is small and reddish in colour.

Potassium

Under extreme K deficiency, there is reduced leaf growth, older leaves develop a yellow brown colour, starting from the leaf tip. Fruits are small, their flesh is pale in colour. The juice is low in vitamin C and acids.

Calcium

In calcium deficient plants, the leaves become pale green having yellow mottling, new leaves which dry from the tip. As the deficiency progresses, the leaves blister and develop a reddish colour. Plant growth is slow and production of slips, suckers and fruit is reduced. The fruits will have internal discolouration and do not mature.

Magnesium

Chlorosis first becomes evident in the older leaves which become light green. Yellow mottling appear which at times coalesces to form an almost yellow strips along the leaf margin. The leaves have a reddish tinge. No fruit production under severe deficiency.

Sulphur

In the early stages of sulphur deficiency, the older leaves show blistering symptoms. As the deficiency progresses, the leaves become light green and in the end, show yellow mottling accompanied with necrosis. The fruit produced under S deficiency exhibit abnormalities in structure and ripening. Unlike normal fruits which ripen from the base upward, the S deficient fruits ripen first at the top. When the fruit is cut in half lengthwise, a vacuole of about one inch in diameter is encountered at the core.

Zinc

In severe zinc deficiency, central leaves bunch together and bend over. Fruit burn may occur from over exposure to the sun. Under high NPK conditions, the Zn deficiency leads to canker formation on the upper surface of the older leaves during the dry season. Small yellow spots appear first, having brown center, which coalesce and finally leaves are completely destroyed.

Copper

The copper deficiency causes curling and twisting of the leaves. The leaves that develop later are typically narrow, light green to yellow, having a thickly waxed surface.

Iron

Development of chlorosis, starting in young leaves, the leaves are generally flaccid, wide and yellow. The old leaves are dry. The fruit is red with a chlorotic crown.

Manganese

The damaged leaves are marbled with light green areas, especially where the veins are situated, surrounding areas with a darker green colour.

Boron

Deficiency symptoms include: orange and yellow discolouration, which becomes brown on only one side of the leaf; minimum leaf growth to two thirds of its normal length and with dry tips. A tendency for the leaf to curl, chlorosis of the young leaves with reddening of the dead borders at the apex, fruits with multiple crowns; formation of a dead suberized tissue between fruitlets, at times accompanied by very small, spherical fruits.

h) Avocado

Nitrogen

The reduction of chlorophyll caused by nitrogen deficiency leads to loss of colour in the leaves. In the early stages, a slight fading of normal colour and slight retardation of shoot elongation occurs. This can later be associated with fruit-set. Later stages, show uniform loss of colour. The dark green leaves turn light green to yellow. Foliage is sparse and leaves are smaller, the plants are also smaller.

Phosphorus

Tree tends to be stunted, lacks vigour and can suffer twig dieback in extreme cases. Small leaves occur much like trees with zinc deficiency, except that the mottling leaf pattern does not occur. Randomly distributed necrotic areas may occur on leaves with severe deficiency. Often the whole grove is not affected; the problem tends to occur in isolated, random groups of trees.

Potassium

The first symptom of deficiency observed in soils lacking potassium is the development of somewhat narrowed leaves of slightly subnormal size. Soon after the leaves reached maturity, they became very deep green. Later, the intensity of green colour decreased. As the leaves aged, the brown specks coalesced into large, irregular, reddish-brown areas of dead tissue between the large veins, along the margins, or at the tip. The trees are badly stunted, and on severely deficient trees, the twigs are very thin and some of them died back during the winter.

Calcium

With calcium deficiency, the young leaves are first affected. They are distorted and have irregular margins. Leaves show spotted/ necrotic areas that fade generally. Necrosis also occurs at the tips of leaves. In extreme cases, defoliation occurs.

Magnesium

Yellowing progresses inwards from the leaf margins and tips towards the main veins, leaving a band of dark green along the main veins. Always appears on older leaves.

Sulphur

Sulphur deficiency symptoms are in many respects similar to those of nitrogen. Young leaves are more sensitive than older ones, in contrast to nitrogen deficiency where all the leaves are affected. Typical symptoms include, the yellow colour and smaller leaf size.

Zinc

Affected leaves show uneven blotchy yellowing between the veins. With a very mild deficiency, leaf size is only marginally reduced and leaf distortion is minimal. A more severe deficiency produces significant leaf distortion and reduction in size, accompanied by the yellowing. With a very severe deficiency, the younger leaves at the end of shoots become yellow, very small and distorted.

Copper

In copper deficiency, terminal growth is affected first. Dieback of growing points is often preceded by shortening of internodes. New leaves are abortive. They almost immediately dry and die back. Older leaves are dull with reddish-brown veins. The colour may spread into the leaves.

Iron

Affected leaves are an overall yellow or white colour with the veins remaining green. Yellowing generally occurs on the young leaves of the summer flush while the leaves of the spring flush remain green.

Manganese

Chlorotic areas develop near the midrib first and then spread. Later, the dark areas become dull and the light yellow areas may become almost grey. In extreme cases, the whole leaf becomes a dull yellowish green.

Boron

The most common symptom found in boron-deficient trees is shot hole of leaves. During the rapid expansion stage of leaves, holes develop due to a collapse of cells giving rise to necrotic areas that abscise in the interveinal regions of the lamina. The yellowish halos around the margins of the holes give this symptom a unique appearance. Another leaf symptom is deformity which occurs when cell expansion on one side of the leaf is retarded, followed by tissue necrosis in that region. This results in a crescent-shaped leaf.

In trees with advanced boron deficiency, leaves yellow in association with reduced growth. Trees showing this symptom have generally lost most of their feeder roots, will flower up to six weeks earlier than healthy trees, and will substantially defoliate during flowering. Reduced growth in trees is accompanied by loss of apical dominance which often stimulates growth from multiple epicormic buds, and the resultant loss of geotropism gives a weeping habit in the tree. In boron-deficient trees deformed fruit may develop during the first six weeks of growth.

Molybdenum

The symptoms of molybdenum deficiency are similar to nitrogen deficiency. Yellow spotting of leaves occurs.

i) Sapota

Nitrogen

Typical chlorotic deficiency symptoms on older leaves and later spread to younger leaves. The leaf cuticular thickness is less and the epidermal palisade parenchymatous cells are loosely arranged, while spongy parenchymatous cells are compactly arranged. In shoots, the epidermal cell size is reduced and vascular bundles are compact with reduction in the size of xylem vessels.

Phosphorus

Purplish flecks on lamina towards the tip are formed with the manifestation of purplish rusty pigmentation all over fully matured leaves. The leaf thickness is more and epidermal cells are loosely arranged. In shoot cross section, the epidermal, cortex, vascular and pith region cells are compact.

Potassium

In recently matured leaves, irregular chlorotic areas along the margin first appear which later turns to blackish grey. The leaf cuticular thickness, size of palisade and sponge parenchymatous are not much influenced but in shoot, cuticular thickness is reduced. Epidermal and cortex cells are thinner and enlarged.

j) Cherry

Nitrogen

Nitrogen deficient trees are generally low in vigour and productivity. Since shoot length is reduced, trees may produce abundant spurs and flower buds. Leaves are pale green in colour, and may develop a reddish tint late in the season. Older basal leaves typically exhibit the earliest and most severe symptoms. Leaves are smaller and often drop prematurely. Fruit are small and earlier to mature. Nitrogen deficient trees are more susceptible to winter cold injury. The reduced shoot growth and decreased flowering tend to lower production. Fruit are smaller, even when trees are thinned to the same fruit load as a tree with better nutrition.

Phosphorus

Symptoms of phosphorus deficiency begin with weak, slender shoot growth and leaves with a dark green colour that eventually turns bronze or reddish purple. Leaves develop a very characteristic leathery texture. A red colouration appears on petioles and young shoots. Eventually leaf size is reduced, and premature defoliation may occur, beginning with basal leaves.

Potassium

The first symptoms of potassium deficiency are a slight upward curling of leaf margins. The undersides of leaves turn bronze in colour and the margins of leaves eventually become necrotic or "scorched". Leaf curling and scorching is usually most severe on the basal leaves of current season growth, but may also appear on terminal leaves. Shoot growth and leaf sizes are reduced. Symptoms are most severe on trees carrying a heavy crop of fruit, since fruit accumulate relatively large amounts of potassium at the expense of the other tree parts. Leaves may drop prematurely from severely affected trees.

Calcium

Calcium deficient trees exhibit reduced shoot growth due to shortened internodes. Defoliation and twig die back follow this. Symptoms on young trees include light brown to yellow markings on leaves. Leaves may become tattered with numerous holes. Leaf blades often roll inward and upward and may develop large chlorotic areas before they abscise. Fruit will have a poor shelf life, reduced flavour and the tree will have reduced vigour.

Magnesium

Deficiency symptoms include an interveinal browning and necrosis starting first on basal (older) leaves. Browning may start in the middle of leaves or progress inward from the margins. Bright red or yellow colouration may border necrotic areas. Affected leaves may fall prematurely.

Sulphur

Uniformly yellow leaves at the shoot tip characterize sulphur deficiency. Symptoms are similar to those of nitrogen deficiency except for the location along the shoot. Nitrogen deficient plants are more uniformly affected along the shoot.

Zinc

Zinc deficiency is common in cherries. It has often been called "little leaf", because rosettes of small pointed leaves form at shoot tips. Shortages cause reduced leaf size and irregular mottling and chlorosis of leaves. Leaves may drop prematurely. Shoots typically fail to elongate normally. This shortens the inter-node distance between leaves and results in tufts or "rosettes" of leaves at the tip of shoots. Affected limbs may be confined to certain portions of the tree or distributed uniformly throughout the tree. Buds on the previous year's growth may fail to open, resulting in lengths of shoots, which are void of leaves. Fruit size and soluble solids content are severely reduced.

Iron

Terminal leaves on deficient trees turn chlorotic to bright yellow between the veins. Veins remain green and stand out against the rest of the leaf. Although terminal leaves are first affected, symptoms may progress in a basal direction to include older leaves. Tissue along the margins of severely affected leaves may die.

Manganese

Chlorosis starts at the margins of leaves and progresses inward between the main veins. Leaves are typically smaller and shoot growth is inhibited in proportion to the severity of the leaf symptoms. In mild instances, symptoms may be most apparent on spur leaves, though spur and shoot leaves are affected in severe cases. Fruit yields and quality may be reduced severely. Fruit from affected trees are well coloured and firm, but small and lacking in juice. Symptoms tend to be general over the whole tree, although apical leaves often appear less affected. Severe deficiency causes some defoliation and shoot dieback.

Boron

Deficient trees exhibit little shoot growth. Some buds may fail to open in the spring, whereas others may open, then shrivel and die. Shoots may grow for some time, then tips cease growth and die. Leaves are distorted in shape with irregular serrations. Leaves may cup or roll in a downward direction and feel thick and leathery.

k) Apple

Nitrogen

Deficient trees produce short terminal shoots. Deficient apple leaves are pale green to yellow. Colour develops uniformly on the leaf with no patterning or mottling and leaf size is small. Symptoms first appear on older leaves because nitrogen moves out of older tissue into actively growing younger leaves. Leaves tend to drop early in the fall. Twigs are thin. Fruit set may be light with considerable drop. Fruits are smaller and mature early.

Phosphorus

Deficiencies trees exhibit reduced shoot growth and flowering. Bud break may be delayed and fruit set will be poor. Fruit mature early and are often misshapen, russeted, or cracked. Trees with low P will produce fruit that will not store well and in varieties that are prone to pre-harvest drop, drop will be severe.

Leaves on P deficient trees develop a characteristic dark green to purple colour, which is usually most intense in or near the main veins of the leaf. Trees slightly deficient in P may not show purple tints in leaves until late summer or autumn. If the deficiency is severe, symptoms may develop soon after bud break. Leaves are smaller than normal and remain upright, with a narrow angle between the petiole and shoot. As the deficiency progresses, leaves turn light green or yellow and older leaves abscise early. Leaf curling and reddish bark are also obvious.

Potassium

Deficiencies result initially in a yellowing of tissue along leaf margins. That tissue later turns a bronze colour and may eventually die, producing a scorched zone along the edges of leaves as the deficiency progresses. Scorched areas do not extend between the veins of leaves. Because K is mobile, symptoms appear first on older leaves but may affect young leaves in severe cases. Fruits accumulate large amounts of K, so leaf symptoms are more likely and severe as fruit approaches maturity during heavy crop years.

Calcium

Calcium deficient trees rapidly cease shoot growth and root growth. Shoot tips dieback, and viable roots thicken abnormally. Characteristic leaf symptoms develop first near the terminal leaf margins but may extend into interveinal tissue. Affected areas turn chlorotic and eventually necrotic. A purplish tint may also develop, appearing first on minor veins and eventually on larger veins.

Magnesium

Inadequate magnesium causes a yellowing of tissue along leaf margins and between the main veins. Symptoms may initially develop along leaf margins, similar to K deficiency symptoms. As the deficiency progresses, the yellow regions turn brown and die, leaving a christmas tree-shaped green area along the main veins in the middle of the leaf. Magnesium is fairly mobile so symptoms tend to develop first on older leaves, usually beginning around midseason.

Sulphur

Symptoms of sulphur deficiency include a uniform chlorosis. Yellow leaves eventually develop necrotic areas near the margin. Tree growth is reduced if symptoms are severe.

Zinc

Leaves are much smaller than normal and narrow. The tissue between the main veins turns yellow and chlorotic. Shoot tips do not elongate fully, resulting in compressed internode lengths and a tuft or rosette of leaves at the terminal.

Copper

Necrotic lesions developing on the terminal leaves of rapidly growing shoots are the first indications of a Cu shortage. Leaves eventually abscise, and the terminal portions of the shoot may die. Branching occurs the following season from below the point of death, often resulting in "witches broom growth". Trees appear "brushy" and low in vigour.

Iron

Symptoms always start at youngest leaves with yellow green. Later, lemon yellow and pale yellow interveinal chlorosis. Main veins still remain green. At acute deficiency, veins become chlorotic too and necrotic areas appear near leaf margin. Severe deficiency produces symptoms at older leaves too.

Manganese

Symptoms are similar to those of magnesium deficiency, but they appear first on young leaves. Youngest leaves yellow between the veins, while the main veins remain green. In severe cases, symptoms may progress to older leaves on the shoot.

Boron

Inadequate boron results in a reduction in leaf size. No distinct patterns occur on individual leaves. New shoot growth is severely reduced and shoot tips may die back in severe cases. Young twigs may develop a browning beneath the bark, similar in appearance to the "measles" symptoms caused by excessive manganese. Fruit set may be reduced. Fruit may become deformed and develop corky, dry lesions in the flesh.

Molybdenum

Molybdenum in necessary for nitrogen fixation. So symptoms of molybdenum deficiencies are similar to those of nitrogen deficiencies.

Leaves turn pale green to yellow; stems and petioles turn reddish brown.

l) Pear

Nitrogen

Nitrogen deficiency is associated with reduction in vegetative growth and paleness in leaf colour. Early flowering, lesser fruit setting and smaller size of fruit with poor colour.

Phosphorus

Typical deficiency symptoms involve leaf marginal and tip burning, decrease in leaf size, shorter terminal growth, die back of new growth, scaly bark and failure of fruits to mature normally. Phosphorus deficiency is more common in acidic soils. Phosphorus deficiency results in high anthocyanin production resulting in purplish red leaf margins.

Potassium

The most visible deficiency symptoms of potassium are scorching of older leaves, which extends towards younger leaves. Older leaves start yellowing at margins followed by appearance of water soaked areas and necrosis. The scorched leaf tends to fold and results in early abscission. Severely affected trees do not develop properly remain small in size and have poor colouration.

Calcium

The symptoms of calcium deficiency on foliage are peculiar and very specific. The earliest leaf symptoms are chlorosis and upward cupping of young leaves followed by marginal chlorosis. Fully developed leaves may show necrotic symptoms in the later stages. Symptoms on fruit involve diffused, golden peer colour and cracking at the calyx end of the fruit. Calcium deficient trees show lesser roots, and stubby growth.

Magnesium

The most common magnesium deficiency symptoms are the appearance of fading areas between the large veins of older leaves which turn pale or yellow and later turn into necrotic. Frequently, 'V' shaped areas of green appear that are centered towards midribs with their apexes towards tips of the affected leaves. The area outside such a 'V' may be yellow, pigmented or necrotic. The older leaves are affected first. The younger leaves show the symptoms of magnesium deficiency with the progress of season.

Zinc

Pears are known to be more susceptible to zinc deficiency than other temperate fruit species. The most common symptoms of zinc deficiency are little leaf, rosette and leaf mottle.

Copper

Copper deficiency symptoms in pear are similar to those of apples. The common symptoms include necrotic lesions, which may be marginal and have brown orange colour parallel to leaf margin. Death and withering of terminal leaves and current shoot growth from the tip are clearly visible. In severe cases, stunted growth is accompanied by small leaves, absence of fruiting and recurrent die back. Renewal of growth may cause a bushy 'witch's broom' appearance. The bark of twigs and trunk is rough and symptoms are often known as 'exanthema' and 'die back'.

Iron

Iron deficiency is most common in pears. The symptoms appear first on terminal leaves, shoots or on youngest leaves of spurs. Symptoms include, interveinal chlorosis accompanied by green vein. The affected leaves become pale yellow to even whitish with severity of deficiency. In severely affected leaves, brown necrotic lesions appear and results in early abscission. Deficiency also results in stunted growth and considerable die back.

Manganese

The deficiency causes chlorosis of the leaf tissue, beginning at the leaf margin and progressing inward. Tissues adjacent to leaf veins and margins turn green, producing a pattern similar to that caused by iron deficiency.

Boron

Reduced shoot growth with failure of basal leaves to develop. Affected twigs and severely affected trees may die during the winter season, Flower becomes brown in colour, early leaf fall, blossom of affected trees may fail to develop fruits causing a severe reduction in crops. In the case of fruits, shallow depressions on the fruit surface may develop which are most abundant at the calyx end of the fruit and may merge. The depression in the fruits are underlain in the flesh with corky tissues. In addition to this, cracking, splitting and malformation also occur.

m) Peach

Nitrogen

Trees stop growth and the leaves become pale green. Reddish tints gradually appear on leaves and branches are reddish-brown with spots. Some laminae are smaller and often twisted and curved badly. Premature defoliation occurred beginning with older leaves. Leaves turn to an orange or reddish-yellow colour prior to abscission. Many fibrous roots are produced. The root-top ratio of deficient trees is three times greater than the healthy trees.

Phosphorus

The deficiency manifests on older leaves in the form of development of purple colour. The purple colour development starts in May and progresses slowly through June and with the onset of rains in July, it becomes more evident. The purple colour of the leaf changes to bronze and in extreme cases, the tree gives a fired look from a long distance. The upward curling of leaves and breaking of leaf margins are the symptoms associated with phosphorus deficiency. There is premature shedding of leaves too.

Potassium

The middle and older leaves are yellowish green, wavy, forward curled and bent downwards like a sickle. The scorched tips and margins of the leaves are reddish brown to purple, tear and have a ragged appearance. The twigs are thin and blossom and fruits are poor in growth. The shoot tips die and the few fruits are small and misshapen.

The potassium deficiency in peach is expressed by small leaves near the ends of terminals and on lateral shoots. These leaves are usually long in relation to width, lighter green than the older leaves and showed a tendency to fold and roll. In acute K deficiency, the older leaves show crinkling and puckering of the tissue along the midrib. These spots later develop out of the leaf blade. Thus, giving a wavy margin and shot-hole effect to the leaf. The twigs are slender with scant fruit buds on them.

Calcium

Young leaves of terminal growth die back from tips and margins or along midrib. Following extreme deficiency, twigs may die back; most leaves are shed; roots are short, bulbous and die back after making little linear growth and new roots emerge profusely behind dead root tips.

Magnesium

The middle leaves are the most affected and exhibit slight, but easily discernible, intercostal chlorosis. Dark green 'water spot' with purplish-red borders around the dead zone appear on the older leaves of annual shoots. The blotches turn a grayish, green-white or light brown to brown. The leaves fall prematurely.

Sulphur

Tip leaves show a uniformly yellow colour, somewhat resembling iron deficiency, but with Fe shortage, there may be a network of green veins on small leaves. There may be some "burning" of leaves where S is deficient.

Zinc

Zinc deficient plants show mottling of younger leaves, which ultimately become small, narrow, pointed and often had a wavy margin. Defoliation of leaves also occur from the upper mid portion of shoots and the roots show gummy deposits. The 'little leaf' and 'rosette' condition of trees are the important characteristic symptoms of zinc deficiency. Deficient trees produce misshapen pointed small fruits with flat cheeks.

Copper

The foliage of the tree is unusually dark green in older leaves but later, the leaves turn greenish-yellow between the fine veins until finally a green reticulate pattern is left on a whitish-green back ground. The young, long and narrow leaves are much smaller than normal, deformed wavy and have irregular margins. They soon drop, leaving the branches bare as in other fruit trees.

Iron

The deficiency is characterized by an interveinal chlorosis which occurs in April when leaves have almost expanded and trees are making rapid vegetative growth. The new flush, which occurs in July-August also, shows similar chlorosis. Symptoms are evident as interveinal chlorosis of top leaves; the fine veinlets and main veins in the leaf blade remaining green. As the chlorosis progresses, the affected areas of the blade lose green colour, they bleach out and in extreme cases, the leaves may become ivory white; the leaves drop prematurely and eventually the denuded shoots die back.

Manganese

Symptoms of manganese deficiency are chlorosis between the leaf veins, checkered yellow and green or sometimes even grayish yellow except for the green along the veins. In the latter stages, reddish blotches or brown spots appear. There may also be drying and dying back of branches.

Boron

The boron deficient trees produce small, twisted, thick and leathery, malformed and yellowish leaves with corky main veins. Leaves may cup with mottled yellowing. The dieback of twig, rosetting of new shoots, failure of leaf and bud development are also noticed in boron deficient trees. Necrosis of the leaf and stem, bark splitting and distorted bark. The symptoms are similar to cold injury. Fruit shows gumminess, cork towards the center, blighted seeds and poor mealy tasteless in quality.

Molybdenum

In the younger leaves, there is a light greenish interveinal yellowing of the main veins and small veins remain green. The older leaves show a marginal burning due to the accumulation of nitrates under low molybdenum supply.

n) Litchi

Nitrogen

The first sign of deficiency (yellowing) is observed in the older leaves. In cases of severe deficiency, the margins of the leaves may curl, leaves may be small or fail to develop or be shed prematurely. Plant growth is stunted, flowering and fruit set are prevented and branch development become poor in N deficient plants. When the deficiency is very severe, new leaves do not mature and leaf drop occurs at the terminals. Fruit become small with low flesh recovery and reduced eating quality.

Phosphorus

The first symptoms of phosphorus deficiency show as dead patches on the tip and margins (coppery brown colour) of mature leaves that progress towards the midrib. Eventually, the leaves curl, desiccate and shed. Only in extreme cases, some degree of necrosis occurs and leaflets become copper coloured.

Potassium

The leaves start to become yellow, the leaf tips die and at last, the bases of the leaves are also affected. Severe deficiency can restrict shoot and root development. Plants may come under flowering but do not set the fruits and trees may die. Plants are stunted and branch development is poor. Leaflets are small, not much dark green in colour and sun burnt at the edges. Necrosis begins at the leaf tips and spreads along the leaf edges towards the base. In severe cases, leaf drop occurs just prior to the final vegetative flush so that only terminal cluster of leaves remain.

Calcium

Calcium deficient leaves show necrosis around the base of the leaves. The margins of the leaves growing more slowly than the rest of the leaf, causing the leaf to cup downward. This symptom often progresses to the point where the petioles develop but the leaves do not, leaving only a dark bit of necrotic tissue at the top of each petiole. Plants under chronic calcium deficiency have a much greater tendency to wilt than non-stressed plants.

Magnesium

The Magnesium deficient leaves show advanced interveinal chlorosis, with necrosis developing in the highly chlorotic tissue. In its advanced stage, magnesium deficiency may superficially resemble potassium deficiency. In the case of magnesium deficiency, the symptoms generally start with mottled chlorotic areas developing in the interveinal tissue. The interveinal laminae tissue tends to expand proportionately more than the other leaf tissues, producing a raised puckered surface, with the top of the puckers progressively going from chlorotic to necrotic tissue.

Zinc

Deficiency symptoms are normally associated with 'little leaf disease', interveinal chlorosis of young leaves and poor fruit set and yields. Leaves become small and narrow with leaf margins bent upward or downward. Internodal length is reduced drastically and the twig with crowded leaves gives rosette appearance and interveinal areas are pale. Zinc deficiency causes general bronzing or mottling of the leaves, smaller shoots and smaller fruit, small and short leaves with normal colour and occasional curving.

Copper

Copper deficiency is most likely in sandy soils with an inherently low copper content. Often the young leaves roll and die. Shoots may also die when leaf concentrations fall below 6 ppm.

Iron

The leaves loose green colour and turn white and size of the leaf is reduced. The growth of the tree is restricted; yield and quality of the fruit is reduced. In some plants, the leaves dry from tip to downwards. The symptom described is called "Bleaching". The deficiency is common in soils with high calcium content. Hence, the effect is known as "calcium induced iron chlorosis". Small plants may die due to acute iron deficiency. There is a general yellowing of the young leaves, spreading to older leaves. When the deficiency is severe, the branches may die.

Manganese

These leaves show a light interveinal chlorosis developed under a limited supply of manganese. The early stages of the chlorosis induced by manganese deficiency are somewhat similar to iron deficiency. They begin with a light chlorosis of the young leaves and netted veins of the mature leaves especially when they are viewed through transmitted light. As the stress increases, the leaves take on a gray metallic sheen and develop dark freckled and necrotic areas along the veins. A purplish luster may also develop on the upper surface of the leaves.

Boron

Litchi is sensitive to boron deficiency especially when high levels of nitrogen are applied or trees become moisture stressed. Boron is essential for flowering and fruit set and deficiencies cause flower and fruit drop as well as misshapen fruit. Boron plays a similar role to calcium in cell wall stabilization which makes it essential for quality factors such as skin strength, fruit firmness and storage life of litchi. Fruit cracking in litchi is mainly due to boron deficiency. Lusterless leathery leaves with thickened veins are the other associated symptoms.

o) Strawberry

Nitrogen

Strawberry plants show stunted growth and have yellowish-green leaves. The leaves are small and have rigid upright stalk. The plants produce few runners. Older leaves turn bright orange-red to red starting at the margins. The teeth of the leaf margin often turn brown and die before the whole leaf is discoloured. The veins at the leaf bases retain their green colour for sometime. Flowering and fruitification are more or less reduced, and the fruits are small.

Phosphorus

Due to phosphorus deficiency, the leaves turn bluish to dark green, starting from the small veins and leaf margins. The leaves droop and turn deep purplish bronze with reddish margins. The leaf stalks are intensive red. On the undersurface of the leaves, the midribs and veins are violet tinged. Flowering is poor and few fruits develop.

Potassium

Trees are weak with stunted growth. Foliar symptoms started with the appearance of broad brown patches on the upper surface of the leaves. The surface of leaf is puckered. There is a greater concentration of brown spot in the marginal area between the veins and they gradually enlarge and coalesce. Finally, scorching of margins took place, which is followed by premature abscission of leaves. The leaves turn bluish green besides the midrib. The edges of older leaves are reddish brown to purplish, later turning brown and drying. Discolouration and necrosis advances until only a small green triangle remains at the base of the leaf. Potassium deficiency also causes a dark brown discolouration of the base of the leaf which spreads into the leaf stalks. Light brown necrotic lesions then appear in the discoloured zone. Fruit development is poor and the fruits are less palatable.

Calcium

Young leaves may die or only leaflet or portion may be affected; appearance of red-brown colour around the base of leaflet; roots show injury before tips are affected, drying back from tips followed by development of new rootlets behind the dead portion; entire mass of root system is affected.

Magnesium

Small, reddish-purple blotches forming a linear pattern appear between the veins. The blotches merge to form belts between the green veins. Finally, the whole leaf turns purplish red, except for a green band. Yellowish green chlorotic blotches sometimes appear near the main nerve before the leaves turn reddish-purple.

Sulphur

Sulphur deficiency symptoms in strawberries resemble those of nitrogen deficiency except that, one leaflet is usually smaller than the other two. Margins of older leaves develop brownish-black serration tips.

Zinc

Zinc deficient plants exhibit pale green to yellow colour of the leaves. Only the larger veins and a narrow strip along the margins remain green. The leaves grow slowly while the leaf tend to be some what narrow and concave. The deficiency is more pronounced on younger leaves.

Copper

The leaves develop intercostal chlorosis and lighter blotches along the leaf margins and near the middle rib. The leaves are smaller than usual, collapse and wilt.

Iron

The young and youngest leaves are yellow to yellowish-white with green sharply defined nerves. In addition, the margins of older leaves turn brown and die. The veins of the younger developing leaves also become chlorotic, if the deficiency persists.

Manganese

Leaves faint, interveinal chlorosis begin at the margins and later on spreads to the whole leaf.

Boron

The leaves are small, severely deformed and puckered with margins turning upwards. The tip of the leaves turned brown. The plant becomes less resistant to diseases. Reduced root growth also occurs. Sometimes, the death of terminal shoot point leads to the development of laterals, which also die back. The leaves become chlorotic and distorted while the petioles become brittle. The flowers burst and the fruits are malformed. Stolons are short with excessive branching and stubbiness in the roots.

p) Kiwi

Nitrogen

Deficiency symptoms develop first on the older leaves and spread progressively to young leaves until the whole plant is affected. Initially, there is a gradual change in the colour of the leaf from the usual dark-green to light green. As the deficiency becomes more pronounced, the affected leaves may become uniformly yellow. However, even on severely deficient plants, the veins remain conspicuously green, particularly on the older leaves.

Phosphorus

Phosphorus deficiency reduces the plant growth without the development of obvious visible symptoms. The stems of affected plants tend to be much thinner and the size of the leaves considerably reduced. Clearly recognizable symptoms only appear on deficient plants when growth is severely restricted. Symptoms include a pale green interveinal chlorosis of the older leaves which spreads from the leaf tip back towards the point of attachment of the petiole. Reddening of the midrib and major veins on the underside of the older leaves may also occur, the colour being more intense towards the base of the leaf blade.

Potassium

The first sign of potassium deficiency is poor growth at bud break. On severely affected vines, the leaves are small and pale yellow-green with a slight marginal chlorosis on the older leaves. As the deficiency becomes more pronounced, there is an upward curling of the margins of the older leaves, that is, particularly noticeable during the warmer period of the day. This symptom may disappear the following day giving the plant an appearance, which has often been erroneously attributed to a lack of water. Later, the margins of the affected leaves remain permanently curled, and the tissue between the minor veins is often ridged upwards. Also the light green chlorosis, which developed initially at the leaf margin, spreads between the veins towards the midrib leaving a zone of green tissue close to the major veins and at the base of the leaf.

Much of the chlorotic tissue quickly becomes necrotic, turning from light to dark brown as if giving the leaf a scorched appearance. As the leaf scorch becomes more extensive, affected tissue becomes brittle and there is a tendency for it to break away at the leaf margin giving the leaf a tattered appearance. Severe potassium deficiency can cause premature defoliation of the vine, although the fruit will remain firmly attached.

Calcium

Symptoms of severe calcium deficiency develop first on recently matured leaves and spread to progressively younger leaves, the veins at the base of the leaf become necrotic and turn black. As the deficiency becomes more pronounced, the necrosis spreads to the fine veins on the remainder of the leaf; these necrotic areas enlarge and coalesce to form extensive patches of necrotic tissue. As the necrotic tissue dries, the leaf becomes very brittle and there is a tendency for the plant to defoliate.

By this stage, the growing point develops a rosette of very small leaves, dies, causing the axillary buds at the junction of the petioles and the stem to develop. At the same time, the oldest leaves develop symptoms that include an upward rolling of the leaf margin and an interveinal necrosis surround by a zone of chlorotic tissue. On severely deficient plants, the root system is poorly developed and in some cases the root apex dies. Large areas of necrotic tissue may also develop at some distance from the root apex and are likely sites for invasion by root pathogens.

Magnesium

Early symptoms of magnesium deficiency include a pale yellow-green interveinal chlorosis of the older leaves on the current season's extension growth of canes. The chlorosis usually develops at the leaf margin and spreads inwards between the veins towards the midrib, often leaving a relatively wide zone of healthy tissue on each side of the main vein.

In some cases, the margin of the leaf remains green and the chlorosis and subsequent necrosis start some distance from the margin. On these leaves, the necrotic tissue usually forms a regular pattern of isolated patches roughly parallel to the leaf margin giving a distinctive "horse shoe" appearance. A large zone of green tissue is also retained at the base of the leaf close to the point of attachment of the petiole, even on severely deficient plants. Initially, there is no necrotic tissue associated with the chlorosis but as the deficiency becomes more pronounced, the chlorotic tissue turns bright yellow and a marginal or interveinal necrosis may develop. The necrosis may also extend a short distance between the veins towards the midrib.

Sulphur

Sulphur deficiency causes visible symptoms, which resemble those of nitrogen deficiency. These include a severe reduction in growth and the development of pale-green to yellow leaves. An important point of difference, however, is that the symptoms of sulphur deficiency are confined to the younger leaves, the older leaves remaining green and healthy.

Initially, a diffuse pale-green to yellow chlorosis develops near the margins of the younger leaves. The chlorosis then spreads progressively over most of the leaf. A distinctive wedge shaped zone of green tissue is often retained at the junction of the major veins and the midrib. In case of a severe deficiency of sulphur, the interveinal tissue of the young leaves becomes completely chlorotic and, unlike severe nitrogen deficiency, the veins also lose their colour.

Zinc

Symptoms of zinc deficiency are confined to the older leaves, particularly those on canes near the junction of the leaders with the trunk of the vine; while the young leaves remain green and healthy even on severely deficient plants. Moreover, there is no reduction in the size of the younger leaves. Symptoms of zinc deficiency in kiwifruit include a bright yellow interveinal chlorosis on the older leaves with the veins remaining dark is very striking. Severe zinc deficiency can also markedly reduce lateral root development.

Copper

The first sign of deficiency is a uniform light-green chlorosis of the young, immature leaves. Later, the chlorosis becomes more intense in the interveinal area with the primary veins remaining dark green; these affected areas may eventually become bleached. Severe copper deficiency results in the death of the growing tip of the current seasons canes with the tips turning black. Premature defoliation of the affected canes is also common. Poor bud break is associated with low copper concentration in these vines.

Iron

Iron deficiency can be diagnosed by the development of a characteristic interveinal chlorosis of the younger leaves grading from yellow to snow white. The older leaves often remain green and healthy. In mild affected plants, the chlorosis is confined to the leaf margins, leaving a large zone of green tissues at the base of the leaf near the point of attachment of the petiole. In severe cases, whole leaf becomes chlorotic except for the veins that remain green. Eventually, even the veins may lose their green colour.

Manganese

Manganese deficiency produces a light green to yellow interveinal chlorosis that appears first on recently matured leaves, but in severe cases, it may affect almost all leaves on a plant. The chlorosis develops initially at the leaf margins and then spreads between the main veins towards the midrib leaving only a small zone of healthy green tissue on either side of the veins. As the deficiency becomes more pronounced, the zone of healthy tissue recedes even further towards the veins so that eventually only the veins remain green.

Boron

The appearance of small irregular patches of yellow tissue near the centre of the younger leaves is the first sign of boron deficiency. These patches enlarge and coalesce to form an extensive area of yellow tissue on both sides of the midrib. A zone of healthy green tissues is usually retained at the margin of the affected leaves. At the same time, the immature terminal leaves thicken, becoming misshapen and twisted. When the deficiency is severe, stem elongation is restricted due to lack of extension of the internodes, giving the plant a stunted appearance.

q) Datepalm

Nitrogen

Symptoms are overall light green colour and decreased vigour of the palm.

Potassium

Symptoms show first in the oldest leaves of the palms, progressing to newer leaves as the problem becomes more severe. Translucent yellow or orange spots form on leaflets. They may or may not have necrotic spots. Leaflet tips are often necrotic, but the midrib remains alive, and it may be orange instead of green.

Magnesium

Symptoms show up first on the oldest leaves and progress up through the canopy as the magnesium is relocated from older to newer leaves. Broad light yellow banding shows up along the margin of the fronds, with the center remaining green. In severe cases, leaflet tips become necrotic.

Manganese

The symptoms are only found in new leaves, as manganese is not translocated from the older leaves. New leaves emerge chlorotic, weak, reduced in size, and necrotic streaked.

Iron

New leaves will be uniformly chlorotic, often with tip necrosis and reduction in leaf size.

r) Walnut

Nitrogen

Early in the season, leaves may be pale in colour and small sized with reduced shoot growth. In the fall, the leaves senescence early, becoming yellow and defoliation earlier than normal.

Phosphorus

Trees usually show very weak shoots and have sparse foliage with smaller than normal leaves. These leaves are yellow and develop necrosis in irregular areas followed by early defoliation.

Potassium

Potassium deficient leaves begin showing symptoms in early to mid-summer. Leaves become pale and later, the edges of the leaves fold upward and curl in. The exposed undersides of the curled part of the leaflet develop a grayish cast (bronzing). Generally, most of the affected leaves are along the middle of the shoots. Leaf size is reduced as well as shoot growth. Invariably, overall nut size also is reduced.

Magnesium

Symptoms appear in mid- to late-summer after considerable shoot growth. The basal leaves, especially those on vigorous shoots, begin to show a distinct chlorotic margin at the apical and lateral edges of the leaflets, eventually leaving an inverted 'V'-shaped green area along the basal part of the midrib. As the season progresses, the chlorotic areas along the margins become necrotic and dark brown in colour.

Zinc

Zinc deficiency symptoms appear early in the season, especially if the deficiency is severe. The first evidence of zinc deficiency is delayed opening of the buds along the distil ends of shoots. Sometimes the delay can be as long as a month. When they open, the leaves are little, chlorotic and appear in clusters, giving rise to the term "Little-Leaf." Leaves that are slightly Zn deficient are only slightly reduced in size but show scattered chlorotic areas between the veins in blades giving them a mottled appearance. Often portions of the leaflet margin fold upward to form a distinct wavy leaf margin. Nuts along the Zn deficient parts of the shoots are markedly smaller in size than normal.

Copper

Copper deficiency symptoms appear in mid-summer. Scorch develops on leaves near tips of the shoots and defoliation follows. Slight shriveling of the shoots occurs and some small dark brown spots appear on the shoot near the tip. Terminal dieback subsequently occurs in late-summer. Kernels are often badly shrivelled.

Iron

Iron deficiency symptoms appear early in the season. The entire leaf is uniformly yellow with the terminal leaves affected more than the basal. Sometimes, the leaves are so chlorotic that they appear white. Often the basal leaves on shoots are green and only the leaves towards the terminal are chlorotic. In severe cases, the chlorotic leaves may develop scorch and abscise.

Manganese

Manganese deficiency symptoms begin to show in early to mid-summer. The characteristic symptom on the leaflet is chlorosis that develops between the main, lateral veins and extends from the midrib to the margin. The chlorotic pattern produces a "herring-bone" effect. The chlorosis gives the leaves an overall pale or slightly yellow appearance. Generally, leaf size is unaffected unless the Mn deficiency is very severe, in which case leaf size could be smaller and reduced yield.

Boron

Boron deficient trees are stunted with weak shoot growth characterized by short internodes, which gives the tree a bushy appearance. Leaves often are chlorotic and misshapen, sometimes reduced to small, bract like forms. In severe cases, terminal dieback occurs.

2. Vegetable Crops

a) Tomato

Nitrogen

The chlorosis symptoms shown by the leaves are the direct result of nitrogen deficiency. A light red cast can also be seen on the veins and petioles. Under nitrogen deficiency, the older mature leaves gradually change from their normal characteristic green appearance to a much paler green. As the deficiency progresses, these older leaves become uniformly yellow (chlorotic). Leaves become yellowish-white under extreme deficiency. The young leaves at the top of the plant maintain a green but paler colour and tend to become smaller in size. Branching is reduced in nitrogen deficient plants resulting in short, spindly plants. The yellowing in nitrogen deficiency is uniform over the entire leaf including the veins. As the deficiency progresses, the older leaves also show more of a tendency to wilt under mild water stress and senescence much earlier than usual.

Phosphorus

The necrotic spots on the leaves are a typical symptom of phosphorus deficiency. As a rule, P deficiency symptoms are not very distinct and thus difficult to identify. A major visual symptom is that the plants are dwarfed or stunted. Developing a distinct purpling of the stem, petiole and the lower sides of the leaves. Under severe deficiency conditions, there is also a tendency for leaves to develop a blue-gray luster. In older leaves under very severe deficiency conditions, a brown netted veining of the leaves may develop.

Potassium

The onset of potassium deficiency is generally characterized by a marginal chlorosis, progressing into a dry leathery tan scorch on recently matured leaves. This is followed by increasing interveinal scorching and/or necrosis progressing from the leaf edge to the midrib as the stress increases. As the deficiency progresses, most of the interveinal area becomes necrotic, the veins remain green and the leaves tend to curl and crinkle. In contrast to nitrogen deficiency, chlorosis is irreversible in potassium deficiency. Typical potassium (K) deficiency of fruit is characterized by colour development disorders, including "greenback", "blotch ripening" and "boxy fruit".

Calcium

These calcium deficient leaves show necrosis around the base of the leaves. Classic symptoms of calcium deficiency include blossom-end rot (BER), burning of the end part of tomato fruits. The blossom-end area darkens and flattens out, then appearing leathery and dark brown, and finally it collapses and secondary pathogens take over the fruit. All these symptoms show soft dead necrotic tissue at rapidly growing areas, which is generally related to poor translocation of calcium to the tissue rather than a low external supply of calcium. Plants under chronic calcium deficiency have a much greater tendency to wilt than non-stressed plants.

Magnesium

Magnesium deficient tomato leaves show advanced interveinal chlorosis, with necrosis developing in the highly chlorotic tissue. In its advanced form, magnesium deficiency may superficially resemble potassium deficiency. In the case of magnesium deficiency, the symptoms generally start with mottled chlorotic areas developing in the interveinal tissue. The interveinal laminae tissue tends to expand proportionately more than the other leaf tissues, producing a raised puckered surface with the top of the puckers progressively going from chlorotic to necrotic tissue.

Sulphur

Sulphur deficiency shows a general overall chlorosis; while still retaining some green colour. The veins and petioles exhibit a very distinct reddish colour. The visual symptoms of sulphur deficiency are very similar to the chlorosis found in nitrogen deficiency. However, in sulphur deficiency, the yellowing is much more uniform over the entire plant including young leaves. The reddish colour often found on the underside of the leaves and the petioles has a more pinkish tone and is much less vivid than that found in nitrogen deficiency. With advanced sulphur deficiency, brown lesions and/or necrotic spots often develop along the petiole, and the leaves tend to become more erect and often twisted and brittle.

Zinc

In the early stages of zinc deficiency, the younger leaves become yellow and pitting develops in the interveinal upper surfaces of the mature leaves. As the deficiency progresses, these symptoms develop into an intense interveinal necrosis but the main veins remain green.

Copper

Copper deficient leaves are curled, and their petioles bend downward. Copper deficiency may be expressed as a light overall chlorosis along with the permanent loss of turgor in the young leaves. Recently matured leaves show netted, green veining with areas bleached to a whitish gray. Some leaves develop sunken necrotic spots and have a tendency to bend downward.

Iron

Iron deficient leaves show intense chlorosis at the base of the leaves with some green netting. The most common symptom for iron deficiency starts out as an interveinal chlorosis of the youngest leaves, evolves into an overall chlorosis and ends as a totally bleached leaf. The bleached areas often develop necrotic spots.

Manganese

These leaves show a light interveinal chlorosis developed under a limited supply of manganese. The early stages of the chlorosis induced by manganese deficiency are somewhat similar to iron deficiency. They begin with a light chlorosis of the young leaves and netted veins of the mature leaves especially when they are viewed through transmitted light. As the stress increases, the leaves take on a gray metallic sheen and develop dark freckled and necrotic areas along the veins. A purplish luster may also develop on the upper surface of the leaves.

Boron

Boron deficient leaf shows a light general chlorosis. Boron is poorly transported in the phloem. Boron deficiency symptoms generally appear in younger plants at the vegetative stage. Slight interveinal chlorosis in older leaves followed by yellow to orange tinting in middle and older leaves. Leaves and stems are brittle and corky, split and swollen miss-shaped fruit.

Molybdenum

These leaves show some mottled spotting along with some interveinal chlorosis. An early symptom for molybdenum deficiency is a general overall chlorosis, similar to the symptom for nitrogen deficiency but generally without the reddish colouration on the undersides of the leaves. This results from the requirement for molybdenum in the reduction of nitrate, which needs to be reduced prior to its assimilation by the plant. Thus, the initial symptoms of molybdenum deficiency are in fact those of nitrogen deficiency.

b) Brinjal

Nitrogen

Nitrogen deficient plants remain stunted and unbranched and turn pale green. The leaves are markedly smaller in size and stiff in texture. Older leaves gradually start getting bleached from margin inwards to a pale in colour and finally shed prematurely. The fruits are small in size, pale in colour and maturing early. Shortage of N supply at any given growth stage results in flowers with short styles. Number of flowers is also reduced. If nitrogen supply is restricted during flowering and fruiting, yield is considerably decreased.

Phosphorus

Plant growth is retarded due to phosphorus deficiency, but they remained normal green in colour. Leaves are smaller in size and as the deficiency progresses, leaves turn dirty green, often with patches and shed prematurely, exposing bare stem in lower parts of the plant. The branching will be poor. The fruits will be small, pale in colour and mature early.

Potassium

Due to potassium deficiency, growth of the plant is suppressed. There is a marked reduction in the number and size of leaves. Brown patches appear on the mature leaves and small necrotic lesions develop along the veins followed by defoliation.

Calcium

Necrosis at tip and margins of newer immature leaves nearest to the terminal growth with a distorted appearance. Retarded plant growth, terminal buds die, brown to black leathery spots on the underside of fruits.

Magnesium

Symptoms of magnesium deficiency results in stunting of growth. In later stages, lower leaves develop a characteristic interveinal chlorosis from margins inwards in a typical inverted "V" shape. Leaves ultimately turn pale yellow.

Sulphur

Newer leaves become light green to yellowish, leaf veins appear lighter in colour, leaf tips cupping downward stems hard fibrous and spindly.

Zinc

Deficiency of zinc causes chlorosis, spreading systematically towards the younger leaves and the tips and margins are mostly affected. These areas rapidly enlarge with an increase in the intensity of chlorosis. Finally, areas near the leaf margins and between the veins become necrotic, producing a scorching effect and the lower leaves begin to wither away. Flowering is delayed and the flowers fail to set fruits and wither. Little leaf disease caused by the deficiency of zinc is predominantly observed in this crop.

Iron

Deficiency of iron restricts plant growth. Leaves are small, and terminal leaves develop interveinal chlorosis, ultimately turn yellow and drop off. The branching is poor and only a few flowers are produced.

Boron

Deficiency causes severe stunted growth. Leaves remain small, stiff and leathery. Apical bud dies, soon followed by browning and collapses of root system. Flowering is very much affected and branching will take place.

Molybdenum

Interveinal mottling on older leaves, while veins remain light green. Mottling also occurs on the newer leaves as they mature. As deficiency continues, the leaf tip and margin become necrotic with the leaf curling inward.

c) Cucurbits

Nitrogen

The vegetative growth is reduced and a yellowing of the leaves due to a loss in chlorophyll. The loss of the green colour occurs first on the more mature leaves and last on the younger leaves because of the degradation of N-containing compounds in the older leaves and movement of N to the young tissue. The size of the plant canopy is reduced. The smaller plant results in a reduced carbohydrate assimilation. Fruit set is reduced and developed fruit are small and subject to sunburn.

Phosphorus

Phosphorus deficient plants have weak roots and produce small, dark, dull, gray green leaves. The oldest leaf at the base of the shoot, turns bright yellow. However, unlike nitrogen deficiency, the leaf directly above this leaf remains dark green. Brown patches appear between the veins on mature leaves. These become scorched and spread until the leaf dies prematurely. Fruit set is reduced.

Potassium

Potassium deficiencies are first characterized by reduced growth, weak stems, and shoot die-back. Early symptoms may be difficult to distinguish. Leaf symptoms include dull bluish-green discolourations, particularly in the interveinal areas of the leaves, general chlorosis particularly at the leaf tips and margins; then tip burning and marginal scorch on the older leaves with the leaves curling under or upwards.

Calcium

Calcium deficiency symptoms appear initially on the young leaves and the shoot apex. New leaves are distorted with hooked tips and curled margins, which are irregular and may show brown scorching or spotting. Calcium deficiency can also cause roots of the plants to become stunted, translucent, and bulbous with dead tips.

Magnesium

Magnesium deficiency symptoms begin with chlorosis, sometimes with red or orange colouration on the oldest leaves. As symptoms progress, the older leaves shrivel and drop, the veins on the remaining leaves stay green and the interveinal areas have an orange chlorotic mottle.

Sulphur

Symptoms of sulphur deficiency are much like those caused by nitrogen: reduced leaf size, stunted growth, chlorosis, orange to purplish pigmentation and delayed maturity.

Zinc

Zinc deficiency symptoms begin with interveinal chlorosis. The first leaves in the plant remain small, the stems fail to elongate and the leaves appear to be in whorls or rosettes.

Copper

The interveinal areas of the lower leaves become severely scorched and the leaves collapse and wither. Downward curling of leaves and failure of plants to produce flowers and fruits are due to Cu deficiency. Interveinal chlorotic blotches on older leaves, which later become dull green or bronze in colour are also found in Cu-deficient plants. Copper deficiency causes reduction in leaf size and development of necrotic spots on the leaves. The roots turn brown. The deficiency markedly reduces the fruit set.

Iron

A deficiency of iron causes a decrease in the size of chloroplasts, reduced chlorophyll and photosynthesis. Symptoms include yellow or white chlorosis between green veins on younger leaves and poor vigor leading to death of the shoot tips.

Manganese

Symptoms of Manganese deficiency start as interveinal mottled chlorosis, the leaves become yellow and then develop small necrotic lesions in the most chlorotic areas. Reduced leave size and crimping of leaf margins are also possible symptoms.

Boron

The deficiency of boron causes dark green colouration and leathery texture of leaves. The older leaves show brownish-yellow interveinal chlorosis and scorched margins, while the younger leaves are often malformed and cupped with prominent veins. Fruits frequently develop longitudinal corky cracks in the skin. The growth of the young plant is stunted. The number of stomata on the lower side of the leaves is also reduced. The older leaves become stiff and cupped upwards. Boron deficiency delays fruit maturation and reduces the weight of the fruits and their yield.

Molybdenum

Initially, the green colour of the leaves fades, particularly between the veins. Later, leaves can turn yellow and die. In some cases, parts of mature leaves remain green at first, giving rise to a blotchy appearance. Symptoms start first in lower leaves and spread upwards, the younger ones remaining green. Growth might appear normal but flowers stay small.

d) Cabbage

Nitrogen

Pale green colour, particularly on young leaves; older leaves may show reddish to purple colours and will shed prematurely.

Phosphorus

Underside of leaves show purple near veins; leaves have dull purple colour; margins of leaves die.

Potassium

Browning of leaf margins and browning between the leaf veins; curds are soft and poorly developed.

Calcium

Leaves roll up at margins with necrotic edges.

Magnesium

Older leaves show patches of orange, red and purple; later the chlorotic patches brown and drop out.

Sulphur

Bleached creamy white to light yellow colour of entire leaf or leaves, including the terminal buds; creamy white colour of older leaves; dying terminal buds; older leaves die and fall off.

Zinc

Zinc deficiency causes appearance of a yellowish viscous substances in root xylem vessels, which impedes water transport to the aerial parts.

Copper

Heads fail to form, leaves become chlorotic, poor growth.

Iron

Bleached yellow colour between the veins of young leaves; leaf veins remain green; yellow and green intermixed in interveinal tissue; older leaves turn yellow, dry out, and fall off; stunted plant growth; leaves and flowers shed prematurely.

Manganese

Manganese deficiency causes yellowing between veins of young leaves. Leaves gradually turn pale-green with darker green next to the veins, petioles and stems.

Boron

Stems are hollow in the head region; leaves are distorted; leaves are brittle, stiff, along margins; leaves making up the head are unattached.

Molybdenum

Molybdenum deficiency sometimes causes "whiptail" in cabbage. Whiptail results in a deformed growing point causing no head to develop, as well as leaf blades consisting mostly of midribs.

e) Cauliflower

Nitrogen

In cauliflower, nitrogen deficiency causes marked reduction in growth; leaves become small in size. The young leaves are pale green in colour, while the older leaves turned purple. The formation of curd is delayed.

Phosphorous

Phosphorous is important for root growth. A shortage of phosphorous stunts plants growth.

Calcium

Calcium deficiency causes tip burn in cauliflower plants.

Magnesium

Older leaves are the first to show deficiency symptoms, which include blotches of interveinal chlorosis. As the chlorosis intensifies, purple blotches may be seen near the leaf margins.

Sulphur

Early deficiency symptoms appear as blotches of interveinal chlorosis on the youngest leaves and the leaves may become reflexed.

Zinc

Young leaves become small; curd colour will be affected.

Manganese

The deficiency appears in the older leaves with wild grey interveinal mottling and necrotic areas. Young leaves retain their normal colour but become chlorotic between veins, which gradually spread to older leaves.

Boron

Cauliflower is highly susceptible to boron deficiency. Generally, the deficiency symptoms of boron are extremely visible on plants after the curd formation only. Growth is reduced, stems are stiff with hollow core. Leaves are curled, leathery and less in number. Curd formation is delayed which turn dirty pale to brown in colour. In early stages, water soaked areas appear on the stem and curd surfaces. As the plant grows, the stem becomes hollow with water soaked tissues covering the internal walls of the cavity. In advanced stage of deficiency, brown or pink coloured areas are seen on the curd surface and therefore, it is also called as 'brown rot' or 'red rot' or 'browning' of the curd. Sometimes, the stem may become hollow even without brown areas on the curd. The curd affected without browning is bitter. The other symptoms of Boron deficiency are thickening, brittleness and downward curling of lower leaves. The foliage colour first changes to dull green and then greenish yellow at the apical margins of the older leaves. Severe deficiency leads to under developed and smaller leaves. The growing point may die in young stage of plant itself.

Molybdenum

In the early stages of molybdenum deficiency, there is marginal scorching, rolling or upward curling of leaves, and withering and crinkling of leaves. In later growth stages, the deficiency shows up as "whiptail," especially in younger leaves. Cupping and drying of the edges of true leaves. There is distorted curding of cauliflower plants.

The chief symptom of whiptail is a reduction in the amount of lamina, usually with irregular and wavy margins. The lamina may be markedly thick and succulent or convoluted. Leaf colour is generally not affected in whiptail, but some specimens are noticeably blue-green. Deficiency of Mo causes twisting of the youngest leaves and chlorotic spots and blotches in the leaves, with the lamina of leaves restricted partly to the midrib.

f) Beet root

Nitrogen

There is no single characteristic symptom that identifies nitrogen deficiency. However, the foliage typically becomes light green in appearance, then yellow, due to the disappearance of chlorophyll from the leaves. This reduces photosynthesis and subsequently crop yield. In sugar beet, nitrogen deficiency is often observed in patches of stunted plants with small yellow leaves.

Phosphorus

Phosphorus deficiency is therefore rare and only occurs when soils are extremely depleted of this element. It is particularly uncommon in mature plants, but manifests itself in seedlings through stunting and dark-green or purple-red leaf colour. In older plants, the roots can also become stunted, while the tap root may form a mass of fibrous secondary roots.

Potassium

High levels of potassium are required for optimum growth of beet roots. The symptoms of potassium deficiency are often referred to as 'scorch'. Patches of dead tissue occur in leaves, which take on a dull olive green to bronze appearance.

Calcium

The most characteristic symptom of calcium deficiency in cultivated beets is 'tip-burn', with young leaves becoming crinkled and reduced in size, and the growing point developing abnormally so that lateral shoots appear. This deficiency is not caused by inadequate calcium in the soil, but by the plant's inability to take it up.

Magnesium

The symptoms of magnesium deficiency are pale yellow and brittle necrotic areas in leaves. Magnesium deficiency is most likely to occur in hot summers for crops growing on sandy soils.

Sulphur

The symptoms of sulphur deficiency are yellowing young leaves with irregular brown blotches.

Boron

The most frequent nutrient deficiency encountered in cultivated beets is boron deficiency. A deficiency of this micronutrient is also referred to as 'black spot', 'heart rot', 'dry rot' or 'canker'. Boron deficiency results in stunted plants, the deformation and death of the growing point, and slow growth. The crown of the root becomes hollowed and blackened, due to water accumulation. Rough black spots can appear on the roots. In beetroot, this gives the roots a bitter taste. The roots may eventually become entirely discoloured, hollowed or split. The yield and quality of cultivated beet roots can therefore be severely affected. Boron deficiency can also result in corky growths on the shoots and stalks. Roots from boron deficient plants store poorly.

g) Carrot

Nitrogen

Nitrogen deficiency symptoms are common in carrot tops close to harvest. The symptoms are yellowing of old leaves with associated red tints in old and young leaves.

Phosphorus

Phosphorus deficiency in carrots is not common in commercial practice but can be recognised by stunted top growth. Purple tints in the foliage appear with severe deficiency. Purple tints are also related to low temperatures in winter and these should not be confused with phosphorus deficiency.

Potassium

Marginal yellowing and scorching of cotyledons and older leaves are the common symptoms of severe deficiency. Potassium deficiency can easily be confused with water stress of seedlings being established during hot periods in summer and with damage by thrips in spring season.

Calcium

The 'cavity spot' in carrot, a typical calcium deficiency disorder, is induced by over fertilization of potassium. This malady is found under waterlogged conditions where NH_4^+ is mainly responsible for reduced Ca uptake. However, this disorder has little relation to Ca status of soil and is more often associated with particular types of weather. The cavity spot is found to be associated with low Ca content in leaves and roots. Leaf necrosis developed in carrot seedling is also due to Ca deficiency.

Magnesium

The symptom of deficiency is bright yellow patches between the veins of the old leaves, especially close to harvest.

Zinc

New leaves are yellow, small and distorted, giving the tops a rosette-like appearance.

Manganese

Symptoms of manganese deficiency are a general yellowing of all leaves and poor growth. Close examination of affected leaves shows that yellowing is largely between the veins.

Boron

A symptom often attributed to boron deficiency is browning below the skin known as '5 o'clock shadow'.

h) Radish

Nitrogen

In radish, nitrogen deficiency causes marked reduction in growth. The number and size of the leaves are reduced and pale green colour of leaves changes to yellow. Roots are thin, stiff and fibrous.

Phosphorus

Phosphorus deficient plants are shorter in height. Leaves are smaller in size, distorted in shape, and pink tinge appears along the margins and veins. Formation of storage root is delayed.

Potassium

The colour of the potassium deficient leaves changed from green to pale yellow, and brown scorches appear on the leaves at later stages. Violet streaks appear on root, which ultimately spread all over. It causes growth retardation, particularly of the storage organ.

Calcium

The deficiency symptoms of calcium are more marked in the younger leaves than in older leaves of the plant. The characteristic symptoms first appear on the young leaves, which become chlorotic. The leaves are fewer in number and the roots are relatively smaller in size.

Magnesium

The deficiency causes chlorosis on mature leaves, which abscise earlier. The roots are smaller in size, stiff and pale in colour.

Iron

The iron deficiency causes chlorosis in younger leaves. Storage roots are reduced in size and become pale in colour.

Boron

Boron deficiency results in reduced growth and an increased content of ionic thiocyanate both in roots and foliage, decreases chlorophyll content and altered distribution of reducing sugars.

i) Onions

Nitrogen

Deficiencies of nitrogen can lead to small plants and bulbs and early maturity. Nitrogen deficiency is characterized by uniformly light green foliage that becomes increasingly yellow as the deficiency progresses. Thus, symptoms are first visible in the older outer leaves and move progressively toward the newly emerging leaves in the center of the plant as nitrogen is translocated toward the stronger sinks near the growing point. Leaf tips die back until whole leaves are lost. Nitrogen-deficient onion plants tend to have thin, small-diameter, erect leaves. Plants grow slowly and stunted.

Phosphorus

Phosphorus deficiency in onions has been associated with slow plant growth, delayed maturity, and a high percentage of thick necks at harvest. Because P is essential for energy transfer within the plant, all aspects of growth are affected. Deficient plants are initially recognized by the dull green colour of the leaves. Leaf tips then wilt and eventually die without yellowing. The necrosis advances toward the base of the leaves, sometimes leaving islands of green within the yellow or brown necrotic tissue. In the final stage, the dead tissues turn black.

Potassium

Onions are fairly susceptible to K deficiency. Leaves of deficient plants initially become dark green from accumulation of nonstructural carbohydrates caused by reduced protein synthesis. Later, the tips of the oldest leaves begin to wilt, especially on the upper sides. As the necrosis progresses, leaf colouration becomes similar to that of nitrogen deficiency.

Calcium

Young leaves become pale, die back, and eventually die. In some plants, an area of the leaf may die, causing the distal section to fall over. Lack of Ca in young plants may also cause the root system to collapse. Leaf tips die back without any significant growth restriction.

Magnesium

Magnesium deficiency in onion is generally characterized by slow plant growth and even death of weak plants. Older leaves become uniformly yellow along the entire length due to loss of chlorophyll, of which magnesium is a key element. Chlorophyll may fail to form in younger leaves. In chronic deficiencies, chlorosis is followed by tip browning and dieback.

Sulphur

Deficient plants tend to produce fewer leaves, but bulb development is usually normal. New leaves exhibit a uniform yellowing caused by a lack of chlorophyll and may in extreme cases become thick and deformed.

Zinc

Zinc deficient onion plants are stunted with marked twisting and yellow striping of the leaves. Cool, wet weather increases the severity of the deficiency.

Copper

Copper deficiency may cause onion leaves to turn white and twist into spirals (pigtail) or bend at right angles to the plant. Leaf tips may become chlorotic and then die back. Bulbs are less solid than those with adequate Cu and tend to mature earlier.

Manganese

Onions are one of the vegetable crops most susceptible to manganese deficiency. Failure of photosystem II reaction centers results in leaves that are initially light coloured. The outer leaves show striped interveinal chlorosis and these leaves develop tip burn and sometimes curl, with development of a progressive necrosis. Plants may become seriously stunted, and bulbing is often delayed. At maturity, a high percentage of bulbs have thick necks. Drought has been shown to increase the severity of the symptoms.

Boron

Onions are lesser responsive to boron than to many other micronutrients. However, severe B deficiency can cause obvious symptoms. Leaves tend to become stiff, brittle, and dark green. Young leaves may be mottled yellow and green. Older leaves often become chlorotic, with tip dieback and distorted, shrunken areas. Regularly spaced transverse yellow lines near the base of the leaves may develop into a series of ladder like cracks.

Molybdenum

Onions are highly sensitive to molybdenum deficiency, which is most often associated with acidic peats and sandy soils. Poor emergence and death of seedlings may occur in highly deficient soils. As the plant grows, leaf tips die back with zone of flabby, wilted tissue between the necrotic and healthy tissues. This wilting and necrosis progress down the leaf and in severe cases may kill the plant.

j) Lettuce

Nitrogen

Lower leaves yellow, stalks short and slender. Older leaves affected first. Younger leaves remain green. No head form with severe deficiency.

Phosphorus

Stunted abnormal dark green. No specific symptoms to identify purple pigmentation under leaves.

Potassium

Older leaves show chlorosis at leaf tips and between veins. Growth is stunted. Small leaves. Leaves are less crinkled with purple leaf margins.

Calcium

Growth severely stunted. Leaves have open rosette growth. Interveinal white spots near the margins of mature leaves. Leaves more crinkled than normal. Plant growth is also flat. Older leaves curl downwards. Leaves are brittle. Chlorosis and later necrosis in leaf margins.

Magnesium

Symptoms appear first in younger leaves as interveinal chlorosis. Leaves have mottled appearance. Plants are smaller. Margins curl upwards. Yellow spots in interveinal areas. Green margin even in severe cases.

Sulphur

Plants stunted and woody. Plant turns to light green. Leaves are thicker. Anthocyanin discolouration on leaf edges.

Zinc

Upper internodes remain short. Rosette appearance. Growth stunted. Leaf margins show necrotic spots.

Copper

Leaves become elongated and cupped. Interveinal chlorotic blotches. Plants remains small. Leaf margins turn up. Chlorosis along edges. Leaves wilt at a later stage. No heads forms.

Iron

Interveinal chlorosis of young leaves. Whole plant is light green. Growth is stunted and stops. Veins remain green initially. Leaves are less crinkled. Anthocyanin discolouration in young leaves.

Manganese

Entire plant became light green to yellow. Young leaves are smoother. Over all pale appearance. Older leaves may become chlorotic. Veins always remain green.

Boron

Restricted root growth. Plant darker than normal. Young leaves hard and rounded. Brown spots at growth points. Growth point dies with severe deficiency. Poor root development.

Molybdenum

In the early stages of molybdenum deficiency, the plants are chlorotic and stunted. The lamina failed to develop and the leaves remained ovate rather than round as in healthy plants. Chlorosis of the leaves is followed by necrosis of the margins. With severe Mo deficiency, lettuce in the juvenile stage appears fired because of accumulation of nitrates.

k) Spinach

Nitrogen

The plant will eventually become a pale green or yellow colour with stunted growth. The leaf turns light green, then yellow, including the veins. The leaves will then become smaller, thinner as the disorder advances, the leaves will develop dead spots and areas. Growth of the plant will slow and eventually stop if the disorder is not corrected

Phosphorous

Slowed growth and becomes bluish green in the older leaves.

Potassium

Browning at leaf tips of older leaves first, then as the problem becomes worse, the leaves will get brown (dead) spots between leaf veins.

Calcium

The old leaves become wrinkled and curled, while the new leaves also will remain curled and a light green band will occur along green margin.

Magnesium

Older leaves become yellow with dark green veins, pale green colour overall. The deficiency begins with mottled yellow areas on the leaf which develops between the leaves veins.

Sulphur

The plant is usually stunted or reduced growth but there are no dead leaves but general yellowing of young leaves.

Zinc

Younger leaves become yellow and holes appear near the veins of the upper surface of mature leaves. The main veins remain green.

Copper

The leaves are curled and petioles bend downwards. Recently matured leaves show netted, green veining with areas bleached to a whitish gray. The leaves tend to bend downwards with brown spots.

Iron

The deficiency begins in the youngest leaves of the plant and eventually causes the entire plant to become a pale yellow. The bleached areas often become dead spots until the entire plant will become almost white.

Manganese

Deficiency begins with a light yellow colour in the young leaves and in the netted veins of the mature leaves. As the stress increases, the leaves become a gray metallic sheen, dark freckled and dead areas occur along the veins.

Boron

In mature tissue, both the pith and the epidermis of stems are affected and therefore, causing the stem to become hollow or roughened. The leaves of the plant is usually brittle and tend to break easily.

Molybdenum

The entire plant will become a yellow colour similar to nitrogen deficiency but does not have the reddish colour behind the leaf. The leaf becomes a mottled and causes yellowing between the veins but the leaves remain green.

I) Potato

Nitrogen

Nitrogen deficient potato plants generally show a light green to yellow colouring of the whole plant. Nitrogen deficiencies will result in a shorter growing season, which will negatively affect tuber yield. Optimum N levels in the plant depend on the growth stage, but leaf deficiency symptoms can generally be expected when NO_3^- levels in the leaf petioles drop below 0.5 to 0.8 % (on a dry mass basis).

Phosphorus

The symptoms first develop on older leaves showing some necrotic spots and plants are dwarfed or stunted. Phosphorus deficient plants develop very slowly. Plants develop a distinct purpling on the stem, petiole and the under sides of the leaves. Plant remains stunted, darker than normal colour. Lower leaf surface gray-green. Leaflets roll upward severely if deficiency is severe.

Potassium

Symptoms first develops on young and full sized leaves in the case of extreme deficiency. Leaflets become crinkled. Some of the leaves show marginal necrosis (tip burn), and at a more advanced deficiency show interveinal necrosis. As the deficiency progresses, most of the interveinal area becomes necrotic, the veins remain green and the leaves tend to curl and crinkle. Leaves take on a scorched appearance with black pigmentation and necrotic (dead tissue) edges.

Calcium

Tip burn appears on the youngest center leaves including the growing point of the plant. Severe deficiencies cause young mature leaflets to roll upward with chlorosis and brown spotting. Since Ca movement within the potato plant is primarily unidirectional, a continuous supply of Ca is required to prevent Ca deficiency of growing points and for tuber formation.

Magnesium

Magnesium deficiency symptoms appear first on young mature leaves as a slight chlorosis with green veining and brown spotting, which terminate as interveinal leaf scorch. Symptoms are most severe in the oldest leaves, which have green veins and large patches of marginal and interveinal necrosis. Leaflets of severely stressed leaves become dry and turn brown but most remain attached to the rachis.

Sulphur

Leaves show a general overall chlorosis. The veins and petioles show a very distinct reddish colour. The yellowing is much more uniform over the entire plant including young leaves. The reddish colour often found on the underside of the leaves. With advanced sulphur deficiency, the leaves tend to become more erect and often twisted and brittle. Leaflet yellowing is uniform.

Zinc

Zinc deficiency in potato often known as 'fern leaf' or 'little leaf', appears on young developing leaves. Deficient plants show severe stunting and bronzing or yellowing of the foliage, usually around the leaf margins, starting from the tips. Youngest leaves are cupped upwards and rolled to such an extent that the terminal growth resembles that of ferns. Leaves of affected plants are smaller and their upper internodes are shorter.

Copper

An early sign of its deficiency is the development of a uniform light green colour of young immature leaf blades similar to those of molybdenum and manganese deficiency. Upward cupping and inward rolling of the young, Leaves.

Iron

Iron deficiency appears initially as a yellowing of young leaves near the growing point of the plant. With time, the leaves become light yellow to nearly white. During the deficiency stage, blade tips remain green for longer time. Netted green veining occurs when traces of iron are absorbed and translocated along the veins for chlorophyll formation.

Manganese

The first sign of manganese deficiency is yellowing and slight cupping of younger leaves. Pinkish colour develops at the base of younger chlorotic leaves and relatively old leaves show dark to black spots. With increased deficiency, dark to black spotting develops between the veins with a large increase in spotting, appearing along larger veins and the mid rib. These symptoms, darkening and cupping, increase in severity with time. During mild deficiency, upper parts of the plants become somewhat chlorotic but do not develop dead spots.

Molybdenum

The leaves show some mottled spotting along with some inter veinal chlorosis. An early symptom for molybdenum deficiency is a general overall chlorosis similar to the symptom for nitrogen deficiency but generally without the reddish colouration on the undersides of the leaves.

m) Sweet potato

Nitrogen

General symptoms include, a uniform light green chlorosis of the leaves and slow growth resulting in a delayed or sparse ground cover.

The development of nitrogen deficiency symptoms vary according to conditions experienced by the crop. When N is initially adequate during the establishment phase but becomes depleted during crop growth, plants may appear normal or near normal in colour and habit, except for yellowing and premature shedding of older leaves due to remobilisation of N from these tissues. In this case, the oldest leaves become uniformly yellow and slightly wilted. A pale coloured necrosis may spread from the tip or margins, but often the leaf is shed before it develops extensive necrosis. Symptoms of chronic N deficiency include uniformly pale colour, reduced leaf size, loss of the normal colour resulting in a dull appearance of the leaves, thin spindly vines and reduced activity of axillary buds leading to less branching. In severe cases, small purple-pigmented or necrotic spots have been observed on the surface of older leaves.

Phosphorus

Mild to moderate phosphorus deficiency may be difficult to recognise in the field. Mild phosphorus deficiency is often associated with a darker than normal, bluish green colour of the foliage. Young to mature leaves remain dark green at all levels of severity. Obvious symptoms on the older leaves are associated with severe stunting. The first sign of phosphorus deficiency is usually the premature senescence of older leaves. In most (but not all) cultivars, yellowing is preceded by the appearance of purple anthocyanin pigments, producing senescing leaves. Yellowing may spread from discrete interveinal patches, which typically become cleared of anthocyanin pigment , or may be more general, but often affecting one half of the blade more than the other . In this case, the chlorotic areas may appear orange or red due to the overlying anthocyanin pigments. Necrotic lesions subsequently develop in the chlorotic zones, and the necroses spread as irregular patches until the leaf blade is entirely brown and dry.

Potassium

Potassium deficiency causes substantial growth reduction before specific symptoms develop. Yellowing appears on the oldest leaves, while the youngest leaves retain a normal colour, size and texture. If young leaves are normally purple, this pigmentation may be reduced in K deficient plants. The oldest leaves develop a yellow chlorosis in marginal and interveinal zones. Brown necrotic lesions develop within the chlorotic zones and eventually spread to cover the entire leaf blade. Cultivars vary in the extent to which lesions spread predominantly from the margins to interveinal zones. Necrosis associated with K deficiency is usually dark in colour, and the necrotic areas become dry and brittle. The yellowing and necrosis may be preceded by a light green interveinal mottle affecting mature to older leaves, and often most obvious on the leaves of axillary shoots.

Calcium

The primary symptom of calcium deficiency is the development of necrotic tissue on young leaves. The necrosis usually begins along the lateral margins, and extends inward mainly in interveinal tissue. Necrotic tissue is mid to dark brown and brittle. Necrosis is not usually preceded by localised chlorosis, although the leaves may be uniformly paler than normal. The symptom is first observed on young expanding leaves, two or three leaves below the tip but as the disorder intensifies, newly formed leaves are affected and finally the apex may die. Necrotic leaves may be prematurely shed. In some cultivars, leaf drop is spectacular, leaving some length of bare stem below the tip.

A second symptom may develop on older leaves, which is apparently unrelated to the necrosis of young leaves. Necrotic spots appear, which are either clustered along the main veins from the petiole attachment point to about half-way to the margin, or scattered uniformly across the interveinal areas. They are roughly circular with abrupt, irregular edges, and do not seem to accelerate leaf senescence. The necrotic tissue in this case is mid brown, and is not brittle. Calcium deficiency inhibits root growth and in severe cases, root tips may die back.

Magnesium

Magnesium deficient crop will tend to have a pale overall colour. The earliest specific symptom of Mg deficiency is an interveinal chlorosis of older leaves. Typically, the main veins retain a relatively broad margin of dark green tissue, but the minor veins are less well defined, resulting in radial bands of pale tissue between the main veins. However, in some cultivars, the chlorosis is more mottled, composed of isolated patches, or the veins retain little green margin and appear as a green network on a generally pale leaf. Chlorosis first appears on the oldest leaves, but may spread to quite young leaves.

It may be accompanied by upward or downward curling of the leaf margins, or a wilty drooping of the leaf blades. Purple or red-brown pigment may appear on older leaves in conjunction with chlorosis.

Most commonly, pigmentation affects the upper surface of interveinal patches near the leaf tip and margins but in some cultivars, the veins under the leaf may become red. Chlorosis generally progresses to yellowing and necrosis of the oldest leaves. Leaves may become entirely yellow and wilted before browning off, or necrosis may develop on interveinal and marginal tissue without prior necrosis. Most commonly, localised yellowing precedes necrotic lesions, which spread from interveinal zones. The vines of Mg-deficient plants may become thin and twining, with lengthened internodes, a response similar to etiolation.

Zinc

The most distinctive symptom of Zn deficiency in sweet potato is a reduction in the size of young leaves. The leaves are thickened but usually not distorted, and may be as small as 1-3 cm in length. After the onset of this symptom, plant growth is severely limited. In some cultivars, internodes are also shortened, but in others, this occurs to a far lesser extent, than the reduction in leaf size. General chlorosis of the young leaves is usual, but may vary from mild to almost complete bleaching. Increased purple pigmentation of the shoot tips may occur in some cultivars. Mature leaves develop an interveinal chlorosis in which a diffuse green margin flanks the major veins, and usually the minor veins to a lesser extent.

Copper

A number of visible symptoms of copper deficiency have been observed on both mature and young leaves, their extent and order of development varying among cultivars. Conspicuous symptoms on the vines are usually associated with considerable growth reduction. Chlorosis, wilting and drooping of mature leaves may be the first visible symptom of Cu deficiency. The chlorosis is interveinal, with a gradual fading of colour with distance from the main veins. Usually, the minor veins retain less green colour than the main veins, but may be sufficiently well defined to give the chlorosis a mottled appearance.

Chlorotic leaves may develop spots or patches of necrosis, which spread until the entire leaf is dead. In some cultivars, necrotic spots have been observed on mature leaves without prior chlorosis. Initially, the necrotic spots are small, dark and sharply defined. They may be clustered close to the point of petiole attachment, or more commonly scattered over the leaf surface. The necrotic spots are clearly visible on the lower side of the leaf and are often adjacent to minor veins. Subsequently, a yellow chlorosis develops around them, spreading to encompass a number of lesions. Later, the area between the initial lesions becomes necrotic, and eventually the whole leaf dies.

Iron

The distinctive symptom of iron deficiency is chlorosis of the youngest leaves. As the chlorosis is a result of photo-bleaching, leaves in bright sunlight will be more affected than those in shade. Initially, a yellow interveinal chlorosis develops, which is characterized by a sharply contrasting green network of veins. However, as the condition becomes more severe, the chlorosis becomes white and the veins may lose their green colour. If young leaves are normally purple, they become pink as a result of Fe deficiency. While all the leaves of the plant may become affected, youngest leaves generally show the greatest intensity. Severely affected leaf blades become necrotic, the necrosis usually spreading from the tip and margins into interveinal zones. Eventually, the growing points may die, both at the terminal apex and in axillary shoots. Necrotic tissue on expanded leaf blades is generally light brown and soft, but may be dark and brittle on the shoot tips.

Manganese

Usually, the first sign of manganese deficiency in sweet potato is an indistinct interveinal chlorosis throughout the plant, but particularly on leaves of intermediate age. The green zones around major veins are relatively broad and fade gradually toward the interveinal zone. In some cultivars, the chlorosis may appear more general than interveinal. Chlorosis may be accompanied by drooping of the leaves (although they remain turgid), slight puckering of the leaf surface or downward curling of the leaf margins. The drooping is caused by bending of the petiole and is not accompanied by wilting of the leaf blades, nor does it progress to the point of leaf senescence and abscission.

On expanding leaves, small areas between minor veins become pale and sunken, and eventually develop into necrotic spots. While all interveinal zones of the leaf are affected, they do not develop at the same rate. Pits in older leaves tend to be darker and are often concentrated near the base of the midrib and main veins. The young leaves become pale, thickened and brittle, and may curl under or buckle as they expand. When pitting occurs in very young leaves, their subsequent expansion causes the pits to develop into larger holes. Severely affected leaves become necrotic, the necrosis spreading from the tip or lateral margins.

Boron

Boron deficiency affects actively growing tissue, both of the shoots and roots. The first sign is usually a thickening of young leaves. The leaves and stem near the shoot tip are brittle and break easily when bent. Young leaves are usually paler than the older leaves, but the extent and pattern of chlorosis is variable: it may be uniform, or diffusely interveinal, fading gradually with distance from the main veins. In other cases, it may be a less distinct interveinal mottle.

Leaves may become puckered (slightly raised in interveinal zones) and the tip and lateral lobes may curl under. Petioles may twist. Internodes may be shortened, producing a compact habit around the apex. In severe cases, the leaf veins may appear callused or overgrown with rough, whitish or pink tissue. Deeply lobed cultivars may display reduced development of the lateral lobes. At greater severity, B deficiency causes death of the growing points. Fibrous roots of B deficient sweet potato plants become short, stumpy and highly branched, producing the coralline structures typical of B deficiency in many plant species. Storage roots are often short and blunt-ended and may split and overgrow, producing cankers and deformities. The skin may be rough and may wrinkle towards the ends. The cut root exudes less white sap than normal and the flesh may be mottled or corky in places.

Molybdenum

Molybdenum deficiency include a general pale green colour, stunted growth with small leaf size and possible reddening of veins on the young leaves. Older leaves developed silvery patches over interveinal tissue, as a result of the upper epidermis separating from the underlying tissue. Eventually these patches became necrotic, but necrosis did not spread readily from the isolated interveinal patches.

n) Yams

Nitrogen

The primary symptom of nitrogen deficiency is a dull, light green colour of the foliage. The colour may be uniform, but commonly shows a weakly interveinal pattern, with slightly darker green on and around the main veins of the leaf. The size of mature leaves is also reduced. Nitrogen deficient plants tend to have relatively few branches. The oldest leaves on the vine may shrivel and drop off earlier than normal. Often, they turn yellow before they dry up, but sometimes there is little yellowing ahead of the spreading dead tissue. Senescence of the oldest leaves is a common symptom of N deficiency in plants, but is less pronounced in yams. Severely N-deficient yams may become increasingly pale from mature to youngest leaves.

Phosphorus

Moderate phosphorus deficiency causes growth reduction with no apparent symptoms. Foliage may be even darker green than in healthy plants. Tuber yield may be adversely affected before symptoms are visible. As P deficiency intensifies, specific symptoms appear in the young leaves. Initially, the internodes are shortened and leaves are thicker and stiffer than normal, and don't expand to their full size. They may be curled down towards the tip of the blade or gathered along the main veins, indicating uneven expansion. The leaves may be light green to yellow, or develop bleached white zones between the main veins. However, in some cases, the normal dark green colour is retained. The shoot tip may become inactive, which can be observed by the relative maturity of first leaves. As severity increases, leaf blades become necrotic at the tip or margins and the whole blade may dry up. The dead tissue is generally dark and brittle. Patches of brown marginal necrosis may be observed on leaves before the onset of deformities, but do not spread. Phosphorus deficiency strongly reduces the growth of tubers.

Potassium

Young mature leaves usually show chlorosis symptoms most distinctly. In mildly deficient plants, the paler interveinal tissue is usually finely divided by the network of greener veins. However, chlorosis patterns may be diffusely interveinal (gradually fading with distance from the major veins). Premature senescence of the oldest leaves is usually characteristic of K deficiency, but in yams, this may not be evident until the onset of tuber filling. At this stage, the lowest leaves may rapidly develop a light-brown necrosis, spreading from the tips and margins of the blades. The necrosis may be preceded by yellowing, but this is often limited to a yellow strip at the leading edge of the dead zone.

Calcium

Calcium deficiency mostly affects the growth of new tissue at the vine, root and tuber tips. The earliest apparent symptom may be premature inactivity of the vine tips. This is evident by the relative maturity of leaves just behind the tip. The vine is typically thin and internodes are shorter than usual and the tip does not extend far beyond the first open leaves. In severe cases, the tip and youngest leaves may die, or while the tip remains green. The inactivity of the shoot tip may stimulate axillary shoots to develop at the upper nodes, but these generally produce one or two leaves before also becoming inactive. Young leaves may also be paler than normal, with either a uniform light-green to cream colour, or an interveinal pattern becoming gradually greener towards the main veins. On all species, young mature leaves often develop dark necrotic spots, which can be scattered on interveinal tissue, or form a more regular arc around the petiole attachment. Mid leaves sometimes show necrosis on the veins on the underside of the leaf. Root growth is particularly affected by Ca deficiency. Root tips may die, resulting in a cluster of branching just behind the tip. The growing tip of the tuber is affected and tubers may be blunt-ended and short, or have constrictions.

Magnesium

The crop appears light green, with an interveinal pattern of chlorosis most pronounced on young mature leaves. Vine tips and stems generally appear healthy. Leaves are generally a healthy size, and have a normal soft texture and a light sheen. Leaves in full sunlight are likely to be paler than those in the shade, as the yellowing results from photobleaching of the green chlorophyll pigment. The pattern of chlorosis varies with variety. Usually, there is a margin of green tissue either side of the main veins, but the minor veins are less defined. The leaf margin may also remain green. In some cases, the chlorosis is most pronounced in a band on either side of the main veins. In others, the yellowing is more uniform across the leaf, contrasting with the greener network of veins. Older leaves develop necrotic lesions that spread from irregular interveinal spots or from the leaf tip. Young leaves that are severely bleached may also develop scattered necrotic spots, but these rarely spread. Commonly, the surface of the leaf becomes stained light brown over the interveinal regions.

Sulphur

Plants are generally a light-green colour all over but the younger leaves are palest. Veins may be slightly greener than the interveinal tissue, but often the colour is uniform across the whole leaf. Plants grow slowly and leaf size is reduced. The vine tip may become inactive, which may be evident from the maturity of the first leaves and the short distance from the first leaves to the tip. These symptoms are similar to those of nitrogen deficiency, but in contrast to N deficiency, leaves tend to be slightly stiffened and may be shiny. In severely deficient plants, expanding leaves may develop white bleached zones in the interveinal tissue or near the tip. The leaf tips may become brown and the necrosis may spread to the whole leaf blade, producing a zone of dead leaves below the vine tip. Necrotic lesions may also occur on the stem or on veins under the leaf. These symptoms are most commonly observed on mature leaves in the middle of the vine. Commonly, mature leaves develop scattered spots, which appear faintly bruised and stained but generally don't become fully necrotic.

Zinc

Symptoms of zinc deficiency are conspicuous, in the form of smaller than normal leaves that are curled or wavy. When less severe, the first symptom of Zn deficiency may be a mild interveinal chlorosis on mature leaves in the mid section of vines. The network of veins is usually clearly evident. However, as this chlorosis intensifies at greater levels of deficiency, the darker green may be restricted to the main veins and the base of primary branches. Necrotic lesions may develop at the margins or in interveinal patches. In older leaves, a dark-purplish pigmentation may form in a mottled pattern across the leaf or predominantly over the main veins. Young leaves become smaller, generally narrower, and often twisted or curled. They may also become pale or lightly bronzed, or they may retain a dark-green colour. Internodes are usually reduced in length. The young leaves may become necrotic, with lesions usually spreading from the tip. Initially, the size reduction may be difficult to observe, but much more severe symptoms often appear on axillary branches. Young leaves exhibit twisted forms and mottled chlorosis. Older leaves may show distinct interveinal chlorosis and necrotic lesions. Young leaves are reduced in size and develop wavy margins. Interveinal chlorosis is most pronounced in young mature leaves. Tubers of Zn-deficient plants often have longitudinal cracks resulting in necrotic lesions.

Copper

The primary symptom of copper deficiency is the appearance of small, deformed young leaves. These are typically crinkled along the veins and cupped upward or curled under. They often appear pinched at the tip, with a necrotic leaf tip, or have torn away from a dead tip as they have expanded, producing a tattered margin. However, this is not always the case, and in some instances, basal lobes or interveinal zones may be necrotic. Leaves may be a normal, dark-green colour or paler, and are often shiny. Symptoms may develop quite suddenly, progressing from normal leaves to severely reduced leaves and dead shoot tips in only a few nodes of growth. Dead or weak shoot tips stimulate the growth of axillary shoots, which grow with strong stems but very reduced leaves. Increased branching towards the top of the vine may result. The common upward cupping or curling of leaves is associated with very low leaf Cu levels.

Iron

The young leaves become chlorotic, initially yellowish green with darker green veins. Young mature leaves may be the first to show distinctive symptoms and generally display a detailed network of greener veins on a pale leaf. Leaves in full sun are affected more than those in the shade. As the deficiency intensifies, young leaves may become completely bleached white. The bleached leaves tend to become necrotic, with dead tissue spreading from the leaf margins or from scattered spots on the leaf blade. The shoot tip may die in severe cases, but usually the tip, stem and petioles remain alive, even when the blades of young leaves are dead. In all species, the shoot tip may become weak and thin with short internodes. Axillary branches may show the symptoms more strongly than leaves on the main vine.

Manganese

The young expanding leaves first develop a fine interveinal chlorosis in which small spots of tissue in between minor veins appear as pale, translucent spots. These spots may be slightly sunken, giving the leaf surface a rough texture. Some or many spots then develop into necrotic lesions. Necrosis also may develop on the stem wings in the upper part of the vine. The youngest leaves are usually pale, but only develop spots as they expand. The necrosis may cause some deformity of the expanding leaves.

Older leaves also develop dark necrotic spots, but these are usually not preceded by pale translucent spots. Initially, they are only apparent on the upper surface of the leaf. Affected leaves may then become yellow and die, but the dark spots remain evident in the paler necrotic tissue of the dead leaf. Manganese deficiency causes considerable growth reduction before distinctive symptoms appear on the vines. Leaves may develop a rugous texture and mottled colour and may be deformed with asymmetrical expansion or curled with wavy margins. Young leaves may become chlorotic, particularly around the margins, and an increase in branching may produce a bushy growth habit.

Boron

The primary symptom of boron deficiency is deformity of the young leaves. This commonly appears as crinkling along the veins, deflection of the leaf tip, cupping either upward or downward, or twisting all due to uneven expansion of different parts of the leaf blade. Constricted tissue (often along veins or in a particular sector of the blade) may appear paler and may become necrotic generally a dark brown, brittle necrosis in irregular patches. Leaves are typically stiffer than normal.

Leaves are also usually brittle and crack easily when bent. Leaf deformity is by wide, shortened leaf blades with curled tips. Young mature leaves, even those lacking deformity, may be paler than normal, with a weakly interveinal, uneven or mottled pattern. Shoot tips may produce several leaves that fail to expand and then become necrotic. The tip itself may die. Weakness or death of the shoot tip stimulates axillary branches, but these typically show stronger symptoms, with very small deformed leaves, often lacking lateral lobes. The root system of B-deficient plants is often restricted, as B is important for the elongation of root tips. Roots often show a cluster of short, thick branches around the tips of main roots and may have necrotic tips. The tuber's growing tip may be arrested, causing blunt round or pear-shaped tubers, or irregular branching. Brown necrotic areas may occur in the flesh of tubers.

Molybdenum

Molybdenum deficient crops will display symptoms resembling N deficiency, including a general light-green colour, small leaves and a reduced growth rate. The symptoms are probably related to the accumulation of nitrate in the plant tissue, rather than symptoms of poor N nutrition. Mature to older leaves may develop a mild interveinal chlorosis. In older leaves, small dark spots may develop on oldest leaves, which progress to become sunken necrotic lesions. Alternatively, larger dark-grey or brown lesions may form on margins or interveinal tissue of older leaves without prior yellowing, but leaves may subsequently turn yellow and senesce fully.

o) Colocasia

Nitrogen

Nitrogen deficient plants have stunted roots, shoots and suckers. Yellowing of the leaf blade starts in older leaves. As the deficiency progresses, all of the leaf blades turn yellow. Premature death of older leaves often results in fewer numbers of active leaf blades, which reduces growth and lowers crop yield.

Phosphorus

Phosphorus deficient plants have stunted root and shoot growth. Older leaf blades may appear darker green due to greater retardation of leaf expansion relative to chlorophyll (green pigment) reduction. Phosphorus deficiency is also characterized by light-coloured dots on the surface of leaf blades. As the deficiency progresses, areas of the leaf margins begin to yellow and turn brown.

Potassium

Potassium deficient plant is characterized by slower growth, increased tendency to wilt, reduced size of leaf blades, and interveinal or marginal "scorching" a burnt appearance between the veins or around the leaf margins. Potassium deficiency is also characterized by irregularly shaped brown spots in the center of older leaf blades. As the deficiency progresses, the spots may coalesce with the whole leaf turning yellow or brown.

Calcium

Calcium deficient plant is characterized by reduced root and shoot growth. Under mildly deficient conditions, the youngest leaf blade trun yellows between the veins. Under severely deficient conditions, the leaf blades become cup-shaped, with yellow areas between the veins and brown areas around the leaf margins. Finally, leaf blades fail to unfurl and the shoot dies.

Magnesium

Magnesium deficient plant has leaf blades with yellowing between the veins, particularly in older leaves. As the deficiency progresses, the margins of the leaf blades turn brown and die.

Sulphur

Sulphur deficient plant has reduced root and shoot growth. Sulphur deficiency is characterized by reduced leaf size and uniform yellowing of the leaf blades, particularly in younger leaves.

Zinc

Zinc deficient plant is stunted. Excessive zinc uptake can result in toxicity characterized by dead spots between the veins in the leaf blades.

Iron

Iron deficient plant is characterized initially by younger leaf blades having yellowing between the veins. As the deficiency progresses, a uniform bleaching occurs in the younger leaves. In addition, lateral root formation is depressed.

Manganese

Manganese deficient crop has reduced growth. Initially, there is yellowing between the veins of the younger leaf blades and as the deficiency progresses, the leaf blade turns uniformly yellow.

Boron

Deficiency symptoms in crop are characterized by stunted growth of shoots and roots.

p) Cassava

Nitrogen

Under severe nitrogen deficiency, cassava plants become dwarfed with low vigour and chlorotic lower leaves. Chlorosis does not advance inward from the apices and leaf edges, as happens with many other nutritional disorders, but, rather, the entire leaf becomes chlorotic at almost the same time. Close observation shows that the Leaves retain some of their green colouring, giving the leaves a slightly greenish colouring. The upper leaves on the plant are generally much paler than normal. Cassava seems to tolerate nitrogen deficiency better than most crops, although differences have been found among cultivars.

Phosphorus

A deficiency of phosphorus can substantially decrease cassava growth without any clearly visible symptoms. Thus, light and moderate deficiencies can be diagnosed only by foliar analysis, soil analysis, or field experiments. The most notable symptom of acute phosphorus deficiency that is easily recognized in the field is the uniform chlorosis that develops on lower leaves. These leaves cannot maintain their turgor and so hang from the petioles. Plants with a severe deficiency of phosphorus, frequently have thinner stems, shorter petioles, narrower foliar lobes and fewer lobes per leaf than do normal ones.

Potassium

Potassium deficiency is mainly characterized by a marked reduction in plant height; petioles are short and the leaves small. Only in severe cases of deficiency, specific symptoms are observed. Initially, small purple or brown spots appear on old leaves, which curl upward from the foliar margins and, in some cases, downward from the tips. As the deficiency intensifies, chlorotic areas develop on the tips and edges of lower leaves and finally join and give way to a marginal necrosis. These leaves and their petioles grow old prematurely and fall.

Calcium

The plant does not develop a healthy root system, and roots become necrotic and decompose. Calcium deficiency causes curling of the tips and deformation of the youngest leaves.

Magnesium

The characteristic symptom of magnesium deficiency is interveinal chlorosis of the lower leaves, beginning on the tips and edges of the foliar lobes and extending inward between the central and secondary veins. The chlorotic tissue is generally yellow and sometimes pale green. When deficiency of magnesium is severe, chlorotic areas can become necrotic and turn brown or whitish.

Sulphur

Leaves deficient in sulphur are uniformly pale green to yellow, similar to nitrogen deficient leaves. Symptoms normally appear first on upper leaves, but later extend over the whole plant.

Zinc

Zinc deficiency is very common in cassava. In young leaves, it causes an interveinal chlorosis with the following characteristics: leaves initially have a normal green colour, but they develop small, white or light-yellow chlorotic spots between the veins; the shape and colour of these spots vary according to the cultivar. Each new leaf that the plant produces is smaller and more chlorotic and foliar lobes become narrow, take on a light green to white colour and curl upward. When deficiency is severe, the foliar tips necrose. The basal lobes of healthy leaves normally point toward the stem, but under zinc deficiency conditions, these two lobes point outward. Although zinc deficiency mainly affects young leaves, older leaves of some cultivars develop necrotic spots.

Copper

The main symptoms of copper deficiency include a uniform chlorosis and deformation of young leaves. Normally, the foliar tips necrose and leaves curl up. The length of stem internodes does not decrease much, so that plant height remains more or less normal. Leaves of the middle part of the plant frequently are suspended from abnormally long and drooping petioles. When copper deficiency is severe, die-back occurs at the tips of the branches; plants resprout from the basal part of the main stem. Root development may also be markedly reduced, causing copper-deficient plants to become susceptible to drought.

Iron

Chlorosis of the young leaves is a characteristic of iron deficiency; initially, they are a uniform pale green with the veins a darker green, similar to plants suffering from nitrogen deficiency. As iron deficiency intensifies, the veins lose their colour and all of the leaf including the petiole, turns yellow and afterward, almost white. Leaves tend to be smaller than normal, but not deformed; the lobes still have the same length: width ratio as normal leaves. Plant growth also is notably reduced. Cassava tolerates iron deficiency better than other crops.

Manganese

Manganese deficiency is characterized by an interveinal chlorosis of recently expanded leaves. Two characteristics, distinguish this chlorosis from that caused by iron deficiency. First, when the plant is deficient in manganese, the veins and tissue adjacent to them remain green, standing out clearly over the yellow background and appearing like a "fish skeleton". If iron deficiency is involved, only the veins remain green when symptoms first appear, but adjacent tissue is chlorotic and not green. Second, symptoms of manganese deficiency can generally be seen on already expanded leaves but not on the younger ones. On the other hand, when manganese deficiency is severe, these differences may disappear and affected leaves can become totally chlorotic with the young leaves also showing symptoms. As in the case of iron deficiency, affected leaves are not deformed, but they may be smaller than normal; plant height is also reduced.

Boron

Deficiency symptoms appear mainly in young parts of the plant such as shoots and root tips. First symptoms generally are seen on the roots, which show poor development and sometimes death of the root tips. These symptoms are sometimes followed by the emergence of short lateral roots having thickened ends; however, not all cultivars present this symptom. Plants deficient in boron tend to be small because of their reduced internode length. Fully expanded leaves develop numerous small pale gray or brown spots, concentrated mainly on the tips and on the margins of foliar lobes. Severely affected upper leaves are small and deformed and have short petioles. Another symptom of boron deficiency is the development of localized lesions on stems and the upper petioles. Lesions at first exude a sticky brown substance and may later turn into cankers.

q) Leafy vegetables

Nitrogen

Plants are weak and severely stunted and older leaves become uniformly pale or yellow. Reddening of older leaves and stems also observed.

Phosphorus

Plants are stunted, leaves are small and older leaves are dull and lack luster compared to new leaves. Other symptoms can include grey-green or purple colouring of old leaves.

Potassium

Plants are slightly stunted. Yellowing or scorching of old leaves occur at the edges and between leaf veins.

Calcium

Scorching of growing tips and the margins of new leaves. Yellowing between the leaf veins can precede scorching and leaf death.

Magnesium

Mild growth reduction. Yellowing and scorching of oldest leaves between the veins, progressing inwards from margins. Leaves sometimes remain green at the base of the blade.

Sulphur

Yellowing and purpling of areas between veins of new and old leaves. Older leaves show marginal scorching.

Zinc

Leaf size and stem length are restricted and yellowing occurs between leaf veins and can progress into scorching between veins.

Copper

Leaves are dark blue-green colour.

Iron

Yellowing of young leaves between leaf veins.

Manganese

Yellowing between leaf veins (mottling), progressing to scorching.

Boron

Yellowing and scorching occurs between leaf veins at growing points.

3. Spices and Condiments

a) Pepper

Nitrogen

Nitrogen deficiency symptoms are generally characterized by poor growth with pale and yellowish leaves. At first, the lower leaves turn yellowish but the upper canopy of affected plants tends to remain relatively green. In severe cases, leaves of the entire plant show a characteristic yellow to orange-yellow discolouration and the extreme end of the leaf tip becomes necrotic in some instances. Leaf abscission is common in severely affected plants.

Phosphorus

The most striking symptom is stunted growth of the plants. Terminal growth is not so much reduced, but more of restricted lateral growth due to poor secondary branching. Leaf blades of mature leaves become very dull looking, turn bronze coloured, tend to be stiff and show necrosis at the tips in some instances before abscission occurs.

Potassium

Potassium deficiency symptoms can be seen on the distal end of mature leaf blades becoming necrotic, brittle and grey in colour. Necrosis is usually confined to the distal end, while the portion beyond the boundary separating necrotic and live tissues displays a 'V' shape band, which is yellow to reddish brown. This band sometimes occurs without the 'tip burn' symptom.

Calcium

Visually, calcium deficiency is first observed in fresh mature leaves as yellowing or chlorosis which starts on either or both edges near to the petiole end or middle part of the leaf blades. The marginal chlorosis advances inward, followed by necrosis. The proximal and distal ends of the affected leaves are either green or pale green. Tiny pinhead necrotic spots may appear scattered between the main veins on the lower and upper surfaces of the leaves. Leaf abscission occurs before the central portion of the leaf becomes necrotic. Die-back may occur at the growing point. Leaves of the lower canopy are usually more severely affected than those of upper canopy.

Magnesium

Magnesium deficiency symptoms first appear on older leaves and progresses to younger leaves. In the early stage, chlorosis occurs in between main veins. This usually starts from the central proximal half of the leaf. The chlorotic area enlarges to the leaf tip and subsequently towards the leaf margin. A light dramatic leaf fall is often induced, leaving the branches quite bare with only younger unaffected leaves remaining on the plant. The area near the petiole often stays green and gives an arrowhead effect of green tissue penetrating the yellowing areas on the leaf.

Zinc

The initial symptom of zinc deficiency is manifested as interveinal chlorosis on younger leaves. Chlorosis further develops into a network of green veins contrasted against a pale yellow lamina. In young flushes, the terminal growth is retarded with concomitant reduction in intermodal length giving a bunched (rosette) appearance.

Copper

Interveinal chlorosis is the first visible symptom of copper deficiency. The symptom appears on the young leaves. Gradually, the entire portions of the lamina become chlorotic. In advanced stages, the veins also become chlorotic. Large brown necrotic spots develop towards the tip and margin of the leaf. Terminal growth is arrested and a new growth from the basal part of the vine is initiated. Lamina turns bronze in colour. Necrotic spots coalesce forming large black areas near the tips and margins. Inward curling of the necrotic margins can also be observed. Shedding of severely affected leaves leads to sparse foliage on the vine.

Iron

Iron deficiency begins in the younger branches and is characterized by interveinal chlorosis. The chlorosis occurs in between the main veins and smaller veins as well, forming a fine reticulate pattern of green veins contrasting sharply with a pale green or yellow background. The youngest leaves may be completely green or even white. In acute deficiency, the internodal length of terminal shoots and lateral branches is markedly shortened and leaves tend to crowd together at the upper end of the canopy. The berries of affected vines appear pale green to yellow.

Manganese

Manganese deficiency symptoms are more severe in the upper canopy of affected plants. Younger leaves turn chlorotic or yellowish white with only the main veins remaining green. The older leaves produce a characteristic "herringbone pattern" with green veins and the areas between the veins turn yellowish white. At a later stage, small necrotic spots appear and grow in size in the pale areas. When deficiencies are prolonged and severe, berries also show characteristic symptoms.

Boron

Failure of development of terminal bud is characteristic of boron deficiency in black pepper vine. Small branches appear below the terminals, the terminal leaves become abnormal in size and, when mature, interveinal chlorotic patches could be seen on the lamina. The affected leaf becomes thick, brittle and presents mottled appearance with bright orange-coloured mottles generally confined to the upper leaves. The affected leaves are retained on the vine for a long time. Only under severe deficiency condition, leaf abscission occurs. Cracking of the necrotic areas of the laminae also leads to withering of portions of the affected plant.

b) Cardamom

Nitrogen

Nitrogen deficiency may be exhibited by general paling, chlorosis (yellowing) on new leaf, young tillers and also lead to poor green colour development of cardamom capsules. Mild nitrogen deficiency will restrict plant growth. Moderate N deficiency will cause leaves to be light green or yellowish. Severe symptoms include necrosis (tissue death) starting at the tips of older leaves, with the tissue death developing a V-pattern down the midrib toward the base of the leaf.

Phosphorous

The deficiency of phosphorus would lead to general stunting, yellowing, poor rhizome and root development, panicle initiation and flowering. Symptoms appears on the plant as small patches of purplish spots and premature dropping of older leaves. Overall growth of plant was stunted and reduced numbers of suckers are observed. Severe deficiency could lead to panicle stunting and pinkish violet discolouration developed near the raceme stalk and also on the panicle rachis.

Potassium

Potassium deficiency in cardamom could be measured indirectly by judging the plant health. Severe lodging of tillers during the south west monsoon season (i.e., heavy windblown during the June and August month in the plantations) due to the poor tiller strength and improper height of the tiller due to excess nitrogen absorption in comparison with rhizome developed (i.e., easy bending of tillers). The deficiency also characterized by marginal yellowing on the leaves of 10 months old tillers to 2 year old yielding tillers on advancement of the potassium deficiency, marginal scorching and drying of leaf blade along the margin progressing towards the mid-rib portions also occur.

Calcium

Calcium deficiency in cardamom plants leads to the poor tiller emergence, poor husk development on the capsules, weak rhizome formation and also frequent rot incidence in the base of the clump. Deficiency symptoms appeared on young leaves. Root and shoot growth was restricted with no new growth of aerial shoots. The stem was thickened giving a bulb like appearance. The leaf margin turned brown with a golden yellow band underneath. Scattered golden yellow spots appeared on the leaves. Stems are thick and woody, terminal buds may die. Older leaves have normal colour but new leaves may be yellowed.

Magnesium

Magnesium deficiency symptoms are as follows: Stems are elongated, spindly and woody. Lower leaves are thick and firm, developing chlorosis which may be confused with nitrogen deficiency. Roots system is extensively developed but spindly roots.

Zinc

The most visible zinc deficiency symptoms are short internodes (rosetting) and a decrease in leaf size. Cardamom plant could exhibit the zinc deficiency very often, since the cardamom soils are very poor in zinc availability. Zinc is involved in the formation of new tillers, panicles and also for the capsule development. Zinc deficiency in cardamom could lead to the poor development of capsule by showing a characteristic yellowing symptom on immature capsule and forms a abscission layer near the base of the capsule. The symptom on leaf is highly characterized by interveinal chlorosis and reduced leaf lamina on the new tillers. Severe zinc deficiency would also causes reduction in the tiller growth, i.e., the tillers would expand its height at 1-1.5 feet from the base and these tillers will be unproductive.

Copper

Deficiency symptoms of copper include dieback of stems and twigs, yellowing of leaves, stunted growth and pale green leaves that wither easily.

Boron

Deficiency symptoms first appear at the growing points as tip burn. This results in a stunted appearance (rosseting), ill filled capsules due to poor pollination, hollow stems and capsule (hallow heart) and brittle, discoloured leaves and loss of fruiting bodies.

Molybdenum

Molybdenum has a significant effect on pollen formation so fruit and capsule formation are affected in molybdenum-deficient plants. Because molybdenum requirement is very low, most plant species do not exhibit molybdenum deficiency symptoms. Unlike the other micronutrients, molybdenum deficiency symptoms are not confined mainly to the youngest leaves because molybdenum is mobile in plants. The molybdenum deficiency on cardamom plants could lead to poor seed development, pale brown colour formation and capsule weight.

c) Turmeric and ginger

Nitrogen

Pale or yellow green leaves symptoms appear first in the older leaves. Reddish tints gradually appear at the leaf margins spread toward the midrib or central vein. Leaves are small in size. Overall growth is markedly reduced. Deficiency results in reduction of rhizome yield.

Potassium

Symptoms first develop on mature leaves. Reduced growth, shortened internodes, marginal burn or scorch (brown leaf edges), necrotic (dead) spots in the leaf, reduction of lateral breaks and tendency to wilt readily

Iron

The deficiency of iron shows up first in the young leaves of plants, which develop interveinal chlorosis and it progresses rapidly over the entire leaf. In severe cases, the leaves turn completely white.

d) Chillies

Nitrogen

Plant development gradually slows down. Gradual drying beginning at leaf margins. The petioles bend and hang downwards, parallel to the stem. The plant develops few flowers and fruit setting is poor. The fruit receptacle is thin, and the ovary is small. Sometimes, there is no fruit development on the plant at all, and on those plants that bear fruits are deformed.

Phosphorus

The plants display limited growth. The leaves are hard and brittle to the touch. Flower formation is defective. Few flowers develop, and in those that do develop, only one in every four or five develops a fruit. The fruit is underdeveloped with a thin receptacle and very few seeds. The root system is undeveloped.

Potassium

Yellow chlorotic spots appear between leaf veins in the lower leaves. The veins and the areas adjacent to these spots do not change their colour. Later, the chlorotic spots become lighter. There is little fruit setting and not much fruit, which is smaller than usual.

Calcium

Deficiency leads to blossom-end rot, Sunken, dark lesion develops at the distal end of the fruit, Young developing fruit is most vulnerable to calcium deficiency, Stunting and necrosis of the youngest leaves and growing points occurs at the top of the plant and excessive N increases the risk of blossom-end rot.

Magnesium

Marked interveinal chlorosis (appears almost white) and develops on the older leaves and then the middle-aged leaves.

Sulphur

Plants develop chlorosis of the shoot tip as a light green colour progressing down the plant. Small necrotic spots can develop on the tips and margins of the young and recently mature leaves. Plants with advanced deficiency are stunted, upper chlorotic foliage turns light yellow to white.

Zinc

Deficiency appears as small thickened young leaves that are deformed and faint. Interveinal chlorosis and large brown necrotic patches develop on the recently mature to mature leaves in the advanced stage.

Copper

Deficiency initially develops as a faint interveinal chlorosis of the young to recently mature leaves, Young leaves begin to roll up at the margin and advanced symptoms develop as a severe rolling of the young leaves gives a cupped appearance.

Iron

Deficiency occurs in the youngest expanding leaves at the tips of the branches.

Manganese

Interveinal chlorosis or speckling of the young, expanding leaves.

Boron

The deficiency manifests itself very quickly. The lower leaves curl upwards. Growth is stunted. The plant develops a thick and short stem. The apex withers and the leaves become yellow from bottom to top of the plant. There is a reduced production of flowers and fruit setting is poor. In case of severe deficiency, the growing points die and decay and the leaves are misshapen

Molybdenum

Plants stunted and pale. Often show similar symptoms to nitrogen deficiency. Leaves may have marginal necrosis.

e) Clove

Nitrogen

The initial symptom of nitrogen deficiency is characterized by a change of dark green colour of the leaves to pale green. The symptom appears first on the lower leaves with the leaf turning chlorotic then to yellow followed by shedding of the leaves. The symptoms advance from lower to the upper leaves. New leaves formed are smaller in size and the plant growth becomes very much retarded. Nitrogen deficiency is unique with general yellowing of older leaves.

Phosphorus

The plants under phosphorus stress show retarded growth. The older leaves become dull green and small brownish spots appear on the margins and laminae mostly towards the basal part of the leaves.

The adjoining necrotic spots coalesce and develop into bigger round spots to give a burnt appearance to the older leaves. The older leaves are shed resulting in sparse foliage on severely affected plants.

Potassium

Mature and older leaves are the first to exhibit potassium deficiency. Leaf tips and margins turn brown followed by drying up of the tips. The symptoms progress inward until about half blade exhibited a scorched appearance. The emerging leaves are pale green to yellowish green in colour. The affected leaves fall off from the plant at a later stage.

Calcium

Reduced shoot growth with dieback of growing tips is the most conspicuous symptom of calcium deficiency. The young leaves are affected first and hence, they are smaller in size. The leaves later become chlorotic and fall off. Growing points of the plant remain dormant without further growth. Later, they become necrotic and fall off leaving a bare tip. The affected plants fail to put forth new growth.

Magnesium

Because of magnesium deficiency, older leaves become yellowish green due to chlorosis spreading from the margins towards the midrib. The leaves become increasingly chlorotic as the deficient plants putforth newer flushes. In the advanced stages, the developing leaf fails to attain normal size and colour. Necrosis of leaf tips and margins as well as downward curling of laminae are also seen in magnesium deficient plants. The leaf remains chlorotic for about two months prior to defoliation. The plants become stunted in growth eventually.

Sulphur

The newly produced leaves are the most affected by the deficiency. They are pale green to yellow in colour and fail to attain normal size. The older leaves remain unaffected. The young leaves of sulphur deficient plant also show marginal necrosis and a tendency to drop off prematurely. Later, growth produces still smaller leaves with shorter internodes. The dieback of tip and complete cessation of growth generally mark the severe stage of the deficiency.

Zinc

Symptom of zinc deficiency is characteristically observed on new flushes. In addition to the loss of normal green colour, the internodal length gets reduced resulting in rosette condition. The petioles and laminae show a tendency to grow downward giving a sickle-like appearance. Unlike, the case of copper deficiency, the curled leaves is retained on the plant for a longer time. The growth of the plants will be considerably reduced under severe conditions.

Copper

Copper deficiency symptoms are manifested on younger leaves. The older leaves do not show any visible symptoms of deficiency. The newly formed leaves are small, narrow and chlorotic. Loss of turgor, bending of lamina and drying up of the tips are associated with the deficiency. Such shoots are short-lived and soon defoliate giving a barren appearance. The growth of the plant also eventually stops.

Iron

Under iron deficiency, the plants show interveinal chlorosis on younger leaves as the initial symptom of deficiency. The emerging leaves will be yellowish white with the midrib remaining green. The leaf fails to produce chlorophyll at normal rate and is papery white. The leaves shed prematurely ultimately resulting in sparse foliage on the plant. The developing shoots show tip die-back also.

Manganese

The initial symptoms of manganese stress resemble those of iron. The symptoms first appear on young leaves as interveinal chlorosis. Leaves are generally smaller and distorted in subsequent flushes following the initial symptoms. In advanced stages of deficiency, growth of the plant becomes stunted without producing new growth. Curling and cupping of leaves can also be observed on severely affected plants.

Boron

In boron stressed plants, the upper leaves become chlorotic. The leaves will have a hard and brittle appearance with occasional tip and marginal necrosis. New growths show reduced size and abort. The leaves dry off prematurely leaving a barren top.

f) Nutmeg

Nitrogen

Uniform pale green discolouration of the laminae is the initial symptom of the deficiency. Leaves gradually turn yellow with orange tinge and in advanced stages, chlorotic blotches also develop. Premature leaf fall is also associated with nitrogen deficiency.

Phosphorus

The symptoms of phosphorus deficiency is manifested as dark green to bluish green colour of the older leaves which later on turn into purple.

Potassium

Deficiency of potassium induces the development of pale yellowing of the older leaves along the margins and tips. Later, the affected areas become chlorotic. The necrosis gradually progresses toward the midribs covering about three-fourths of the lamina. In the advanced stages, the affected leaves are generally shed.

Calcium

Thick, brittle and smaller-sized leaves with blunt ends develop as the initial symptoms of calcium deficiency. Gradually, light orange yellow chlorosis develops along the leaf margins and tips. The affected leaves become crinkled.

Magnesium

Development of pale yellow discolouration of midrib main veins and adjacent areas at the basal portion of the older leaves are the initial symptoms of magnesium deficiency. Later, the entire lamina turns yellow and necrotic. The affected leaves show outward cupping at the margins and tips.

Sulphur

Symptoms of sulphur deficiency first appear on the young leaves as uniform lemon yellow discolouration. The leaves of the previous flush turn pale yellow. The affected leaves later develop tiny purplish brown necrotic spots with yellow halo, which later turn to characteristic bird's eye or shot holes. Defoliation and dieback follow this.

Zinc

Zinc deficiency symptoms appear as pale yellow interveinal chlorosis. Leaves become abnormally long and narrow. Veins turn dark green presenting a network of dark green veins against a yellow background. Characteristic mottling is also observed.

Copper

The deficiency symptoms appear on younger leaves as light yellow chlorosis. Later, the symptoms become conspicuous with the development of orange yellow spots. Affected leaf margins and tips show inward cupping. In the advanced stages, the leaves become severely distorted and malformed resulting in defoliation.

Iron

The initial iron deficiency symptoms appear as uniform straw-yellow discolouration of the lamina of young leaves. Necrosis of the margins and tips follows as the deficiency advances. New leaves formed are reduced in size.

Manganese

The first visible symptom of manganese deficiency is the development of interveinal chlorosis. Areas along the lateral veins become sunken giving a crinkled appearance. Water-soaked necrotic areas soon develop along margins and tips, which progress inwards covering about three-fourths of the lamina. The affected areas dry up and wither away.

Boron

In nutmeg, chlorosis of the tips of the terminal leaves is the initial symptom of boron deficiency. The terminal leaves become thick, brittle and reduced in size. Later on, bright orange yellow patches develop in between the lateral veins. In advanced stages, necrosis develops at tips and margins of the leaves. Cracking of the necrotic regions results in tearing off of the affected areas.

g) Fennel

Nitrogen

Pale or yellow green leaves symptoms appear first in the older leaves. Reddish tints gradually appear at the leaf margins spread toward the midrib or central vein. Leaves are small. Overall growth is markedly reduced.

Phosphorus

Slow plant growth, leaf colour may intensify, browning or purpling in foliage in some plants, thin stems and loss of lower leaves.

Potassium

Symptoms develop on young leaves in the case of extreme deficiency. Reduced growth, shortened internodes, marginal burn or scorch (brown leaf edges), necrotic (dead) spots in the leaf, reduction of lateral breaks and tendency to wilt readily.

Magnesium

Symptoms appear on the older leaves. Older leaves may fall off, or they develop a triangular arrowhead at the base of the leaf, near the stalk. Reduction in growth, marginal chlorosis, interveinal chlorosis (yellow between the veins) in some species, eventually the whole leaf becomes yellow.

Sulphur

Stunted plant growth, leaves turn to pale yellow. The leaves show a general overall chlorosis. Under severe deficiency, the yellowing is much more uniform over the entire plant including young leaves.

Iron

The deficiency of iron shows up first in the young leaves of plants, which develop interveinal chlorosis and it progresses rapidly over the entire leaf. In severe cases, the leaves turn completely white.

Manganese

Symptoms occur on new growth. Leaves appear mottled, light green to yellow green in between the veins.

4. Plantation Crops

a) Coconut

Nitrogen

Nitrogen deficiency reduces the chlorophyll content of the leaves. In the early stages, the crown of the palm loses its dark green glossy appearance and turns to a pale green colour. In more advanced stages, the older leaves will turn golden yellow and may be prematurely dropped. The yellowing of the younger leaves gradually intensifies. The yellowing of the leaf follows a typical pattern. It starts at the tip of the leaf and the leaf-lets, progressing inwards along the mid-ribs, the margins remaining pale green. There is no sharp division between the greener and the yellow parts of the leaves. New leaves will not grow out to their full length and older leaves will drop prematurely, reducing the number of leaves in the crown and giving it an umbrella-like shape. Young nuts may fall off prematurely and the number of female flowers per inflorescence is reduced. In severe cases, whole bunches may abort. Yields will gradually be reduced. The stem of the palm gradually tapers to a so-called "pencil point" condition and the crown consists of a reduced number of leaves being shorter than normal.

Phosphorus

Generally, phosphorus deficient palms have a dark green foliage, slightly darker than normal. Only in severe cases, leaves may turn yellow before dying prematurely. Rosetting of the leaves with shorter petioles are also observed.

Potassium

The first symptoms begin in the older leaves. They are characterized by yellowing of the leaflets, followed by necrosis. Yellowing starts at the tip of the leaflets, progressing along the margins towards the base. The tip of the leaflet withers and becomes necrotic, necrotic spots also appearing in the yellow part of the leaflet, mainly along the margins. In severe cases, the yellow spots coalesce giving the leaves a scorched appearance. In advanced stages, only a strip along the mid-rib remains green, widening towards the base of the leaflet. This is also the pattern in the entire leaf, the margins yellowing and the green part along the midrib widening towards the base.

In advanced stages, the oldest leaves wither and drop prematurely. The youngest leaves remain green. The number of inflorescences and the number of nuts produced gradually decrease. The growth of the palm is reduced, the trunk narrows and the internodes are short.

Calcium

The first symptom appears on younger leaves and even on the spears, with the occurrence of round yellow spots, becoming brown in the centre. These spots appear isolated at first, become coalescent, and finally dry out. Severe distortion of leaflets and leaves also occur. In addition to the easily visible symptoms on the leaves, abnormalities also occur in the root system of all coconut varieties. Root systems had a low number of primary roots, and most of them will rot at severe deficiency including secondary and tertiary roots.

Magnesium

The first deficiency symptom is the inter-vascular yellowing of the older leaves. The petiole of the leaf and the mid-ribs of the leaflets remain green. In the early stages, a small margin on either side of the mid-rib of the leaflets also remains green. Gradually, the yellowing proceeds to younger leaves and the older leaves turn to a yellow-orange and even to orange colour. The youngest leaves remain green. The yellowing of the leaves particularly occur on those patches of the leaves being exposed to sunshine, the shaded parts rarely showing chlorosis. This shade effect is more accentuated in Red and Yellow Dwarfs than in Talls or hybrids. Old leaves may assume a bronzed appearance and dry. In severe cases, the leaflet extremities show necrosis and turn to a characteristic reddish-brown. Translucent spots may appear on the yellow parts of the leaf.

Sulphur

In sulphur deficient palms, all leaves tend to become chlorotic, even the young ones. Sulphur deficient seedlings became yellow much later than nitrogen deficient palms. The mature leaves progressively lost their green colour from the margins of the leaflets towards the mid-rib, so that the green seemed to form a haze on a yellow background. The rachis of the leaves are weak and pliant.

On the older palms, the total number of living leaves is strongly reduced and the leaves are of a yellowish-green colour. Sulphur deficient palms also have a marked tendency to retain their dead leaves. There is a premature bending of the leaves above the normal abscission layer and consequently, a considerable number of dead leaves are hanging along the stem.

Nut production is severely reduced, the nuts being small, producing a poor quality of copra. The endosperm is of normal thickness but on drying, it collapses to a thin rubbery copra. This copra is difficult to break and may clog machines. It does not express its oil freely. It has a low oil content. The copra has a capacity for absorbing moisture more readily than normal copra, leading to a rapid deterioration. Higher sugar contents are connected with lower oil contents, pointing to a breakdown in fat synthesis in the maturing nut.

Zinc

Zinc deficiency is characterized by formation of small leaves wherein, the leaf size is reduced to 50 per cent. Leaflets become chlorotic, narrow and reduced in length. In acute deficiency, flowering is delayed. Zinc deficiency will also lead to button shedding.

Copper

The clear change of the colour of the younger leaves in the upper region from green to yellow, most of the petioles are in arc-shape, eventually losing turgidity. Distinctly, the lamina along the midrib remains green. A close-up of copper-deficient leaflets showing the premature dry up and necrosis of the tips of leaflets and change in colour from dark green to yellow from the tips towards the leaf petiole, while the lamina along the midrib remains green.

Iron

Iron deficiency usually appears on palms growing in poorly aerated soils or those that have been planted too deeply. Water logged soils and deep planting effectively suffocate the roots and reduce their effectiveness in taking up nutrients such as Fe. The main symptom of iron deficiency is chlorosis or yellowing between the veins of new leaves Uniform chlorotic new leaves as the deficiency progresses, the tips become necrotic and leaf size reduced.

Manganese

Emerging leaflets are chlorotic with longitudinal necrotic streaking. Leaflet tips become necrotic, withered and curled giving the leaf a frizzled appearance commonly called "frizzletop." On coconut palm, necrotic leaflet tips fall off and the leaf has a singed appearance.

Boron

The length of the petiole and the number of leaves on the crown are reduced. Sometimes, the petiole gets thickened, abnormally elongated and the tip of the tubular leaf becomes whipped. Leaflets show hook like symptoms on single side/both sides with or without a brown solution oozing out through the hooked leaflets. The hooking can be seen in various types and shapes. Fan like and boat like appearance of leaves are also associated with boron deficiency.

The leaflets are asymmetrically arranged so that the space between the leaflets is abnormally greater than that of the normal one. Sometimes, the bracts of the petiole become hardened/ thickened. Petiole without any leaflets having tip necrosis is also common deficiency symptom due to boron. Longitudinal or transverse cracking along the petiole with or without gummosis is also observed.

Boron deficiency also adversely affects the production and productivity of coconut. The older and matured leaves looked healthy without any malformation. However, the younger leaves on the crown exhibited the deficiency symptoms like fasciations, hooking of leaves, crowding of leaves, etc. Crowding of leaves around the apex resulted in a choking of the crown. This situation is normally referred to as 'crown choking' disease of coconut. The extent of degree of each symptom depends on the time and stage of the plant growth cycle at which the deficiency occurs.

Button shedding is also associated with boron deficiency. The inflorescences are aborted without any visible external symptoms. Necrotic inflorescence with branched spikelets having male flowers only and abnormally big inflorescence with a few female flowers are also observed in a few cases. Bulging is observed at the base of certain nuts. Some bunches showed hen and chicken symptoms. Cracking is observed on the husk and shell of the nuts.

Molybdenum

The deficiency symptoms are manifested by chlorotic leaf blade, Small slender leaves, rosetted plants and occurrence of whip tail.

Chlorine

Chlorine deficiency is characterized by yellowing and/or orange mottling of the older leaves, with a drying-up of the outer edges and tips of the leaflets - a symptom very similar to that of potassium deficiency. Chlorine deficient leaves are droopier and the leaflets may be folded slightly similar to palms suffering from drought. Chlorine deficient palms have a lower leaf production and a lower number of leaves growing slower than normal. Chlorine seems to have less influence on the number of nuts produced than on the size of the nuts.

b) Arecanut

Nitrogen

Plants are stunted and generally yellowing of lower leaves and older leaves show golden yellow colour.

Potassium

Symptoms first appear on older leaves and later spread to younger leaves. Translucent yellow or orange spots develop on leaflets. The tree appears yellow, the trunk become slender with few short leaves. Leaflets with necrotic areas along the margins which later wither off.

Phosphorus

Intra-veinal chlorosis of older leaves. The chlorosis spreads throughout the leaf followed by marginal necrosis and defoliation. Stem is slender without fresh growth and stunted growth of roots.

Calcium

Defoliation of young leaves with necrosis along the veins. The stems become woody and short with yellowish tips. Roots are stubby and dry.

Magnesium

Magnesium deficiency includes; Broad light yellow band along the margin of older leaves, the center of the leaf remaining green, In severe cases, leaflet tips may become necrotic. Older leaves become bronzed and show dry appearance and leaflets show necrosis and turn to reddish brown with translucent spots.

Sulphur

Slight chlorosis of leaves, lack of plant growth with slender stem.

Zinc

Young leaves show interveinal chlorosis and yellowish white spots on leaves.

Copper

Copper deficiency includes: Reduced growth and dark green colour of leaves, twist malformed leaves and new leaves shriveled, busy growth.

c) Oil palm

Nitrogen

Nitrogen deficiency is expressed in uniformly pale, yellow green leaflets and a sharply reduced growth rate. Midrib tissues become bright yellow. Unlike magnesium deficiency, the symptoms are equally pronounced on both upper and lower rank pinnae. Nitrogen deficiency may also be caused by poor drainage.

Phosphorus

In oil palm seedlings, phosphorus deficiency causes the older leaves to become dull and assume a pale olive green colour while in adult palms high incidence of premature desiccation of older leaves occurs. Phosphorus application increases the bunch production rate, bunch weight, number of female inflorescences and thereby the sex ratio.

Potassium

The most typical and widespread form of K deficiency is known as "confluent orange spotting". The first signs of K deficiency are pale green spots on the pinnae of older fronds. In a more advanced stage, the rectangular spots become orange-yellow and transmit light when held up to the sky. Later, the tips of leaf pinnae start to dry up. In very severe cases, entire older fronds may dry up.

Magnesium

Severe magnesium deficiency symptoms are most striking and have been named as 'orange frond'. While the lower most leaves are dead, those above them show a gradation of colouring from bright orange on the lower leaves to faint yellow. The youngest leaves do not show any discolouration. The most typical magnesium deficiency symptom is the shading effect in which the shaded portion of the leaflet will be dark green while the exposed portion of the same leaflet is chlorotic.

Zinc

Zinc deficiency appears as small, narrow white streaks on lower and mid-crown fronds. A different condition that produces blotchy leaf symptoms has also been identified as Zn deficiency.

Copper

Copper deficiency appears initially as whitish yellow mottling of younger fronds. As the deficiency intensifies, yellow, mottled, interveinal stripes appear and rusty, brown spots develop on the distal end of leaflets. Affected fronds and leaflets are stunted and leaflets dry up.

Iron

Iron deficiency is very rare in oil palm and occurs where soil pH is very high (i.e., more than 7.5). It is easily identified, as symptoms appear first on the youngest fronds, which appear droopy and show diffuse blotchy yellowing and white freckles.

Manganese

Manganese deficiency shows as a yellowing of interveinal areas. In contrast to Mg deficiency, the symptoms are found on young rather than on older fronds. The symptoms are equally pronounced on upper (sun exposed) and lower (shaded) rank pinnae.

Boron

Boron deficiency is expressed in a range of leaf symptoms. However, in all cases, the distal end of leaflets at the tip of the frond are most affected. Pinnae are misshapen, stiff and brittle. "Hook leaf" is one typical symptom of B deficiency.

d) Cocoa

Nitrogen

Restricted growth of tops and roots and especially lateral shoots. Plants become spindly with general chlorosis of entire plant to a light green and then a yellowing of older leaves which proceeds toward younger leaves. Older leaves defoliate early.

Phosphorus

Restricted and spindly growth similar to that of nitrogen deficiency. Leaf colour is usually dull dark green to bluish green with purpling of petioles and the veins on underside of younger leaves. Younger leaves may be yellowish green with purple veins.

Potassium

Older leaves show interveinal chlorosis and marginal necrotic spots or scorching which progresses inward and also upward toward younger leaves as deficiency becomes more sever.

Calcium

From slight chlorosis to brown or black scorching of new leaf tips and die-back of growing points. The scorched and die-back portion of tissue is very slow to dry so that it does not crumble easily. Boron deficiency also causes scorching of new leaf tips and die-back of growing points.

Magnesium

Interveinal chlorotic mottling or marbling of the older leaves which proceeds toward the younger leaves as the deficiency becomes more severe. The chlorotic interveinal yellow patches usually occur toward the center of the leaf with the margins being the last to turn yellow.

Sulphur

Older leaves become yellowish green and the stems become thin, hard, and woody. Some plants show colourful orange and red tints rather than yellowing. The stems, although hard and woody, increase in length but not in diameter.

Zinc

In some plants, the interveinal chlorotic mottling first appears on the older leaves and in others, it appears on the immature leaves. It eventually affects the growing points of all plants.

Copper

Leaves at top of the plant wilt easily followed by chlorotic and necrotic areas in the leaves. Leaves on top half of plant may show unusual puckering with veinal chlorosis.

Iron

Starts with interveinal chlorotic mottling of immature leaves and in severe cases, the new leaves become completely lacking in chlorophyll but with little or no necrotic spots. The chlorotic mottling on immature leaves appear near the bases of the leaflets.

Manganese

Starts with interveinal chlorotic mottling of immature leaves and it is indistinguishable from that of iron. On fruiting plants, the blossom buds often do not fully develop and turn yellow or abort. As the deficiency becomes more severe, the new growth becomes completely ceases.

Boron

From slight chlorosis to brown or black scorching of new leaf tips and die-back of the growing points similar to calcium deficiency. Also the brown and black die-back tissue is very slow to dry so that it can not be crumbled easily. Both the pith and epidermis of stems may be affected as exhibited by hollow stems to roughened and cracked stems.

Molybdenum

Older leaves show interveinal chlorotic blotches, become cupped and thickened. Chlorosis continues upward to younger leaves as deficiency progresses.

e) Cashew

Nitrogen

Deficiency symptoms of nitrogen always appear first in older leaves, stem becomes thin and growth is stunted. The leaves are so smaller than those of other plants. Colour of the leaves change gradually from dark to pale green and to yellow, coincided with stunted growth.

Phosphorus

Phosphorus deficiency reflects on stunted growth with dull-green foliage. Lower leaves are withered and fallen.

Potassium

Potassium deficiency symptoms may develop within 2 months after planting. Lower leaves turn yellow, starting at the apex and along the margins showing necrosis symptoms spread rapidly from lower part of leaf towards the top.

Calcium

Cupping of young leaves with marginal scorch and death of growing points are the deficiency symptoms of calcium.

Magnesium

Interveinal chlorosis and yellowing of lower leaves are the characteristic visible symptoms of manganese deficiency. The symptoms spread rapidly from the lower leaves towards the upper leaves.

Sulphur

The deficiency has been characterized by pale green to greenish yellow discolouration of younger leaves which later turn to uniform yellow. Small necrotic spots appear on the affected leaves which coalesce to become larger necrotic areas. Shedding of affected leaves and growth retardation also occurs in the deficient plants.

Manganese

Interveinal chlorosis of younger leaves is the first visual symptoms of Mn deficiency. In some cases, brownish patches can also be observed near leaf margins. Leaf development will also be poor. In severely affected plants, chlorosis may cover the entire lamina except the major veins and laterals more conspicuous. Dwarfing and reduced leaf productions are also due to deficiency of Mn.

Boron

Boron deficient trees fail to expand young terminal leaves which are distorted and remain rolled and the growing points die. The deficiency causes poor seed development. It's deficiency causes 'gumming' in nuts. The kernel quality is also reduced to considerable extent.

f) Rubber

Nitrogen

The main indication of nitrogen deficiency is a pale yellow-green colour of the leaves. The whole leaf becomes first yellow-green in colour. In young unbranched tree, the symptoms will first appear on the older leaves in the lower storeys, and only when very severe, the deficiency affect the upper and younger storeys on the tree. In old branched tree, nitrogen deficiency results in a marked retardation of growth, particularly by reducing the size of the canopy. The deficiency symptoms are likely to be more pronounced on leaves exposed to full sunlight than on leaves in the shade of the canopy.

Phosphorus

The distinguishing symptom of phosphorus deficiency is a bronzing of part of the under surface of the leaf. The bronzing is often largely restricted to one half of the leaf and is usually first seen at the distal end of the lamina; the leaf tip frequently dies back. Pale bronze tints are sometimes found on the under surface of very young not fully expanded leaves.

In young unbranched tree, the symptoms are usually first observed on leaves in the middle and upper storeys, and considerable defoliation can occur. The yellowing and purpling of the oldest senescent leaves on the tree just before abscission.

In old branched tree, deficiency symptoms would be expected, on leaves exposed to full sunlight, and not on shaded leaves. Shaded leaves on weak, lower branches of the canopy often show a general yellowing and bronzing prior to the defoliation, death and fall of the branch due to self- pruning.

Potassium

The characteristic symptom of potassium deficiency is the development of a marginal and tip chlorosis, which is followed by a marginal necrosis. In young unbranched plant, marginal chlorosis usually appears first as a yellow mottling. Later, the mottled areas coalesce forming a fairly distinct marginal band of chlorotic tissue. This band quickly becomes necrotic, the necrosis first appearing at or near the leaf tip.

On leaves from older branched trees, the chlorosis usually commences at the tip and along the leaf margins but with the rapid development of the chlorosis over the leaf. It frequently found that only the tissue at the base of the leaf, and adjacent to the midrib, remains green. Sometimes one half of the leaf will be more affected than the other.

Calcium

In young branched tree, the symptoms are observed on the leaves on the upper or younger storeys on the plant and in severe cases, the growing point itself may die back. There can be considerable reduction in leaf size.

In old branched tree, the symptoms of white papery scorch associated with calcium deficiency, have been observed only on leaves in the shade of the canopy and not on leaves exposed to full sunlight.

Magnesium

The characteristic symptom of magnesium deficiency is the development of a chlorosis in the interveinal areas on the leaf. Generally, the chlorosis appears to spread inwards from the leaf margins, giving a herring-bone pattern.

In cases of severe deficiency, the chlorosis which is generally a deep yellow colour is often followed by interveinal necrosis and sometimes by marginal necrosis. In cases of mild deficiency, the "herring-bone" pattern may be seen but the chlorosis is of a dull yellowish green. There is usually only little reduction in leaf size, although when the deficiency is extremely severe, there can be considerable defoliation and marked reduction in tree growth and leaf size.

Sulphur

The first indication of the sulphur deficiency is very similar to that of nitrogen deficiency, namely a gradual uniform yellowing of the entire leaf, which is much reduced in size. Later, a tip scorch develops which results initially in the cupping of the leaf and later affects the whole of the distal end of the leaf.

Zinc

The characteristic feature of zinc deficiency is that the lamina becomes very much reduced in breadth relative to its length. Frequently, the lamina may become twisted and the margins appear wavy or undulating. There is also a general chlorosis of the leaf, similar to that observed in fairly severe cases of manganese deficiency, the midrib and main veins remaining dark green in colour.

Copper

The first sign of copper deficiency is wilting of the leaf margin at the tip, with subsequent upward cupping of the leaf tip. The wilting of the margin develops into a very pale brown scorch which often spreads down the lamina from the tip; finally the lamina is shed. Frequently, a large number of leaves are shed at the same time, leaving behind a bare stem. After such defoliation, the apical growing point usually dies and the new shoots which develop from axillary meristems give rise to multiple branching. Under moderate deficiency, leaf cupping, scorching, leaf abscission, and death of the apical meristems will again occur. In severe cases, the lamina on the new side shoots fail to expand properly, and the new side shoot maybe covered with numerous dead and shrunken petioles.

Iron

The first leaf symptom of iron deficiency is a general leaf chlorosis resembling that of manganese deficiency. With increase in severity of the deficiency, the entire leaf assumes a pale lemon yellow to white colour. There is considerable reduction in leaf size.

Manganese

The typical deficiency symptom is an overall paling and yellowing of the leaf with bands of green tissue outlining the midrib and main veins. As the deficiency becomes more severe, the leaf becomes progressively more yellow, but the green midrib and main veins still remain evident. No necrosis of leaf tip or margin occurs. Little reduction in leaf size occurs even when the deficiency is severe, and it is frequently found that the symptoms are more often observed on the large leaves of a tree than on the small leaves. Manganese deficiency symptoms will usually be found on shaded leaves on low branches in the tree.

Boron

Boron deficient leaves are distorted, reduced in size and somewhat brittle. The leaf deformation does not follow any consistent pattern. There is no loss of colour, and occasionally the veins may appear to be wider than normal.

In young unbranched tree, the first sign of boron deficiency is observed in the younger, upper storeys on the plant, which will not be separated by any discrete internode. The individual storeys cannot be distinguished, and consequently there appears to be a continual production of new leaves, resulting in the 'bottle-brush' appearance of the stem. When the deficiency is severe, the apical meristems may die and axillary meristems develop near the top of the stem.

Molybdenum

Molybdenum deficiency is characterized by the development of a very pale brown scorch around the leaf margins, particularly in the region of the leaf tip.

g) Coffee

Nitrogen

Leaves rapidly becoming pale green; emerging leaves uniformly pale green with a dull green sheen. Entire plant becoming pale green, with sparse vegetative growth; leaves becoming yellow-green at advanced stages; whitish veins may be present in lower leaves.

Phosphorus

Phosphorus deficiency symptoms include irregular, yellowish spots appear distributed over the entire leaf blade; the chlorotic spots may have a red or purplish tint. These symptoms first appear on the older leaves. In acute conditions, these spots coalesce and leaf readily falls off the branch.

Potassium

The deficient plants show first small brownish spots near the spatial margin. Gradually, the spots coalesce and a dark brown, almost black, necrotic area is formed at the tip of leaf on the adjacent margins. The central tissue is little affected. The leaf abscises easily. Potassium deficient plants are easily prone to anthracnose disease.

Calcium

Younger leaves show chlorotic symptoms on their margins all around the blade. On both the sides of the central vein, the tissue remains green. Sometimes, cupping appears as a result of unequal development of the veins and the leaf blade. The tissue cracks and suberous tissue is formed. The veins assume dark colour due to necrosis in the phloem. In young plants, the apical bud dies under acute deficient condition. Yellow areas develop between the veins spreading from the margins towards the centre. Later, the colour of the patches changes to bright orange and they tend to coalesce.

Magnesium

Chlorosis occurs between the main vein and the secondary veins. A narrow band of green tissue exists around the yellow area. The chlorotic spots gradually expand towards the edges of the leaf. In acute conditions, a dark necrotic spot appears at the tip of the leaf.

Sulphur

The blade of the younger leaves show yellowish green colour. Chlorosis appears as a band, which comprises the main vein and extends towards the middle of the blade. The lower surface of the leaf is much lighter in colour than the upper one.

Zinc

In zinc deficient plants, the new leaves are small, narrow and distorted. The veins form a green net work against a pale background of yellowish tissue. In acute conditions, the young leaves easily break and have a leather-like appearance. The branches show reduced number of leaves. Mostly of the leaves might fall. As a consequence of the shortening of the internodes, a rosette of small, narrow, pointed leaves develop at the extremities. Die-back of the branches sometimes occurs.

Copper

In plants suffering from copper deficiency, the young leaves gets 'S' shaped appearance along the main vein followed by inward curling. Due to lack of proper development of the veins, a puffed appearance may be observed. Necrosis follows disploration of the affected leaves.

Iron

Iron deficiency is frequent in the nurseries and also in older trees, especially when an intensive production of young leaves is in progress. In mild shortage, the leaf blade becomes pale green and the veins retain the normal dark green colour.

Manganese

Young leaves are first affected. The internodal regions are light green and later show numerous little yellow spots. These spots coalesce later giving rise to larger yellow areas. The older leaves show necrosis at the tip and on the margins. The necrosis progresses slowly towards the basal region of the leaf.

Boron

Due to the higher demand for boron in the actively growing regions, a striking symptoms of deficiency is the death of terminal buds and such dead branches and top of the plant remain attached to the plant for some time. Later, a few branches develop below the dead terminal bud giving the appearance of a fan. Leaves are usually narrow and twisted with irregular edges. Internodes are shortened.

Molybdenum

Yellow-green irregular spots may occur near the margins of the leaves deficient in molybdenum. The leaf curls downwards till the lateral edges touch each other. At later stages, the leaves lose the yellow colour and visible necrotic spots appear.

h) Tea

Nitrogen

Yellowing of leaves. Retardation of root and shoot growth and formation of small leaves with short internodes.

Phosphorus

Phosphorus deficiency is characterized by dull bronzing of recently matured lower leaves with or without brown spotting and loss of glossiness. The stems become thinner; the internodes of shoots become shorter and finally shedding of leaves. Sometimes leading to the death of plants.

Potassium

The deficiency symptoms appear in the following sequence: Pronounced marginal necrosis on mature leaves; premature leaf fall, leaving a crown of young foliage at the top, development of thin and twiggy wood and ultimately the death of plant.

Calcium

Calcium deficiency is characterized by brittle old leaves and the young leaves with discoloured lamina edges becoming dark brown.

Magnesium

Magnesium deficiency is characterized by the yellowing of old leaves and an inverted '*V*' shaped yellow tint between the veins. There is premature leaf fall from the affected bushes.

Sulphur

Sulphur deficiency of tea is known as "tea yellows" and is characterized by a progressive degeneration of tea bush.

i) The early stage: When the leaves are of normal size and texture but are mottled with yellow.

ii) The intermediate stage: When the leaves are yellow, brittle and very much reduced in size and are scorched on the margins and tips.

iii) The late stage: When all leaves are shed leaving a small leaf at the tip of shoot. The internodes are very much shortened and axillary buds grow into very small shoots with yellow leaves.

iv) The final stage: When the bush is totally defoliated, the terminal buds of the shoot die and gradually die-back sets in, ending in the death of the entire bush.

The most striking symptom is the "net veining" in the younger leaves where the leaf blade takes a striking yellow colour and the veins down to the finest branching stand out prominently dark green.

Zinc

Zinc deficiency in tea manifests by way of unevenly developed leaves becoming brittle, with chlorotic greenish yellow mottling. The deficient leaves become sickle-shaped, sometimes the margins become wavy, the internodes become shorter and the shoots present a bunchy appearance. The affected leaves are smaller and are arranged in rosettes.

Copper

Due to copper deficiency, young leaf tips become white with narrow and twisted leaves. The growth of internode is depressed due to lack of adequate copper. The deficiency causes poor fermentation, which reduces the quality and flavour of tea.

Iron

There is inter-veinal chlorosis or yellow stripping in young leaves. In extreme cases, scorching of the leaf tips and margins may occur.

Manganese

The manganese deficiency symptoms are exhibited by the mature leaves and they develop pale yellow patches along the margin of the leaf. These extend towards the midrib resulting in interveinal chlorosis and the affected leaves show an irregular mottling with the veins and surrounding areas standing out green. This is followed by the appearance of reddish brown necrotic spots within the chlorotic patches close to the leaf margin.

Boron

The first sign of boron deficiency is the restriction in the growth of terminal bud, which becomes dormant. The leaves become dark green, thick leathery and crinkled. The growing points ultimately die and as a result of the loss of apical dominance, many axillary buds try to grow but these also die back. Cluster of small shoots fills the upper axils; corky streaks may develop on the stem rather like elongated lenticels.

5. Flower Crops

a) Rose

Nitrogen

The leaves of a nitrogen deficient rose show an overall yellow or light green colour. The older leaves are more affected than the newer growth because nitrogen is mobile within the plant and will move from the older foliage to newer leaves when in short supply. This older foliage may turn clear yellow and ultimately fall off. Slower growth, stunted plants and reduced stem length and diameter are also symptomatic of nitrogen deficiency. Flowers of the darker coloured cultivars may appear several shades lighter than normal and the buds of any variety may fail to open.

Phosphorus

The first symptom of phosphorus deficiency is an overall stunting of the leaves and shoot growth. Later, the older leaves may lose their luster, becoming dull green and drop off. Root development is reduced, which results in poor flower production and slow-to-open buds on weak stems. A slight purpling of the underside of the midrib can develop in some cultivars

Potassium

Deficiency results in slow growth, shorter than normal. Weak flower stems and small short deformed flower buds. Tip and marginal leaf burn beginning with the older leaves. Potassium deficiency may contribute to the production of "blind shoots" (new stems with no flower).

Calcium

Calcium deficiency causes abnormal root growth causing the tips to shorten, get brittle and eventually die. Young leaves are distorted, older foliage become dull green and may curl down at the margins. Calcium deficiency can also contribute to weak stems and in the cut flower industry is associated with a condition known as 'black tip' on some red rose varieties.

Magnesium

Deficiency can cause stunting and yellowing between the veins on the foliage. These symptoms show up on the older. This yellowing can eventually cause purplish or brown dead spots that can eventually cover the entire leaf.

Sulphur

Sulphur deficiency causes young leaves to develop light green to yellowish in colour. Plants may be spindly and smaller than normal.

Zinc

Zinc deficiency symptoms are similar to those caused by copper deficiency. However, zinc deficiency can also the lateral shoots that develop after the terminal bud dies to stunt causing "little leaf" symptoms.

Copper

Copper deficiency symptoms include distorted young leaves with yellow tips. Ultimately, the growing point dies, and short, stunted, lateral shoots develop.

Iron

Iron deficiency results in interveinal chlorosis of the younger foliage (yellow foliage, green veins). If the condition worsens, the newly formed leaves may remain small and can become completely pale yellow to white. Iron deficiencies can be induced by high levels of manganese, zinc, or phosphorus. Manganese deficiency can cause pale mottled leaves and interveinal chlorosis of young foliage.

Manganese

Main symptoms starts with interveinal chlorosis of young leaves. Next, necrotic tissue may drop out giving the leaf a ragged appearance.

Boron

Boron deficiency can cause small, thickened, curled, scorched leaves and death to the terminal bud. Death of the terminal bud causes lateral buds to develop contributing to "witches broom effect". Boron deficiency can cause "bullheads" (flowers with shortened petals, that are abnormally thick and have the margins roll in).

Molybdenum

Molybdenum deficiency can cause stunting and lack of vigour. It can also cause malformed leaves with marginal scorching and cupping.

b) Chrysanthemum

Nitrogen

The entire plant becomes lighter green, but the effect is most noticeable on the older foliage. Gradually, the oldest leaves lose their green colour, and on most plants they become yellow. However, a red or bronze colouration of the lower foliage occurs with some plants. The leaves dry and turn brown under severe deficiency and do not drop. The leaves are smaller and are borne rather stiffly, in a somewhat upright position on the stem. The stems are more slender and woody than on plants having an adequate supply of nitrogen. With some plants, the stems are shorter, but in the case of the rose, there are fewer shoots, although some long-stemmed flowers are produced. The flowers are smaller and may lack well developed colour.

Phosphorus

A purplish colouration, developing first on the underside of the petiole and spreading to the main veins of the leaf.

Potassium

In potassium deficiency, the margins of the leaves of the older foliage become yellow, and the chlorosis progresses toward the midportion of the foliage as this deficiency increases in severity. Necrotic areas or spots may appear on the margins. The older leaves may drop in very extreme cases of deficiency. Sometimes, chlorotic or aerotic spots or blotches appear in areas between the veins of the older leaves near the margin.

Calcium

Calcium deficiency is the development of short clubby roots. Cambial activity of the roots occurs, but division of cells at the tip is reduced, giving rise to short, thickened roots. The stems may be rather drastically reduced in length and in severe cases, the youngest foliage may develop chlorosis followed by death of the stem tip.

Magnesium

In pronounced magnesium deficiency, the lower leaves becomes yellow in blotchy irregular spots between the veins rather quickly. The remaining portion of the leaf is almost a normal green.

Sulphur

The upper foliage becomes chlorotic and the veins are usually more yellow than the surrounding tissue.

Copper

Dull green leaves, chlorotic veins. Veinal chlorosis produces an inverse "netting". Margins of leaves become chlorotic. Affected leaves wilt during the day, outer margins turning upward. Flowers are small, reflex and soft.

Iron

The root growth is reduced or killed, chlorosis of the new leaves will appear rather quickly on some plants, the severity depending upon the extent to which the root system has been damaged. As iron deficiency becomes more intense, necrotic areas appear on scattered portions of the yellowed leave and the affected foliage may drop.

Manganese

In manganese deficiency, the top foliage becomes yellow between the veins, and eventually the entire leaf turns chlorotic. The chlorotic areas are rather small and scattered throughout the leaf blade, while the smallest veins remain green, giving a netted appearance to the leaf.

Boron

The new foliage is thick or leathery and quickly becomes chlorotic. The rose flowers are usually malformed. The stem tip dies, giving rise to growth of shoots immediately below which in turn die at the tip, and a "witches broom" effect is observed.

c) Carnation

Nitrogen

Deficiency of this nutrient limits plant growth, produces thin stem and stiff leaves which curves at the tip, particularly during winter and general yellowing of mature leaves occurs besides disturbing flower initiation.

Phosphorus

Deficiency of phosphorus reduces growth and produces stunted plants. The leaves become narrow and the tips dry off and turn yellow. In severe cases, the whole plant turns yellow and dies prematurely.

Potassium

Deficiency of potassium in carnation is characterized by scorching of leaf blade. In severe cases, white spots develop on foliage just below the flower; flowers become malformed and lose brightness; yellowing of calyx occurs. Deficiency of potassium also reduces flower yield, grade, stem strength and longevity.

Calcium

Deficiency of calcium produces thin and weak stems, small flowers, production of many sleepy flowers, appearance of crescent shaped necrotic lesions at the leaf tip and calyx scorch.

Magnesium

The typical symptom of magnesium deficiency is interveinal discolouration in older leaves. The leaf area between the veins turns pale yellowish-green, while the veins remain deep green.

Boron

Boron deficiency symptoms on carnation show on the youngest foliage. Leaves are yellow and twisted and the terminal bud sometimes is dead. Axillary shoot proliferation at upper nodes is common, resembling the "witches broom" caused by virus infections in other crops. Flower buds are abnormal and die before opening. The new foliage turn pale brown with a reddish-purple band and partial yellowing of the leaf base.

d) Gladiolus

Nitrogen

Nitrogen deficiency results in pale green colouration of foliage and reduction in the number of florets per spike and number of spikes per corm.

Phosphorus

The symptoms of phosphorus deficiency are dark green foliage and purple colouration in the lower leaves.

Potassium

The deficiency results in shortening of spike length, reduction in the number of florets, general yellowing of older leaves and yellowing between the veins of younger leaves. In case of severe deficiency, the older leaves show marginal leaf burn. Delay in flowering is also the cause of potassium deficiency.

Calcium

Calcium is an important element for producing good quality flowers. Mild deficiency results in a disorder called 'topple', which causes breaking over of gladiolus in the vase after most of the florets have opened. Severe calcium deficiency results in symptoms similar to physiological 'bud-rot'. The affected florets do not open normally, petals curve inwards and there may be break down of petal tissue with water soaking apparent. In acute deficiency, the leaves emerge from sheathing leaves in a soft, gelatinous condition, initially light-coloured but later turn brown. Some horizontal cracking may also appear.

Magnesium

Symptom of magnesium deficiency is yellowing between veins, which appears first in the older leaves. It also causes delay in flowering.

Copper

Soft spike, weak, soft stem and flaccid leaves characterize the disorder caused by copper deficiency.

Iron

Iron deficiency is manifested first as yellowing (chlorosis) between the veins of new leaves and as the deficiency becomes more severe, the whole leaf becomes yellow

Manganese

In case of manganese deficiency, the leaf veins tend to remain green when the leaf is yellow. The deficiency is also expressed by yellowing of veins of new leaves.

Boron

Deficiency of boron causes horizontal cracking, starting from the margin of the leaf and extending towards the midrib. In some cultivars, there is less cracking but translucent areas develop between the veins. Boron deficiency causes reduction in size and thickening of young leaves and edges sometimes become thin, cracked and colourless. In older plants, the cracking occurs near the base of larger leaves.

e) Orchids

Nitrogen

When nitrogen is deficient, the plants are stunted and matured too early. Older leaves turn yellow and drop off.

Phosphorus

Phosphorus deficiency leads to stunting, but the leaves, instead of turning yellow, become dark green. Severe reduction in the growth is observed.

Potassium

Deficiency results in dwarfness with the edges of the leaves frequently scorched and dead. Reduction in growth and drying of foliage is often observed.

Calcium

Due to deficiency of the calcium, the new growths are stunted and distorted. The discolouration of leaves starts from the margin and extends the mid-rib. It develops due to loss of chlorophyll. The veinlets become tinged with purple coloured, while the rest of the tissues in the affected area generally change from greenish yellow to dark brown or black. Mature leaves turn completely black and dry off. In advanced stage, the tip of the plant and tips of upper leaves die back and the young pseudobulbs turn black.

Magnesium

The deficiency of magnesium causes chlorosis between veins of older leaves and plants do not thrive.

Sulphur

The deficiency affects the root system.

Iron

The deficiency of iron causes yellowing of younger leaves.

Manganese

Deficiency of manganese causes chlorosis between the veins.

f) Snapdragons

Nitrogen

Nitrogen deficiency causes stunted plant growth and an overall yellow-green colour of the foliage. Few side branches are produced and stems tend to be thin, hard and wiry.

Phosphorus

Phosphorus deficiency also caused stunted plant growth and the young leaves become very dark green in colour. In some cases, purpling, especially of the undersides of older leaves. Young leaves may have the tips recurved downward and inward toward the central stem.

Potassium

Potassium deficiency causes symptoms in snapdragons somewhat different from those in most plants. The initial response is an interveinal chlorosis of the young leaves similar to the classic iron deficiency symptom. This is followed later by the development of necrosis of tips and margins of the older leaves on the plant.

Calcium

Calcium deficiency in snapdragon first affects the developing root system, because calcium is an element essential for the formation of primary cell walls in meristematic areas.

Magnesium

The symptoms of magnesium deficiency in snapdragons are more typical of the classic pattern. Interveinal chlorosis and necrosis appear first on the older leaves of the plant, while the young, newly developing leaves remain relatively healthy in appearance. Tips of the older leaves have been reported to curl downward and tips of younger leaves to curl upward.

Sulphur

Sulphur deficiency causes pale, yellow-green upper leaves in which the main veins are lighter than the rest of the leaf blade. As the deficiency continues, the symptom pattern progresses downward. A fine interveinal chlorosis may be observed on the older leaves, while the young leaves are more uniformly yellow-green.

Iron

Iron chlorosis of snapdragon appears first as interveinal yellowing of the youngest leaves and progresses to nearly complete loss of all green colour from these leaves. Small axillary shoots that develop on the main stem may be almost completely white.

Boron

Lack of boron inhibits meristematic activity and a common symptom will be death or abortion of the terminal growing point (in snapdragons, the terminal flower spike) followed by the growth and development of bypassing shoots from axillary buds further down the stem.

g) Cyclamen

Nitrogen

Small, pale (chlorotic) foliage and, with extreme deficiency, weak growth.

Phosphorus

Dark green, stiff leaves rich in anthocyanins, especially in petioles and on lower leaf surfaces.

Potassium

Smaller leaves with necrotic dotted margins on older foliage; necrotic areas increase with deficiency severity, flower scapes are shorter than normal.

Calcium

Flexing down of leaf and flower stalks and development of macroscopic brown streaks on edges of young leaves; roots remain short; tubers internally glassy with vessels partially brown.

Boron

Young leaves thickened and irregularly curled, flower buds remain small and dry on short peduncles thickened near the base.

Iron

A pale green to greenish-yellow in new foliage.

6. Forest Trees

a) Teak

Nitrogen

During the initial stages, yellow patches appeared towards the margins of older leaves. Later, the entire lamina turned pale yellow. In the acute stage of deficiency, the entire seedling appears severely chlorotic and these leaves gradually dried prematurely.

Phosphorus

Symptoms appears first on older leaves as purple bronze patches which later changed into yellow chlorotic patches. New leaves are pale in colour. Gradually, the bronze patches extended towards the entire leaf resulting in premature defoliation.

Potassium

The symptoms are first manifested on lower leaves. The leaves have chlorotic tips at the beginning. These chlorotic areas gradually spread through the margin upwards. Necrosis progressed from the lower part of chlorotic leaves.

Magnesium

The older leaves produced small necrotic areas during the initial stages of deficiency. Characteristic chlorotic pattern between the veins is also observed. However, the midrib and the veins remained green. At the acute stages, the chlorotic areas developed into necrotic regions.

Sulphur

Sulphur deficiency symptoms first appeared as discolouration of the terminal leaves from dark green to pale green. The symptoms gradually advanced from the margin inwards. At the moderate stages of deficiency, only the regions close to the midrib appeared green. Later, necrosis set in and at the acute stage, the entire leaf developed chlorotic. The affected leaves are yellowish in colour.

Zinc

The veins and veinlets of recently matured leaves became more prominent. Gradually, white chlorotic spots appeared on the midrib and veins of the leaves. Later, whole leaf became chlorotic and necrosis started from the tip of the leaf. Development of bronze coloured spots is also observed on lower leaves and later they merged together to form patches. At the acute stage, margins of older leaves became necrotic and premature leaf shedding occurred. In some cases, the newly formed leaves are extra ordinary big in size and are drooping.

Copper

During the early stages of deficiency, yellow patches appeared on the margins of older leaves. Later, the veins and veinlets of matured leaves became green and very clear (vein clearing) and the interveinal area remained yellow as in Fe deficient leaves. As the plant growth advanced, the entire leaf became chlorotic. Gradually, necrosis developed from tip of the leaves. In between, white chlorotic mottling appeared on younger leaves and the affected leaves lost their softness and they became crisp. Some of the leaves produced were oblong in shape. In the advanced stage of deficiency, the new leaves formed are wrinkled in shape and the growth of leaves are arrested.

Iron

Gradually, the entire leaf became yellow and the green colour is restricted only to the veins. This particular pattern of leaf with green veins and yellow interveinal area (interveinal chlorosis), expressed by the newly opened leaves is found to be a characteristic symptom of Fe deficiency. Later, most of the leaves of the plant became yellow and at the acute stage, the yellow leaves became necrotic and the whole leaves died off. Another characteristic symptom observed due to the deficiency of Fe is cupping of leaves. The margins of the leaves especially the upper and immediate below started to curl outwardly and attained the shape of a cup. Development of rusty patches on older leaves and the formation of multistems are also noticed in Fe deficient plants.

Manganese

Leaves became completely chlorotic. As the deficiency progresses, the whole leaves become chlorotic. But at later stage, lower leaves remained chlorotic and the upper leaves became green.

Boron

The new leaves developed are small in size and they are found clustering on the top. Both the upper and lower leaves are very brittle. All the leaves are yellow in colour and at the acute stage, leaves became necrotic leading to premature leaf fall.

Molybdenum

Molybdenum deficiency results in yellow patches on the tip and upper margin of the older leaves. Later, these patches gradually spread to the margins of the leaves and became necrotic. Marginal scorching is also due to the deficiency of Mo. In some plants, the leaves became bluish green in colour, the newly matured leaves expressed vein-clearing symptoms and the interveinal area remained green.

b) Eucalyptus spp.

Nitrogen

Deficiency symptoms first appear on older leaves. These become pale yellow in colour. There may be yellow spots on the leaves and chlorosis. In some cases, the veins become reddish. The colour depends on the degree of deficiency and on the species involved. There may also be premature leaf drop.

Phosphorus

Phosphorus deficiencies show up first in the older leaves which take on a purple colour, often with dry tips and margins.

Potassium

Deficiency symptoms appear in older leaves first. The main symptom is dead tissue on the leaf margins spreading to the leaf tip.

Calcium

Deficiency symptoms show up first on new leaves and include necrosis (dead patches) and dieback. Plants may be stunted and generally have poor root development.

Magnesium

Deficiency symptoms show up first on older leaves. The main symptom is chlorosis starting from the tip and edges of the leaves and progressing inwards. There may be a green "V" along the midrib. Some older leaves may be reddish in colour.

Sulphur

Deficiency symptoms include chlorosis with pale veins and stunted growth.

Zinc

Deficiencies appear on young leaves first. Symptoms include chlorosis and crinkly leaves with necrotic (dead) patches. There may be dieback of the tips.

Copper

Deficiency symptoms include wilting and dieback of the terminal shoots and failure of the plant to grow straight. The stem or trunk may be twisted and growing almost horizontal in some cases.

Iron

The main symptom of iron deficiency is chlorosis particularly between the veins (which remain green). New growth is pale yellow.

Manganese

Chlorosis is the main symptom of manganese deficiency. There is also likely to be cupping of the leaves and necrosis (dead patches on the leaves). Young leaves are the first to be affected.

Boron

Symptoms of deficiency vary depending on the species involved but the most common symptom is dieback of the young shoots. New leaves may have yellowish edges.

c) Pinus spp.

Nitrogen

Branching of trees and foliage which turns a uniform yellowish green, becoming yellow under severe conditions.

Phosphorus

Thin crown, grey green colour. Needles may be short, the tips yellow and sometimes fused. Within the stand, growth is very uneven and there will be a range of symptoms.

Potassium

Chlorosis occurs mainly in the lower lateral branches, the younger upper branches remaining unaffected, while the leading shoot remains apparently healthy.

Calcium

Tip dies back, side branches sprout giving a hedged, flattened appearance to the top of the tree. There may be abundant cone production and copious resin bleeding near terminal bud.

Magnesium

Needle tips appear golden yellow as if dipped in paint. In some seasons, yellowing occurs on stem needles.

Sulphur

Overall yellowing of trees with yellowing most pronounced at base of needles. In severe cases, lesions and resin bleeding on stems.

Zinc

Terminal needles are greatly reduced in length and clustered at end of shoot (rosetted).

Copper

Symptoms not specific but there may be dark blue-green foliage, distorted branches and bushiness. Prostrate growth in extreme cases.

Manganese

Paleness of the foliage, especially towards the top of the trees, and most recent growth extremely short, somewhat curved.

Boron

Terminal shoots and leader die back with bright orange-red colouration. Buds fail to flush and the main stem forks and becomes deformed. Pith is black or dark brown. Young pith is flecked with brown; older pith is pitted with dark areas. Resin bleeding from bud.

7. Medicinal Plants

a) Geranium

Nitrogen

Nitrogen deficiency slows down the growth and development of plants. The plants appear stunted with light green lower leaves, while the upper leaves remain green. With prolonged nitrogen deficiency, yellowing or reddish pigmentation of older or lower leaves occurs depending on cultivars. This is followed by browning and death of leaf margins.

Phosphorus

In early phosphorus deficiency, plants appear darker green than normal with reduced growth. As the deficiency progresses, the older, lower leaves develop irregular spots of red or purple pigmentation along the leaf margins, which eventually turn brown to dark brown and die. In most cases, lack of phosphorus delays flowering in plants.

Potassium

The leaves of potassium deficient plants are small and darker green than normal plants. These symptoms are followed by a sudden development of irregular necrotic tissues along the margins and tips of lower, older leaves. The browning eventually covers the whole leaf, giving the leaf a scored appearance. Geranium leaves can be potassium deficient for several weeks before necrotic symptoms develop.

Calcium

Lack of calcium generally affects the growing points. The symptoms first appear on the root tips as a black necrosis. When deficiency is severe, cell death occurs on the entire primary and secondary roots.

Magnesium

Normally, magnesium deficiency symptoms appear in lower and older leaves with greenish-yellow to yellowish -green chlorosis developing along the leaf margins and tips, which progresses inward between the leaf veins. As the deficiency persists, necrosis develops between the veins and the leaves curl downward.

Sulphur

Initially, uniform light greenish-yellow chlorosis develops on young and/or mature leaves, but rarely on lower, older leaves. As the symptoms advance, uniform chlorosis spreads to the rest of the leaves.

Zinc

Zinc deficiency severely affects plant growth. Young and recently matured leaves develop marginal cupping, veinal chlorosis and necrosis. Some plants develop purple/pinkish pigmentation. Shoot and root growth is reduced.

Copper

Initially, young and maturing leaves appear stunted and misshapen with pointed margins. Overall, the plants appear "Bonsai-like". This is followed by impaired flower development that includes reduced size, premature abscission, or abortion. Sudden death of tissue with symptoms similar to localized tissue dehydration develops on recently matured leaves as a result of poor xylem tissue development.

Iron

Iron deficiency symptoms develop on young leaves and shoots. Generally, young leaves develop interveinal chlorosis from the base, but in some species, it develops from the tips. Over time, the chlorosis intensifies and the pattern becomes less interveinal. Even the stems appear chlorotic.

Manganese

Young and recently matured leaves develop chlorosis followed by a necrotic stippling on recently matured leaves. Drastic reduction of shoot and root growth is common. Flowering is strongly inhibited.

Boron

Boron deficiency develops on the shoot and root meristems and also on young leaves. The symptoms first develop on the roots three to four days earlier than the shoots. Overall, the roots are thick and short. Primary roots develop thick, swollen root tips with numerous short secondary roots developing close to the tips, resembling 'witches broom' growth. Meanwhile, foliage becomes darker and glossy. Young and recently matured leaves become thick, leathery, and brittle with severe distortions. Loss of apical dominance is a common symptom in the roots and shoots.

b) Marijuana

Nitrogen

The lower leaves become yellow between the veins until the entire leaf is eventually yellow. It wilts and dies. The symptoms spread upwards affecting all sun leaves and then the smaller and younger leaves as well. Plant growth is inhibited. The plants remain short and stunted if nitrogen deficiency is severe in the growth phase. During flowering, early deficiency can lead to a significant decrease in yield.

Phosphorus

In young plants, phosphorus deficiency slows growth. The plants are underdeveloped with a poor root zone and small bluish-green leaves. In flowering plants, the leaves turn reddish-purple. This generally starts at the leaf tips and spreads towards the leaf stem. Stems may also redden. In acute cases, leaves turn necrotic and fall off.

Potassium

Potassium deficiency sometimes resembles deficiency of nitrogen in the first stages of leaf deterioration. But the difference is that the leaves also develop rust-coloured spots. Eventually, the affected leaves wilt and drop. In severe cases, the stems become brittle.

Calcium

Hollow stems found on harvested plants is a sign of inadequate calcium.

Magnesium

The lower to middle leaves become yellow between the veins until the entire leaves are eventually yellow. Rusty spots appear dotted throughout the leaves. The leaf edges appear necrotic and curl up. Some leaves develop a whitish tinge. Symptoms spread throughout the plant by chronic deficiency.

Sulphur

Sulphur deficiencies are usually displayed by yellow new growth.

Zinc

Plants that have white areas on leaf tips or intermittently between leaf veins.

Iron

Iron deficiencies turn leaves on shoots pale while veins remain dark green.

Boron

New growing shoots appear burnt, brown, or grey.

Molybdenum

Plants with a molybdenum deficiency express yellowing leaves in the middle of the plant.

❑❑❑

Part IV :
REMEDIAL MEASURES

Chapter 22

Application of Nutrients

A. Methods of Fertilizer Application

The method of application of fertilizers is an essential component of good agricultural practices. The amount and timing of nutrient uptake depends on various factors, such as crop variety, planting date, crop rotation, soil and weather conditions. For good agricultural practices, one should choose the timing and the quantity in such a way that as much as possible of the nutrients is used by the plants. For optimum crop use efficiency and minimum potential for environmental pollution, one must apply the nutrients as near to the time the crop needs them as is practical. This is of particular importance for mobile nutrients such as nitrogen, which can easily be leached out of the soil profile, if they are not taken up by the plant roots.

All primary and secondary nutrients should be incorporated immediately after application in regions where intense rainfall is expected, to avoid losses due to run-off and erosion. Different soils react differently with fertilizer application. Similarly, the N, P, K requirements of different crops are different and even for a single a crop the nutrient requirement are not the same at different stages of growth. Nutrients to be used by plant must be placed in such a manner that they can be dissolved by the moisture in the soil. The rates and distance that plant food element can move within the soil depend on the chemical nature of the material that furnishes the nutrients and character of soil. Different methods of fertilizer application are as follows viz.,

A. Application of solid fertilizer

B. Application of liquid fertilizer

C. Application of gaseous fertilizer

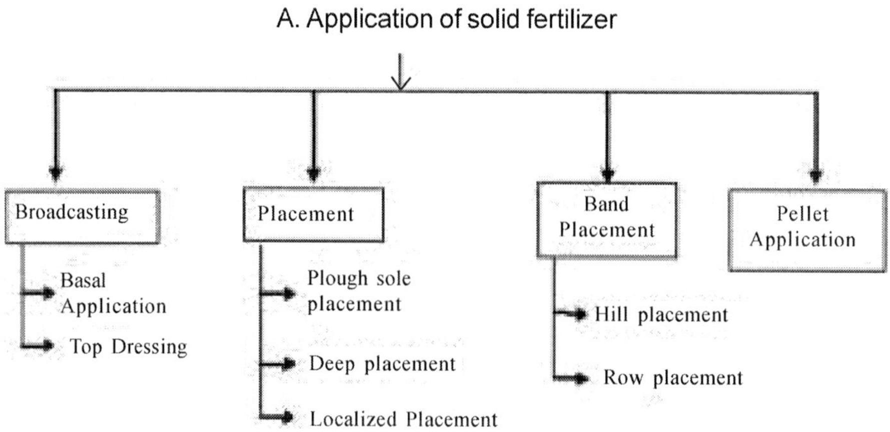

Time of application

The time of application depends upon various factors viz., the nature of fertilizer, soil type and the difference in nutrient requirement and nature of field crops. The time of application includes i) Basal application and ii) Top dressing.

i) Basal application

a) **Before preparatory tillage:**

Bulky organic manures, green manures, soil amendments and soil conditioners are applied before preparatory tillage for thorough mixing with the soil.

b) **Basal dressing:**

Application of manures and fertilizers before last ploughing/ puddling or before sowing or planting.

c) **At sowing or planting:**

Concentrated organic manures, readily soluble and higher mobile fertilizers, slow release fertilizers, starter dose of N fertilizer to legume crops and fertilizer for specific nutrient deficient soil are applied during this time.

ii) Top dressing

It is the application of manures and fertilizers to the established crop within crop duration. Top dressing may be done to the soil or to the foliage. Split application of nitrogen and potassium is done throughout the cropping period to increase the fertilizer use efficiency.

A. Application of fertilizers in solid forms

1. Broadcasting

The fertilizer is spread by hand with the main objective of providing uniform and evenly distribution of whole of fertilizer over the entire area of the field. This method is known as broadcasting. The fertilizers are generally applied in soil with last preparatory tillage just before planting. There are two types of broadcasting method of fertilizer application depending on the time of application viz., (i) Broadcasting at planting and (ii) Top dressing.

i) **Broadcasting at planting:**

The fertilizers are broadcasted to soil just before planting and fertilizers are incorporated to the soil by ploughing followed by planting. All the P fertilizers, 1/3 or 1/4 of nitrogenous and potassic fertilizers are applied as broadcasting.

ii) **Top dressing:**

Applying fertilizers in the standing crop is known as top dressing. By this method, the nutrients mainly nitrogen is readily available for growing plants. In addition to N fertilizers, K fertilizers are also being applied as top dressing to reduce the losses and to increase the use efficiency.

2. Placement of fertilizers

It is a method of applying fertilizer in the soil before sowing or after sowing the crops. It has been found that placement of phosphatic fertilizer has increased its efficiency than broadcasting. The best method of placement of fertilizers is determined by several factors such as nature of the crop to be grown, soil type, soil fertility status, fertility type and farmers preference. Placement method includes (a) Plough sole placement (b) Deep placement (c) Sub-soil placement.

a) **Plough sole placement:**

The fertilizers are placed in the plough sole after opening the furrow by plough and the furrows are covered immediately. This method is generally recommended to dry soil where there is moisture only in the plough - sole layer and in problem soils where a problem of fixation.

b) **Deep placement:**

The ammoniacal nitrogenous fertilizers (eg. Ammonium sulphate) or ammonia forming nitrogenous fertilizers (e.g. urea) is placed deep in reduced layer to check denitrification. The N remains in NH_4^+ form and is available to the crop during active vegetative period.

c) **Sub-soil placement:**

This method is adopted mainly in humid and semi-humid regions where soils are mostly acidic in reaction. The phosphatic and potassic fertilizers are placed by this method to reduce the fixation of nutrients.

3. Localized placement

It is a method of placing fertilizers into the soil close to seed or plant. The root of young plants can get nutrients from the fertilizers. This method reduces the fixation of P and K. Localized placement or spot application includes different methods like

a) drill placement,

b) band placement,

c) side dressing,

d) pellet application,

a) **Drill placement:**

In this method, the fertilizer is applied at the time of sowing by means of a seed-cum-fertilizer drill. This places fertilizer and the seed in the same row but at different depths. Although this method has been found suitable for the application of phosphatic and potassic fertilizers in cereal crops, but sometimes germination of seeds and young plants may get damaged due to higher concentration of soluble salts.

b) **Band placement:**

In this method, the fertilizers are placed in bands on oneside or both sides of the row at about 5 cm away from the seed or plant in any direction. There are two types of band placement.

i) **Hill placement:**

The fertilizers are placed close to the plants in bands on one side or both sides of wide spread plant such as cotton, castor, cucurbitaceous crop and fruit crops. Nitrogenous and phosphatic fertilizers are applied by this method.

ii) **Row placement:**

The fertilizers are placed on one or both sides of the row in continuous bands by hand or a seed drill. This method is practiced in row crops (e.g. sugarcane, potato, tobacco, cotton and vegetables) with wide spaces between rows. Higher rates of fertilizers are possible with row placement than hill placement. For applying small amount of fertilizers, hill placement is usually most effective.

c) **Side dressing:**

In this method, fertilizers are applied along the side of a row or around the plant and mixed them with soil. There are two types viz., i) Spot placement and ii) Basin placement.

i) **Spot placement:**

The fertilizers are applied in spaces in between the plants and mixed them with soils. This method is employed for applying fertilizers in vegetable crops.

ii) **Basin placement:**

The fertilizers are placed in shallow basin already made around the base of the fruit trees. The applied fertilizers are mixed with the soil by spade.

d) **Pellet application:**

Small pellets of convenient size are made after mixing the nitrogenous fertilizers, particularly urea and applied one to two inches deep between the rows of paddy crop. The pellets are deposited in soft mud of paddy field. Application of fertilizers by this method reduces the loss of N through leaching or by run-off of water. The main advantages of placement of fertilizers are:

- When the fertilizer is placed, there is minimum contact between the soil and the fertilizer, and thus fixation of nutrients is greatly reduced.
- The weeds all over the field can not make use of the fertilizers.

- Residual response of fertilizers is usually higher.
- Utilization of fertilizers by the plants is higher.
- Loss of nitrogen by leaching is reduced.
- Being immobile, phosphates are better utilized when placed.

Significance of solid fertilizer application

- Since phosphates generally move only short distance from their points of placement, so for better availability, phosphates must be placed in the zone of root development for being utilized by plant.
- The placement of water soluble phosphorus in bends tends to reduce contact with the soil and result in lesser fixation than in broadcast application.
- In contrast to phosphorus, the nitrates salts are mobile and move vertically or horizontally within the soil as the water moves. In fine textured soil, the movement of N is restricted.
- Potassium salts are less mobile than the nitrates but are more mobile than P.
- In general, N and K carrying fertilizers are more readily soluble than the P material. Therefore, they cannot be safely concentrated in as large amount near the seed or the roots of plants because of the danger of salt damage.
- Likewise, a reduction of soil moisture increases the concentration of the soil solution. Therefore, relatively large amount of fertilizers placed too near the seed or seedling roots are likely to cause injury during the dry periods, practically when such periods occur soon after the application of fertilizer.
- It is desirable to divide the total requirement of fertilizer nitrogen into several parts to be applied from time to time during growing season. Fertilizers rich in potassium should be placed in a band to the side and below the seed or transplant.

B. Liquid Fertilizer Application

Water-soluble fertilizer can be applied with spray cans or applied with sprinklers or furrow irrigation. Small amounts of liquid fertilizer applied to young vegetable plants at the time of transplanting is called a starter solution. Some plants require micronutrients such as zinc and iron that they cannot get through the soil, but their leaves can absorb them. Water-soluble fertilizer is usually sprayed on leaves when they first show signs of nutrient deficiency and in periods of drought when the soil is to dry to absorb the nutrients. Following are the common methods of applying liquid fertilizers

B. Application of liquid fertilizers

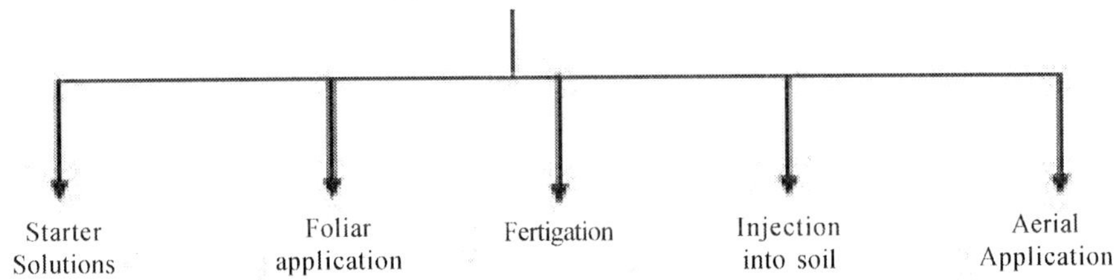

a) Starter solutions

It refers to the application of solution of N, P_2O_5 and K_2O in the ratio of 1:2:1 and 1:1:2 to young plants at the time of transplanting, particularly for vegetables.

Starter solution helps in rapid establishment and quick growth of seedlings.

The advantages are viz.,

- The nutrients reach the plant roots immediately, and
- The solution is sufficiently diluted so that it does not inhibit growth
- Starter solution stimulates growth of young plants

b) Foliar application

- It refers to the spraying of fertilizer solutions containing one or more nutrients on the foliage of growing plants.
- Several nutrient elements are readily absorbed by leaves when they are dissolved in water and sprayed on them.
- The concentration of the spray solution has to be controlled, otherwise serious damage may result due to scorching of the leaves.

- Foliar application is effective for the application of minor nutrients like iron, copper, boron, zinc and manganese. Sometimes insecticides are also applied along with fertilizers.

c) Fertigation

This is the most common method for the application of liquid fertilizer solutions and dissolved solid water soluble fertilizers. The concentrated fertilizer solution is injected into the irrigation water, mixes with it and enters the soil during irrigation. This is the most efficient method for the fertilization of irrigated crops, especially for those irrigated with drippers, micro- jets and micro- sprinklers. A range of injection accessories is available for fertigation.

d) Injection into soil

Liquid fertilizers for injection into the soil may be of either pressure or non-pressure types. Non-pressure solutions may be applied either on the surface or in furrows without appreciable loss of plant nutrient under most conditions. Anhydrous ammonia must be placed in narrow furrows at a depth of 12-15 cm and covered immediately to prevent loss of ammonia.

e) Aerial application

In areas where ground application, is not practicable, the fertilizer solutions are applied by aircraft particularly in hilly areas, in forest lands, in grass lands or in sugarcane fields etc.

C. Application of fertilizers in gaseous form

Anhydrous ammonia is typically stored in a liquid form, most commonly under pressure, and to a lesser degree, under refrigeration. Anhydrous liquefied ammonia is applied by subsurface injection. The ammonia quickly vaporizes, but is captured by several components in the soil including water, clay and other minerals.

B. Foliar Nutrition

Foliar nutrition is recognized as an important method of fertilizer application, since foliar nutrients usually penetrate the leaf cuticle or stomata and enter the cells facilitating easy and rapid utilization of nutrients. It is advantageous over soil application in specific occasions in certain crops. For instance, in calcareous soil, deficiencies of iron, zinc and boron is inevitable due to precipitation of zinc and iron as insoluble carbonates and calcium-boron antagonism. In soils with high indigenous phosphorus supply, zinc and iron deficiency are common. Some of the · other nutrient antagonism commonly encountered are copper-molybdenum, copper, iron, potassium-magnesium etc. On such occasions, soil application of zinc sulphate or borax may not be useful. So foliar application is the best option.

The ability of plant leaves to absorb water and nutrients was recognized approximately three centuries ago. The application of nutrient solutions to the foliage of plants as an alternative means to fertilize crops such as grapevine agriculture was noted in the early 19th century (Gris, 1843). Following this, research efforts were made to try and characterize the chemical and physical nature of the plant foliar cuticle, the cellular physiology and structure of plant leaves as well as focusing on potential mechanisms of penetration by foliar sprays.

Aerial plant surfaces are generally covered by a hydrophobic cuticle and very often possess modified epidermal cells such as trichomes or stomata. The outer surface of the cuticle is covered by waxes that may confer a hydrophobic character to the plant's surface. The degree of hydrophobicity and polarity of the plant surface is determined by the species, chemistry and topography which are also influenced by the epidermal cell structure at a microscopic level.

A transverse section of a typical angiosperm leaf consists of a cuticle that covers the upper and lower epidermal cell layers enclosing the mesophyll. Leaves differ in their structure between species but generally consist of palisade parenchyma near the upper epidermis and spongy parenchyma between the palisade layer and the lower epidermis. There are large intercellular spaces among the mesophyll cells, especially in the spongy parenchyma. The epidermis is a compact layer with sometimes two or more layers of cells and the principal features, related to nutrient and water transport, which characterize the epidermis are the cuticle and the stomata.

Cuticles

The cuticle covering aerial plant parts is an extra-cellular layer composed of a biopolymer matrix with waxes embedded into (intra-cuticular), or deposited onto (epi-cuticular waxes), the surface. On the inner side, a waxy substance called cutin is mixed with polysaccharide material from the epidermal cell wall, which is chiefly composed of cellulose, hemicellulose and pectin in a ratio similar to that found in plant cell walls. Therefore, the cuticle itself can be considered as a 'cutinized' cell wall, which emphasizes the compositional and heterogeneous nature of this layer and its physiologically important interaction with the cell wall underneath determine the degree of polarity and hydrophobicity. Differences in surface polarity and hydrophobicity in relation to variable growing conditions, plant species, varieties and organs can be expected and these will have an influence on the effectiveness of foliar sprays.

Cuticular permeability

The cuticle consists of three layers, namely the epicuticular wax layer (EW), the cuticle proper (CP) and the cuticular layer (CL). The EW layer is the outermost and most hydrophobic component of the cuticle. The CP that lies beneath the epicuticular waxes contains mainly cutin and/or cutan and is by definition free of polysaccharides. The CL is located under the CP and consists of cutin/cutan, pectin and hemicelluloses that increase the polarity of this layer due to the presence of hydroxyl and carboxylic functional groups. The middle lamellae(ML) and pectin layer is situated beneath the CL.

A gradual increase in negative charge from the epicuticular wax to the pectin layer creates an electrochemical gradient that may increase the movement of cations and water molecules. The intra-cuticular waxes limit the exchange of water and solutes between the plant and the surrounding environment, while the epicuticular waxes influence the wettability, light reflectance and surface properties of the plant organ.

Point of deliquescence

The processes of hydration and dissolution of a salt are determined by its point of deliquescence (POD) which is a physical property associated with a compound at a given temperature (Table 1). Deliquescent salts are hygroscopic substances (i.e. capable of trapping water from the surrounding environment) and will dissolve once a critical relative humidity threshold has been attained. The point of deliquescence is defined as the relative humidity value at which the salt becomes a solute. Thereby, the lower the point of deliquescence of a salt is, the sooner it will dissolve upon exposure to ambient relative humidity. When ambient relative humidity is higher than the point of deliquescence of the foliar applied compound, the substance will dissolve and will be available for absorption by the leaf.

Table 1. Point of deliquescence (POD) of various salts

Compound	POD (%)
$CaCl_2 \cdot 6H_2O$	33
$Ca(NO_3)_2 \cdot 4H_2O$	56
$MgCl_2 \cdot 6H_2O$	33
$Mg(NO_3)_2 \cdot 6H_2O$	56
$MgSO_4$	90
$Zn(NO_3)_2 \cdot 6H_2O$	42
$ZnSO_4$	90
KCl	86
KNO_3	95
K_2SO_4	98
$K_2CO_3 \cdot 2H_2O$	44
K_2HPO_4	92
KH_2PO_4	95
NH_4NO_3	63
Ca-propionate $\cdot H_2O$	95
Ca-lactate $\cdot 5H_2O$	97
Ca-acetate	100
$FeCl_3 \cdot 6H_2O$	44
$Fe(NO_3)_3 \cdot 9H_2O$	54
$Mn(NO_3)_2 \cdot 4H_2O$	42
$MnCl_2 \cdot 4H_2O$	60

(Schonherr, 2002)

Effect of the environmental factors on foliar absorption of nutrients

- Light
- Temperature
- Humidity

Light

Chemical ion uptake by leaves can be directly influenced by light as a result of physical and chemical changes in the cuticle and also from the direct involvement of light on the energy and metabolite availability on the uptake and assimilation of foliar-applied nutrients.

The amount and composition of synthesized waxes and their arrangement on the surface is directly influenced by light including photosynthetically active radiation, as well as by UV-B radiation. The thickness of the cuticle and the amount of cuticular waxes in various plant species was found to be higher on those grown under high rather than low light intensities and the development of secondary wax structures is increased by higher light intensities. Foliar penetration of chelated Fe preferentially occurs during the night and therefore, foliar applications during the late afternoon are recommended.

Temperature

Temperature can influence foliar absorption through its effect on the rate of drying of the spray application; the nutrient solution physico-chemistry; as well as its impact on leaf cuticles; and on plant metabolism, ion uptake and assimilation. The most immediate effect of high temperature is increasing the drying rate of the spray droplets which will directly reduces absorption. However, increased absorption by leaves over prolonged high temperatures has been recorded in several species. This may be the result of the high temperature during leaf development altering the amount and composition of synthesized waxes and their arrangement on the surface.

Humidity

As with light and temperature, humidity can affect multiple processes that ultimately influence the rate of foliar absorption of aerial applied fertilizers. The key processes affected by humidity are

- The reaction of the foliar application during aerial transport and once deposited on the plant surface
- The effect of humidity of leaf cuticle structure and stomatal function and
- The effect of humidity on leaf metabolism and transport processes. In the short term, the effect of humidity on nutrient absorption by leaves is primarily related to the rate of drying of droplets during aerial transit to the plant surface and their persistence once deposited on the plant surface. High relative humidity favors uptake as it delays rapid solution drying which can lead to

crystallization on the leaf surfaces. Additionally, high air humidity causes the swelling of the cuticular membrane which favours the absorption of hydrophilic compounds. Therefore, humidity at the time of foliar application affects the velocity of penetration by two independent mechanisms:

- Swelling of the cuticle; and
- Over longer time frames (days to weeks), the amount and composition of synthesized waxes and their arrangement on the surface is influenced by relative humidity which consequently may alter the rate at which foliar applications can penetrate the plant surface. Post absorption humidity can have general effects on plant responses to foliar nutrients by affecting xylem and phloem transport processes.

Time of application

For efficient absorption of foliar fertilization, leaf stomata's should be open and temperature should not be too high to cause burning of plant foliage. In the afternoon when air temperature is low (after 2–3 P. M.), is the best time for foliar fertilization. Another factor that may affect foliar fertilization is a windy day, which can drift the spray solution. Hence, windy days should be avoided for foliar spray. There should be at least 3 to 4 hours for the applied nutrient to be absorbed by plant foliage. Hence, there should not be rain for at least 3 to 4 hours after application of the nutrient solution. When applying a nutrient solution as a spray, some sticking material should be added to the solution to stick the spray drops to plant foliage.

Foliar application should be made when the plant is not in water stress, either too wet or too dry. Nutrients are best applied when the plant is cool and filled with water (turgid)). Applications that are misapplied or too late in the season may not be effective. The most critical times to apply are when the crop is under a given nutrient stress. Stress periods occur during periods of active growth activity. This is likely when the plant is changing from a vegetative to a reproductive stage. The amount of fertilizer and water volume used in foliar nutrition is given in Table 2.

Table 2. Amount of fertilizers and water volume used in foliar spray of macro and micronutrients

Nutrient element	Chemical formula	Kg per 500 liter of water
N	$CO(NH_2)_2$	3-5
N	$(NH_4)_2SO_4$; NH_4NO_3; $(NH_4)_2HPO_4$; NH_4Cl; $NH_4H_2PO_4$	2-3
P	H_3PO_4; others see N above	2-3
K	KCl; KNO_3; K_2SO_4	1.5-2.5
Ca	$CaCl_2$; $Ca(NO_3)_2$	1.5-2.5
Mg	$MgSO_4$; $Mg(NO_3)_2$	3-10
Fe	$FeSO_4$	3-6
Mn	$MnSO_4$	1-2
Zn	$ZnSO_4$	1.5-2.5
Cu	$CuSO_4$	0.5-1
B	Sodium borate	0.25-0.5
Mo	Sodium molybdate	0.1-0.15

(Fageria and Barbosa Filho, 2006).

The uptake speed of single nutrients is considerably vary. Cations penetrate membranes faster than anions. Plant leaves can absorb all main nutrients and microelements with the speed listed below (Table 3).

Table 3. Rates of Nutrients Absorption into Plant Tissue

Nutrient	Time for 50% Absorption
Nitrogen (as urea)	1/2 - 2 hours
Phosphorus	5 - 10 days
Potassium	10 - 24 hours
Calcium	1 - 2 days
Magnesium	2 - 5 hours
Sulphur	8 days
Zinc	1 - 2 days
Manganese	1 - 2 days
Iron	10 - 20 days
Molybdenum	10 - 20 days

The uptake speed of ions in a plant affects the efficiency of foliar nutrition. The observed nutrients by a plant are distinguished by varying mobility. Regarding immobile nutrients, it is therefore necessary to repeat sprays at the time when the plant needs it the most. A range of nutrients which are slowly received and relatively immobile in plants can be extremely efficient in a form of foliar nutrition and can be a precaution to ensure or eventually eliminate their deficiencies. The foliar nutrition can prevent soil nutrient overloading and lower the risk of environmental threats. During the foliar nutrition, nutrient efficiency can reach up to 85%, whereas application of fertilizers through soil only 30 - 60% of efficiency depending on nutrient type. The efficiency of foliar nutrition is dependent on concentration and the dose of solution which cannot be too high to prevent leaves from burns. Regarding macrobiogenic elements, average 2% solutions are recommended, regarding microbiogenic elements, an optimal concentration is from 0,1 to 0,5 %.

II. Types of Fertilizers Materials/Additives/Water

Fertilizer Materials

Not all fertilizers are suitable for use as a foliar spray. The primary objective of a foliar application is to allow for maximum absorption of nutrients into the plant tissue. Therefore, foliar fertilizer formulations should meet certain standards in order to minimize foliage damage. Qualifications for fertilizer materials follow:

i) Low salt index: Damage to plant cells from high salt concentrations can be considerable, especially from nitrates (NO_3^-) and chlorides (Cl^-).

ii) High solubility: Needed to reduce the volume of solution needed for application.

iii) High purity: Needed to eliminate interference with spraying, solution compatibility, or unexpected adverse effects on foliage.

Nitrogen

Urea is the most suitable nitrogen source for foliar applications, due to its low salt index and high solubility in comparison to other nitrogen sources. Urea has been shown to stimulate absorption of other nutrients by increasing the permeability of leaf tissue. However, the urea utilized in foliar sprays should be low in biuret content.

Secondary and Micronutrients

Foliar application of these nutrients (secondary: calcium, magnesium and sulphur; micronutrients: zinc, manganese, iron, copper, boron and molybdenum) can be highly effective, but because of difficulties associated with leaf tissue absorption and translocation of some of these nutrients (notably calcium magnesium, iron, boron and molybdenum), choosing the correct fertilizer sources for these nutrients becomes very critical. Chelate sources, while valuable for soil application, have been shown to be generally unfavorable for foliar application because most chelating agents have a molecular size too large to be effectively absorbed by leaf tissue. However, organic chelating agents (including citric and malic acids, amino acids, phenolic acids, glucoheptonate and glucosylgycine) have been shown to enhance secondary and micronutrient foliar absorption.

Micronutrients should be applied according to need as determined by a tissue test, and should always be applied along with nitrogen in the solution. Combinations of certain nutrients may pose solubility problems, especially where nutrient solutions are combined with fungicides and pesticides. Generally, unless compatibility with fungicides and pesticides is known, nutrient sprays should be applied separately. Urea is compatible with most pesticides, exceptions being lime, sulphur, Karathane, Glyodin, Phygon, Phosdrin and Sevin. Magnesium sulphate is not compatible with arsenicals or copper sprays. Manganese solutions should not be mixed with phosphate, iron sulphate. Lead arsenate should not be mixed with any nutrient solution.

Additives

Agents added to the foliar fertilizer solution which buffer the pH of the solution (preferably between pH of 5.0 and 6.0) and provide for quick and uniform coverage of the spray droplets are highly recommended. Foliage burn is caused by a high concentration of fertilizer salts (i.e., nitrate and chloride) rather than low pH in the fertilizer solution. Low pH foliar solutions have been shown to increase the absorption rate of fertilizers. Leaf and stem tissues can inhibit initial nutrient absorption by means of waxy substances in the cuticle (outer layer of plant cells), pubescence (fine hairy growths) and drooping leaf angles. To achieve maximum nutrient absorption via foliar applications, a fine mist application with spreading and wetting agents is desired (Figure 1). These agents provide quick wetting of plant tissue and more uniform coverage with increased spray retention by reducing the surface tension of the spray droplets. Effective foliar applications depend on maximum absorption of soluble nutrients, avoiding losses due to evaporation and/or runoff as much as possible.

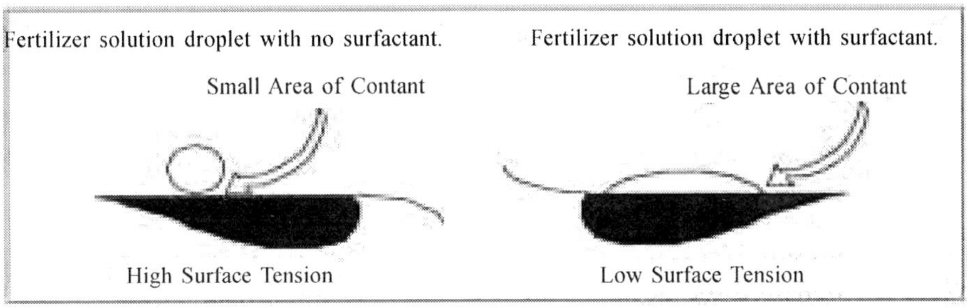

Figure 1. Illustrates the beneficial effect surfactants can have in enhancing foliar fertilizer adsorption.

There are currently available a large number of adjuvants (additives) encompassing a wide range of properties and uses. Adjuvants (broadly defined as any substance added to the spray tank, separate from the pesticide formulation, that improves a pesticide's performance) may be grouped into two categories: i) activators and ii) special-purpose adjuvants.

i) The activator category consists of the following wetter-spreaders, stickers, (e.g., builders and extenders), emulsifiers (dispersants and suspending agents), emulsifiable oils (activators), and plant penetrants (translocators).

ii) Special-purpose adjuvants would include foliar nutrients, compatibility agents, drift retardants, foam retardants, buffers, inverting agents, soil penetrants, stabilization agents (UV filters), feeding stimulants, washing agents, sinking agents, and protectant binders.

Mechanism of absorption

The penetration of the spray solution or nutrient solution occurs through the trichomes and stomata, and then diffuses through cuticle, the layer of polymerized wax which occurs on outer surface of the epidermal cells of leaves. After penetration in the cuticle, further penetration take place through fine, thread like semi-microscopic structures called ectodesmata. This extends through the outer epidermal cell wall, from the inner surface of the cuticle to the plasma membrane. When the substance reaches plasma membrane of an epidermal cell, it will be absorbed by a mechanism similar to those which operate in root cells. The spraying should be done until the foliage attains complete wetting.

Foliar Spray Nutrition is Important in following cases

- When the absorption of plant nutrients is disturbed by weeds, poor aeration, low soil temperature, frequent rainfall etc.
- When fertigation is not practiced due to rainy climate
- When the root system suffers mechanical injuries or pest damage.
- Uptake of nutrients through foliage is considerably faster than roots.
- Foliar spray is also the method of choice when prompt correction of nutrient deficiencies is required.
- When rapid recovery is required under adverse conditions such as drought, disease or insect attack, foliar sprays are much more than a cure to poor growing conditions.

Suggestion for effective foliar nutrition

- Have the deficiency symptoms diagnosed by local specialist / adviser is checked before spraying.
- Do not exceed the recommended solution concentration.
- Adjust concentration according to type of spray used (Particle size of the spray droplets)
- Do not spray during hot, windy periods and against the breeze.
- Wherever necessary, neutralize spray solution with lime.
- Use only clear filtered solution
- Repeat spray as and when recommended,
- Avoid spray during misting to prevent washing away of the spray.
- Double check the calculations used for preparing the final solution.

- Do not expect a foliar spray to leave residual effect tor next crop.
- Do not store spray solution, particularly of iron. Use freshly made solutions.
- Use wetting agents to improve effectiveness of spray @ one ml/ litre of spray fluid.
- Before buying a product, examine clearly its nutrient makeup and composition and whether its meets the growing demand.

Precautions for foliar fertilization

Hardness or amount of salts contained in the water used to make the solutions affects the degree of concentration of which fertilizer may be applied with safety. The 'harder' the water, the greater the concentration of the fertilizer chemicals can be used. Hydrated lime can be used in soft water to enable the absorption more concentrated solutions.

Merits of foliar application

- Application rates are much lower than the soil application.
- Uniform distribution is easily obtained
- Response to the applied nutrient is almost, immediate, therefore deficiencies can be corrected during the growing season.
- Suspected deficiencies can be more easily diagnosed with spray trials.

Demerits of foliar fertilization

- Nutrient demand often is high, when plants are small and leaf surface is insufficient to foliar application.
- Leaf burn may result if salt concentration is in excessive, it may be too late to correct the deficiency and still obtain maximum yield.
- There is little residual effect from foliar sprays.
- Extra costs may be required when more than one spray may be needed.
- Boron cannot be applied with $CaCl_2$

- $CaCl_2$ can be mixed with pesticide sprays.
- $CaC1_2$ is more effective when used in low concentrations.
- When combined with pesticide, the solution be acidified to lower the pH to about 6, High pH may be one of the reasons for the occasional leaf injury caused by concentrated sprays of $CaCl_2$.
- Copper deficiency can also be readily corrected with Bordeaux mixture.
- Effectiveness of foliar sprayed iron can be increased by addition of 0.1 citric acid (1% $FeSO_4$ + 0.1% citric acid).
- Epsom salt sprays ($MgSO_4$) on compatible with most pesticides
- Application of manganese sulphate to the soil to correct Mn deficiency is not effective on alkaline or heavy limed soils.
- Urea should not be mixed with cyprex. Dodine, Karathane, Dikar etc,
- Higher rates of Zn may causes severe injury to shoots, buds, leaves or fruits.
- Addition of Mn to zinc foliar spray provides better response.
- Addition of urea improves Zn absorption.
- Foliar application of B, Cu, Mg, Mn, and Zn for controlling deficiencies of these elements in have advantages over soil application. Those advantages are high effectiveness, rapid plant responses, convenience elimination and reduction of toxic symptoms. The disadvantages of foliar nutrition are that the effects or sprays are temporary and are not transmitted in to the next year. Thus, the annual sprays are necessary.
- Storage of $FeSO_4$ solution should be avoided otherwise ferrous Fe will be oxidised to ferric and precipitated.
- Zinc sulphate can be mixed with phasolone or monocrotophos and $ZnSO_4$ should not be mixed with phosphatic fertilizer
- Generally 400 litre /ha at spray fluid is recommended for hand operated sprays.
- Ammonium sulphate is unsuitable for foliar application. Because, the solution is acidic and if the solution is neutralized by lime water, ammonia volatilises causing N loss.

- If the concentration of fertilizer solution is high, it may cause salt burn. The concentration of N fertilizer, such as urea, usually does not exceed 2%. The concentration of H_3PO_4 should be very low i.e. 0.4 % or less. Otherwise, it will injure the plants.
- Sulphate salts are recommended for foliar spray at 0.5-1 % and slightly vary depending upon the crop and local condition. Foliar spray of single nutrients can be done depending upon the deficiency. Sometimes, same plant may exhibit deficiency of more than one element. In such cases, it does not mean that each compound can be added at 0.5 %- 1.0% concentration. In such situations, total concentration should be within this level, obviously, in such cases, the individual nutrient is applied at one third of the total concentration.

C. Fertigation

Fertigation is defined as the application of solid or liquid mineral fertilizes via pressurized irrigation systems, thus forming nutrient-containing irrigation - water. Although the practice of fertigation only started commercially in the mid - 20th century, there is evidence that the concept of irrigation with dissolved nutrients was well known in the past. Liquid ammonia was probably the first commercially produced liquid fertilizer, but in modern fertigation, the use of ammonia as the nitrogen source is negligible. The increased production of ammonia and other nitrogen solutions led to the development of direct application of these solutions into the soil by mechanical means i.e. through drills, knives and injection tubes etc.

Need for fertigation

- Maximizing water and fertilizer use efficiency
- Delivers water and nutrients simultaneously to the root zone
- Applies water and nutrients accurately and uniformly
- Increases the availability and the uptake of nutrients
- Reduces nutrient losses from leaching, volatilization and/or fixation
- Maximizes water and nutrient productivity
- Maximizing crop and soil productivity
- Minimizing production costs

Fertigation concepts

Equipment

The irrigation system consists of the irrigation tubing (main line, laterals and drip lines) and the control head (filters, pressure gauge, check valves and pressure regulator). The fertigation equipment is an integral part of the control head, consisting of various fertilizer application units (by-pass tank, venturi or pumps). Fertigation can also be performed by dissolving the fertilizers in a specific water volume (tank system).

The incorporation of the fertilizers into the irrigation water can be carried out in different ways. The selection of the method and equipment depends on the form of fertilizer to be used (solid or liquid), the available power source and size of the plot and cost (Table 4). Fertigation can be achieved by using single or multiple nutrient fertilizers in their solid or liquid form. An essential prerequisite for the use of solid fertilizer in fertigation is its complete and fast dissolution in the irrigation water.

Table 4. Fertilizer application units in fertigation

Fertigation method	venturi	Fertilizer tank	Fertilizer pump
Description	Sucks the stock nutrient solution by differential pressure and dilutes it into the irrigation water	Water passing through the by-pass tank dissolves the solid fertilizer	Energy-driven device sucks the stock nutrient solution and dilutes it into the irrigation water
Fertilizer form	Liquid *	Solid or liquid	Liquid*
Fertilizer injection	Proportional	Quantitative	Proportional
Automation	Yes	No	Yes
Cost	Low	Medium	High

*Use liquid solutions or prepare a solution by dissolving solid fertilizers

TYPES OF WATER SOLUBLE FERTILIZERS

- Liquid/Fluid fertilizers: Fluid or liquid in which the plant nutrients are in true solution. These are solutions which contain one or more plant nutrients.

- Suspension fertilizers: Fluid fertilizers that have solid nutrients dispersed throughout the fluid.
- Water soluble solid fertilizers: Fertilizers that are in solid state plant nutrients are fully water soluble.

Salient Features of Water Soluable Fertilizers

- Nutrients are 100 per cent water soluble.
- Acidic in nature.
- They correct and maintain the soil pH.
- Nutrients are applicable in smaller quantity.
- Nutrients are in readily available form.
- Free from Na and Cl salts.
- Apply in precise and uniform form.
- Correct placement in root zone.
- There are no chances of fixation of P and K.
- No soil deterioration.
- Application as per physiological stages of crop.
- Improve up take of nutrients present in soil.
- Efficiency is more than 90 per cent.
- Improve and fasten the uptake of nutrients in soil.

Solubility of fertilizers

Fertilisers have different solubilities and therefore, need different amounts of water to dissolve and should be completely mixed before being injected. This is where agitation plays an important role in the effectiveness of injection. Agitation is easier in vertical tanks as there is a smaller surface area at the base of tank. Good agitation and a fine particle size results in a quicker dissolve rate, but the maximum concentration that's able to be dissolved does not change (Table 5).

Table 5. Solubility rating of various fertiliser products

Product	kg / 100 L@20°C	Product	kg / 100 L@20°C
Ammonium nitrate	192	Calcium nitrate	60
Ammonium sulphate	75	Magnesium sulphate	71
Mono-ammonium phosphate MAP	37	Magnesium nitrate	71
Liquifert P (Tech MAP)	20	Soluble boron	9.5
Mono-potassium phosphate MKP	12	Zinc sulphate	44
Potassium chloride	34	Liquifert K (KCI)	20
Potassium nitrate	8	Liquifert N (Urea)	25
Potassium sulphate	10	Urea (water temperature 5°C)	45

If applying two chemicals at once, test the compatibility with each other and with the irrigation water as a precipitate (sediment) may form. Urea, muriate of potash, potassium nitrate and chelated trace elements are generally considered safe to mix. However phosphates, sulphates, calcium, magnesium and trace elements can create problems. When this happens, the following can occur:

- Precipitates may settle to the bottom of the tank or block filters and emitters.
- Precipitates may also form if the water is hard (i.e high in Ca and Mg or contains carbonate).

Therefore, the trace elements to avoid are copper, zinc, manganese and iron sulphates. These cannot be mixed with calcium nitrate, MAP, MKP and always use chelated forms of trace elements if mixing products (Table 6).

Table 6. Product compatibility chart

	Urea	Ammonium nitrate	Ammonium sulphate	Mono-ammonium phosphate MAP	Mono-potassium phosphate MKP	Potassium nitrate	Potassium sulphate	Potassium chloride	Calcium nitrate	Magnesium sulphate	Soluble boron
Urea		YES	YES	YES	YES	YES	YES	YES	YES	YES	YES
Ammonium nitrate	YES		YES	YES	YES	YES	YES	YES	YES	YES	YES
Ammonium sulphate	YES	YES		YES	YES	YES	YES	YES	NO	YES	YES
Mono-ammonium phosphate MAP	YES	YES	YES		YES	YES	YES	YES	NO	NO	YES
Mono-potassium phosphate MKP	YES	YES	YES	YES		YES	YES	YES	NO	NO	YES
Potassium nitrate	YES	YES	YES	YES	YES		YES	YES	YES	YES	YES
Potassium sulphate	YES	YES	YES	YES	YES	YES		YES		YES	YES
Potassium chloride	YES	YES	YES	YES	YES	YES	YES		YES	YES	YES
Calcium nitrate	YES	YES	NO	NO	NO	YES	NO	YES		NO	NO
Magnesium sulphate	YES	YES	YES	NO	NO	YES	YES	YES	NO		NO
Soluble boron	YES	YES	YES	YES	YES	YES	YES	YES	NO	NO	

Nutrient application

Fertigation allows nutrient needs to be supplied according to the physiological stage of the crop and the site- specific requirements (Crop type and cultivar, growing system, yield target, soil type and climatic conditions). Nutrients status should be monitored regularly in the plant tissue (leaf and petiole analysis), in the soil and in the irrigation water and leaching solutions (Table 7).

Table 7. Typical N: P: K ratios in a fertigation programme according to the physiological stage of the crop (eg: Tomato)

Physiological stage	Ratio of supplied nutrients			Observation
	N	P_2O_5	K_2O	
Planting-establishment	1	1	1	High P and K for best rooting and plant establishment
vegetative	2	0.5	1.5	High N and K for best plant growth and leaf development
Fruit set-fruit ripening	1	0.3	3	High K for best fruit quality (size, colour, flavor)Reduce N to avoid lush growth, soft fruit and pest problems
Fruit ripening-harvest	1	0.1	2	High K for maintenance

Fertilisers and Irrigation Water

The pH of the irrigation water has to be near to neutral and its electrical conductivity (EC) to be within acceptable limits that are not well defined, but a value of around 1 dS/ m is acceptable. Addition of fertilisers to the water raises its EC and changes its pH. The objective is to have a fertigation solution somewhat acid and with a low EC.

The optimum pH of fertigation water is 5.5 - 6.0. An anticipated EC of the fertigation solution is calculated by measuring the EC of the irrigation water prior to addition of the fertiliser solution and adding it to the estimated EC of the fertiliser solution. The measured EC of the fertigation solution collected from the dripper should be within 10% of the calculated value. A similar EC in both solutions is normal and if the EC of the drainage solution is more than 20% higher that of the fertigation solution, there is a risk of salinization of the growth medium.

The fertilisers should not precipitate as insoluble salts by reacting with the ions present in the irrigation water. Typically Mg and Ca are the ions present in the water that can precipitate with phosphates or sulphates from fertilisers. The solubility of different fertilisers varies considerably, and the temperature also has a big effect on the solubility. If the user makes tank solutions from several products, the presence of other salts already in the solution can lower the solubility. This restricts the concentration of stock solutions, but generally 10% solutions can be made quite safely from many of these products.

Advantages of Fertigation

- In drip fertigation, the amount and form of nutrient supply is regulated as per the need of the critical stages of plant growth.
- Saving of fertiliser applied, due to better fertiliser use efficiency and reduction in leaching.
- Optimisation of nutrient balance in soils by supplying the nutrients directly to the effective root zones as per the requirement.
- Reduction in labour and energy cost by making use of water distribution systems for nutrient application.
- Better yield and quality of products obtained.
- Timely application of precise amounts of fertilisers directly to the roots zone, this improves fertiliser use efficiency and reduces nutrient leaching below the root zone.
- Ensures a uniform flow of water and nutrients.
- Improves availability of nutrients and their uptake by crop.
- Safer application method, as it eliminates the danger affecting roots due to higher dose.
- Soil and water erosion are prevented.

Disadvantages of Fertigation

- Both the components (drip and water soluble fertiliser) are very costly.
- Maintenance of drip irrigation is difficult. There is possibility of theft and rat infestation.
- Good quality water is very essential.
- Clogging of emitters may cause a serious problem.
- It needs water soluble fertilisers, the availability of these types of fertilisers is limited.
- Adjustment of fertilisers to suit the need is not easy.

- Infestation of insect pest and diseases increases.
- Area under micro irrigation is now increasing mainly because of subsidy in micro irrigation, if subsidy is withdrawn, the area under micronutrient may also reduce.

D. Chelates

The word chelate derives from the Greek word "chel", meaning a crab's claw, and refers to the pincer-like manner in which the mineral is bound. Some chelating chemicals are shaped like a letter 'C' and surround the mineral with just one molecule. This type of binding is called a "complex". When two molecules of the same material surround a mineral it is known as a chelate. It is important to note that some minerals, like boron or molybdenum, have only one chemical bond and are thus limited to forming only a complex. Strictly speaking, boron and molybdenum cannot be chelated minerals but they are often presented and sold in the market as chelated minerals. Chelates need a "glue" to hold the protective shell in place.

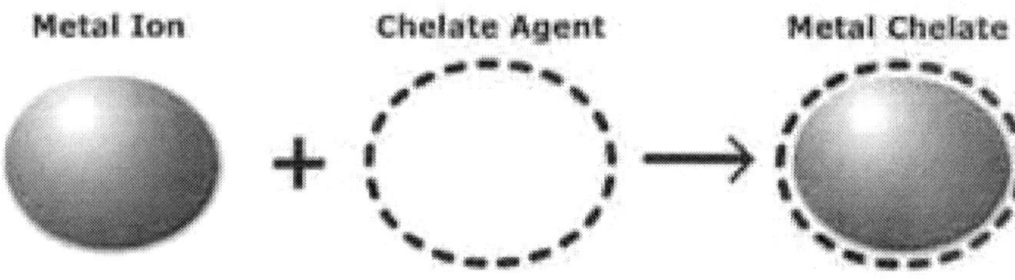

Chelates are molecules with a neutral charge, which is very important. Micronutrients normally have an electrical charge on them. For example, calcium and magnesium are both +2 charge. Soil is generally negative in charge, especially clay soils. This means that calcium and magnesium will likely react with the soil and be insoluble and not available for plants to use. Moreover, since they are the same charge, calcium and magnesium compete with each other for entry into the plant. Pores on the leaves of plants (also known as stomata's) are negatively charged, so positively charged molecules trying to penetrate the plant get tied up at the stomata entrance thus slowing absorption. But the interaction with chelates is very different. The neutral charge of chelated minerals allows chelated minerals allow them to enter the stomata unimpeded.

Chelation

Chelation is a term that describes an encapsulation process. The idea about chelation was initiated by Horner *et al.* (1934). Chelation is a natural process. In order to prevent absorbed nutrients from precipitation resulting from the interaction of nutrients, such as iron forming precipitation with phosphorus, upon entering plant cells cationic nutrients will immediately form chelates with organic acids such as citric acids, malonic acid, and some amino acids. This chelation process will then enable the nutrients to move freely inside the plants (Figure 2).

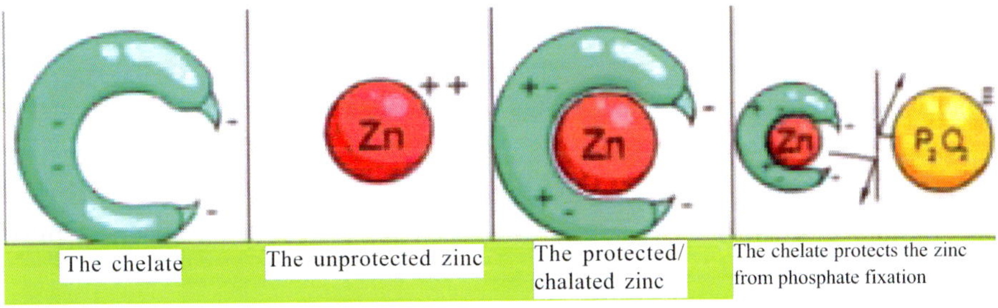

Figure 2. Chelation process

Chelation in soil increases nutrient availability to plants. Organic substances in the soil either applied or produced by plants or microorganisms are the natural chelating agents. The most important substances having this nature are Hydroxamate Siderophores, Organic Acids and Amino Acids. Under Iron deficient conditions, microorganisms will produce siderophores to overcome the iron starvation.

Micronutrients in soils are interacts with soils for clays have negative surface charges that strongly attract and immobilize the positively charged metal ions. Thus, clay effectively lock up the micronutrient. In a chelated form, however, the negative charge of the chelated metal micronutrient allows movement of the micronutrient in the soil solution for rapid, effective and efficient uptake by root system. The chelate is unaffected by the charges on the clay surface. When chelated nutrients are applied to the soil, it diffuse to the root cell where metal ions are absorbed leaving the chelating agent to rediffuse and mobilize further metal ions from the solid phases of soil.

Stability of chelates

The chelates are very stable. The stability of a chelate is affected by number of rings formed, the metal ion and the solution pH. Chelate stability also varies depending on the specific chelating agent used. For four micronutrients, the following order of stability of some of the major chelates has been calculated by Norvell (1972).

Fe:EDDHA > DTPA> CDTA > EDTA > EGTA = HEDTA > NTA

Cu: DTPA > HEDTA = CDTA > EDTA = EDDHA > EGTA =NTA

Zn: DPTA > CDTA = HEDTA = EDTA > NTA > EGTA

Mn: DPTA = CDTA > EDTA = EGTA = HEDTA > NTA

Generally, bonds are stronger with metals in approximately in this order:

$Fe^{+3} > Cu^{2+} > Zn^{2+} > Fe^{2+} > Mn^{2+} > Ca^{2+} = Mg^{2+}$

Except in soils of very high pH, iron chelates tend to be more stable than those of copper and zinc, which are inturn, more stable than those of manganese. Hence, iron is more strongly affected by the chelating agents than the other micronutrients. Until pH ranges from 6.2 to 6.5, DTPA is a good fertilizer choice. In cultivation systems with higher pH in the root environment EDDHA or EDDHMA chelate is recommended (Table 8).

Table 8. pH and chelates

Chelate formula	Effective from pH
Fe EDDHMA	3-11
Fe EDDHA	3-10
Fe DTPA	1.5-7.5
Fe HEDTA	1.5-7
Fe EDTA	1.5-6.5
Mn EDTA	3-10
Zn EDTA	2-10
Cu EDTA	1.5-10
Ca EDTA	5-10
Mg EDTA	6-10

Characteristics of a chelate applied to soil

- Should not be easily replaced by other polyvalent cations in the soil.
- Must be stable against hydrolysis.
- Has to be resistant to microbiological decomposition.
- Should be soluble in water
- Not easily precipitated by ions or colloids in soils
- Must be available to plants either at root surface or within the plant.
- Must be non damaging to plants at concentrations required to prevent deficiencies.

Practical consideration of chelates application

- Chelated micronutrients can be mixed with other inorganic NPK and other fertilizers.
- Albio is a aminoacid chelate. Since aminoacids are the basic building block of protein formed in all living things, the chelation of minerals with aminoacid provides a tremendous advantage in the efficiency of absorption and translocation of minerals within plants. These chelated minerals are marketed under the trade name metalosated.
- Chelated micronutrients are costlier than the ordinary inorganic micronutrient salts.
- Chelated nutrients are most effective, smaller amounts are required to correct the deficiency than the inorganic micronutrient salts.
- Chelated nutrients should not be mixed with liquid fertilizers that contain ammonia. Due to high pH of ammonia, the metal forms precipitates, because the (OH) groups compete with the donor groups of the chelating agent for metal ions.

Chelated nutrients are more plant available than complexed nutrients and complexed nutrients are more plant available than mixtures and uncomplexed nutrients.

Types of chelating agents

The two type of chelating agent are viz., Natural chelating agents and synthetic chelating agents.

Natural chelating agents	Synthetic chelating agents
Citric acid,	EDTA,
Tartaric acid,	DTPA,
Glycine,	CDTA,
Gluconic acid,	EDDHA,
Malonic acid, Fulvic acid.	HEDTA

Organic acids and amino acids such as citric acid and glycine are also naturally occurring chelating agents. Citric acid is one of the organic acids commonly used as chelating agents. Other naturally occurring organic acids such as malonic acid and gluconic acid also play an important role in plant mineral nutrition.

Glycine is the simplest amino acid with a molecular weight of 75. Chelates of glycine with cations such as iron, zinc, and copper. The chelates usually contain 2 moles of ligand (glycine) and one mole of metal as demonstrated in the following Figure 3.

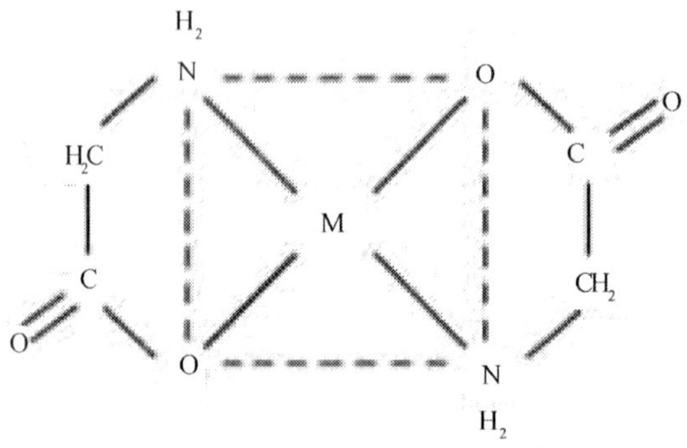

Figure 3. Glycine of chelate

One of the most common forms is Ethylene DiamineTetraacetic Acid (EDTA), which has been on the market for many years. EDTA is a large synthetic molecule that binds very tightly to minerals and resists chemical interactions desirable characteristics for chelates used in the soil.

Chelating agents as well as minerals contribute to the stability of chelates. A strong chelating agent may bond the mineral too strongly and make it unavailable to plants. On the other hand, a weak chelating agent may not be able to protect the chelated minerals from chemical reactions with other compounds and thereby reduce their availability to plants. Combination of chelating agents can improve product stability and broaden product effectiveness.

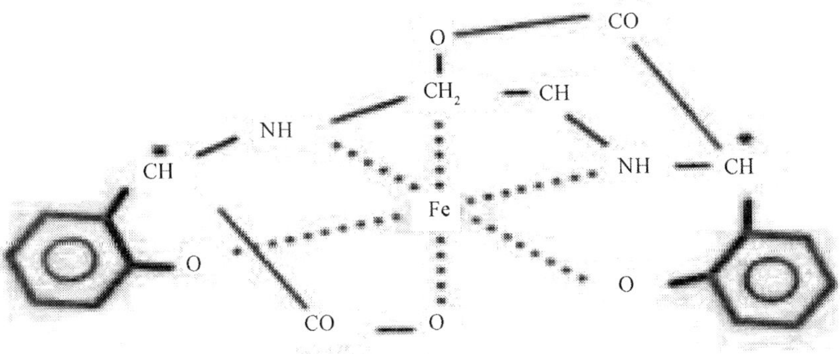

Figure 4. EDDHA chelates with metal

Chelated compounds are more stable than non-chelated compounds. Therefore, metallic chelates are widely used in agriculture as micronutrient fertilizers to supply plants with Iron, Manganese, Zinc and Copper. The most common chelates used in agriculture are EDTA, DTPA and EDDHA (Figure 4).

Foliar Absorption Mechanisms

In between the leaf epidermal cell wall and cuticular membrane is a layer of pectin. The cuticular permeability of mineral nutrients has most commonly been studied using radio labeled isotopes. These studies indicate that cuticular penetration is largely a diffusion process and that the use of certain facilitators increases the rate of diffusion. The permeability has been shown to differ depending upon the size of molecules. Larger molecules such as EDTA, DTPA and EDDHA penetrate at a slower rate compared to natural chelating agents. Cell membranes do not have the capacity to absorb synthetic chelates. For the mineral to be absorbed into the cell, chelates must release the mineral, leaving a vacancy within the chelate molecule and display charges that must be satisfied. For example, EDTA has a very high affinity for calcium. As a result, the synthetic chelate will scavenge existing free calcium from the surrounding environment, including cell walls and membranes. This chain of events has the potential to cause the collapse of the cell walls and the leakage of cell contents, leading to phytotoxicity effects.

Chelated nutrients are also useful for foliar application. Plant leaves have a waxy coating that prevents them from drying. The wax repels water and inorganic substances making it difficult for inorganic nutrients to penetrate into the leaf. However, organic molecules can penetrate the wax.

The organic coating around the chelated nutrient allows it to penetrate through the wax into the leaf. Once in the leaf, the chelate releases the nutrient so that it can be used by the plant (Figure 5).

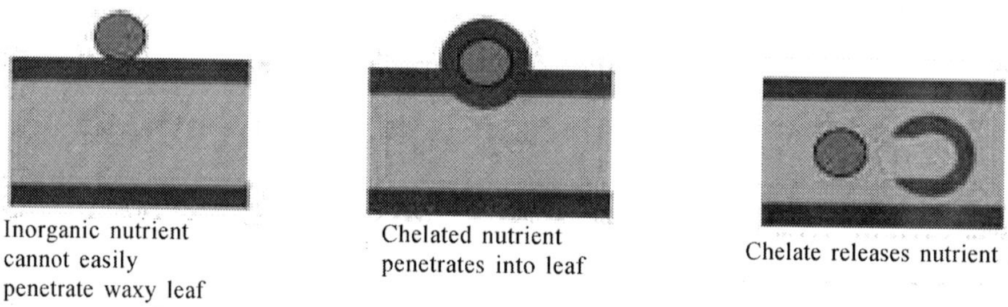

Figure 5. Foliar absorption of chelates

Role of Chelates

- **Increase the availability of nutrients.**

  Chelating agents will bind the relatively insoluble iron in high pH soil and make it available to plants.

- **Prevent mineral nutrients from forming insoluble precipitates.**

  The chelating agents of the metal ions will protect the chelated ions from unfavourable chemical reactions and hence increase the availability of these ions to plants.

- **Reduce toxicity of some metal ions to plants.**

  Chelation in the soil may reduce the concentration of some metal ions to a non-toxic level. This process is usually accomplished by humic acid and high-molecular-weight components of organic matter.

- **Prevent nutrients from leaching.**

  Metal ions forming chelates are more stable than the free ions. Chelation process reduces the loss of nutrients through leaching.

- **Increase the mobility of plant nutrients.**

  Chelation increases the mobility of nutrients in soil. This increased mobility enhances the uptake of these nutrients by plants.

- **Suppress the growth of plant pathogens.**

  Some chelating agents may suppress the growth of plant pathogens by depriving iron and hence favour plant growth.

Advantages of Chelates over Tradition Forms

The chelated forms of micronutrients have a number of advantages over more traditional forms of trace elements such as oxides and sulphates:

- Much lower quantities are necessary as compared to inorganic compounds, because they are completely assimilatable by crops. Chelates are thus cost effective even though they are a little more expensive.
- Chelates are much more easily absorbed by plant roots or leaves because chelates are of organic nature. The chelation process removes the positive charge from the micronutrients allowing the neutral or slightly negatively charged chelates to slide through the pores on the leaf and root surface more rapidly. Since these pores are negatively charged, positively charged micronutrients would normally be 'fixed' at the pore entrance would be difficult to be assimilated by plants. When neutral chelated micronutrients are used, there would be no such restriction barriers.
- Chelates are more easily translocated within the plant as their action is partly systemic.
- Chelates are easily assimilated within the plant system.
- The chances of 'scorching' of crops while using chelates is less because they are organic substances.
- Under alkaline conditions, chelated iron, zinc, manganese and copper is a better way to provide micronutrients to a crop.

- Chelates are compatible with a wide variety of pesticides and liquid fertilizers, as chelates do not react with their components. Most chelates can be mixed with dry mixes and liquid fertilizers.
- Chelates are not readily leached from the soil as they adsorb on to the surface of soil particles.
- The application of micronutrient chelate fertilizers to soil or as foliar spray have their own advantages. Some advantages of foliar application are the lower application rates needed, immediate effect on the crop, feasibility of uniform application even when using small amounts and no build up of harmful levels.
- Thus, they serve as a good supplement to soil application as well as a mid-course correction and practical method for tree crops. On the other hand, advantages of soil application include: application of deficient micronutrient right at the beginning of the crop cycle with generally no injury to plants, provide residual benefits to crops that follow the current one, no sprayer needed and skilled labour not essential. In case of soil application, it is important to have adequate water supply to wash the chelates down into the soil up to the extent of the spread of the root system.
- Iron chelates in commercially produced soluble fertilizers are degraded by light. Therefore, proper storage of fertilizer stock solution in opaque containers should be followed to check degradation of iron chelates. Solid forms of the fertilizer are not influenced by light.

References

Abaye A.O. 1996. Potassium fertilization of cotton. Virgina Cooperative Extension, Virgina State University. pp. 1-1-8.

Ahlawat, I. P. S., Gangaiah, B., and Zahid, M. A. 2007 Nutrient management in chickpea. Pages 213-232 In: Chickpea Breeding and Management. S. S. Yadav, B. Redden, W. Chen and B. Sharma,(eds). CAB Int. Wallingford, UK.

Akehurst, B. C. 1981. Tobacco, Longman Publications, London.

Alexander, K.C and Arulraj, S. 1995. Sugarcane production Manual. Coimbatore. p. 129.

Alloway, B. J. 2008. Micronutrient Deficiencies in Global Crop Production. Herdelberg, Germany, Springer Science, Berlin.

Anderson, W.B., 1982. Diagnosis and correction of iron deficiency in field crops - An overview. *J.Plant Nutr.,* 5: 785-95.

Anonymous, 1980. Monograph on cotton, TNAU, Coimbatore.

Anonymous 1991. Nitrogen deficiency in oil palm. Symptoms and correction. *Oléagineux,* 46: 247-250.

Anonymous 1992. Potassium deficiency in oil palm Symptoms and correction. *Oléagineux,* 47: 587-591.

Anonymous. 2014. AESA based IPM - Cucurbitaceous vegetable crops. Department of Agricultural and Co-operation, Govt. of India, New Delhi.

Arabhanvi, F., Pujar, A.M and Hulihalli, U.K. 2015. Micronutrients and productivity of oilseed crops - A review. *Agri. Review,* 36: 345-348.

Arnon, D. I., and Scott. P. R. 1939. "The Essentiality of Certain Elements in Minute Quantity for Plants with Special Reference to Copper." *Plant Physiol.* 14: 371–375.

Arora,C.L, Brar, M.S. and Takkar, N.1985. Nutrient indexing of irrigated groundnut and its response to zinc and boron application in some villages of Ludhiana. *Fert. News.*, 30: 36-39.

Asher, C.J. and Lee, M.T. 1975. Nutritional disorders in ginger. Department of Agriculture, University of Queensland, St. Lucia, Australia.

Asher, C.J., Edwards, D.G. and Howeler, R.H. 1980. Nutritional Disorders of Cassava. Department of Agriculture, University of Queensland, St Lucia, Australia.

Balakrishnan, K. 2000. Chemo and anatomical detection of nutrient deficiency in crops. In: ICAR Summer short course on "Diagnosis and correction of nutritional and physiological disorders in crops" held between June 12-20 at Dept. of Crop Physiology, TNAU, Coimbatore.

Balakrishnan, K. Anbu, S and Azhakiamanavalan, R.S. 1998. Diagnosis of manganese deficiency in fruit crops. *Micronutrient News,* 12: 1-4.

Balakrishnan, K. Anbu, S and Azhakiamanavalan, R.S. 1999. A key to identify nutritional disorders in horticultural crops and their control measures. In: Souvenir XX Flower, vegetable and fruit show, held at Govt. of Pondicherry. Dept. of Agriculture,pp. 62-66.

Balakrishnan, K. Anbu, S and Azhakiamanavalan, R.S. 2000. Diagnosis of iron deficiency in fruit crops. *Micronutrient News,* 14: 1-3.

Balakrishnan, K. Anbu, S and Azhakiamanavalan, R.S. 2000. Potassium and fruit quality. *Agro India* (Feb.): 20-21.

Balakrishnan, K. and Azhakiamanavalan, R.S. 1998. Diagnosis of magnesium deficiency in fruit crops. *Micronutrient News,* 12: 2-3.

Balakrishnan, K. and Azhakiamanavalan, R.S. 1998. Diagnosis of sulphur deficiency in fruit crops. *Micronutrient News.* 12: 1-3.

Balakrishnan, K. and Azhakiamanavalan, R.S. 1999. Calcium for fruit crops. *Micronutrient News.* 13: 1-3.

Balakrishnan, K., Rajangam, J. and Thangaraj, T. 2001. Molybdenum. *Agro India (July)* 10-12.

Balakrishnan, K., Venkatesan, K., Thandapani, V. and Sambandamurthi, S. 1994. Diagnosis of zinc deficiency in fruit crops. *Micronutrient News.* 8: 1-4.

Balasubramanian, T.N. 2001. Influence of climate on soil fertility In: Advance in Integrated management system for sustaining crop productivity and soil fertility, C.R. Chinnamuth et al (eds) TNAU, Coimbatore.

Balasubramanian, V 1998. CREMNET Technology brief No.2 IRRI, Philippines

Barak, P. and Helmke P.A., 1993. The chemistry of zinc. In zinc in soils and plants. (ed.) A.D. Robin. Dordecht: Kluwer Academic Publishers.

Barber, S.A. 1995. Soil Nutrient Bioavailability: A Mechanistic Approach, 2nd Ed. John Wiley & Sons, New York.

Barker, A. V. and Pilbeam. D. J. 2007. Handbook of plant nutrition. Taylor and Francis. New York.

Barnard, R O, Cilli. G.E.B and Kotze. J.M. 1991.Deficiency Symptoms in Avocados. Department of Microbiology and Plant Pathology, University of Pretoria, Pretoria. South African Avocado Growers' Association Yearbook. 67-71.

Bates, T.E., 1971. Factors affecting critical nutrient concentrations in plants ad their evaluation: A review. *Soil Sci.,* 112: 116-30.

Bennet, W.F. 1993. Nutrient deficiencies and toxicities in crop plants. APS. The American Phytopathological Soc., St. Paul, MN, USA.

Bergmann, W. 1992. Nutritional Disorders of plants: Visual and Analytical Diagnosis. Fischer, Jena, New York.

Bergmann, W. and Neubert, P. 1976. Plant diagnosis and Plant analysis. Gustav Fischer Verlag Jena, New York.

Bhargawa, B.S. 2002. Leaf analysis for nutrient diagnosis, recommendation and management in fruit crops. *J. Indian Soil Sci.* 50: 352-373.

Bhatt, J.G., M.R.K. Appakuttan E. and Nathan, A.R.S. 1982. Leaf reddening in hybrids of cotton. *Comm. Soil Sci. and Plant Analy.* 13: 151-156.

Bhatt, R. and Sujatha, S. 2004. Crop management. In: Arecanut (Eds: Balasimha, D. and V. Rajagopal), CPCRI, Kasaragod, pp. 76-102.

Black, C. A. 1992. Soil fertility evaluation and control. Lewis Publication, London.

Blarney, F.P.C., Edwards, D.G. and Asher, C.J. 1987. Nutritional Disorders of Sunflower. Department of Agriculture, University of Queensland, St Lucia, Australia.

Bolle-Jones, E.W. and Ismunadji, M. 1963. Mineral deficiency symptoms of the sweet potato. *J. Exptl. Agri,* 31: 60-64.

Bould, C., Hewit, E.J and Needham, P. 1983. Diagnosis of mineral disorders in plants. Vol.I. Principles. HMSQ, London.

Brajendra, Vishwakarma, A.K., Sreedevi, B and Sharma, M. 2011. Diagnosing nutritional disorders in field crops and their management. Narendra Publishing House New Delhi.

Brown, J.C., 1961. Iron chlorosis in plants. *Adv. Agron.,* 13: 329-69.

Brown, P. H.. Welch, R. M and Cary, E. E. 1987. "Nickel: A Micronutrient Essential for Higher Plants." *Plant Physiology* 85: 801–803.

Cakmak, I and Römheld V. 1997: Boron deficiency induced imparments of cellular functions in plants. *Plant Soil,* 193: 71-83.

Chanda, K. L and Shikhamany S.D. 1999. The Grape- Improvement, Production and Post Harvest Management. Malhotra Publishing House New Delhi.

Chapman, H.D. 1966. Diagnostic criteria for plant and soils. University of California, Division of Agricultural Science, Riverside.

Chen, Y. and Barak, P. 1982. Iron nutrition of plants in calcareous soils. *Adv. Agron.* 35: 217-240.

Cheong, S.P. and Ng, S.K. 1977. Copper deficiency of oil palms on peat. In: Earp, D.A. and Newall, W. (eds.) International Developments in Oil Palm. The Proceedings of the Malaysian International Agricultural Oil Palm Conference, Kuala Lumpur, 14-17 June 1976, pp. 362-370.

Choudhary, O.P and Yaduvanshi, N.P.S. 2016. Nutrient management in salt-affected soils. *Indian J. Fert.* 12: 20-35.

Chundawat, B.S. 1998. Sapota. Agrotech Publishing house, Udaipur, India.

Clark, R.B. 1982. Mineral nutritional factors reducing sorghum yields: micronutrients and acidity. Sorghum in the Eighties: In:- Proceedings of the International Symposium on Sorghum. International Crops Research Institute for the semi-Arid Tropics, 2 – 7 November 1981, Patancheru, Andhra Pradesh, India, 179-190.

Cooke, G.W. 1985. The role and importance of potassium in the agricultural system of the humid tropics. Proceedings of Colloquium on Potassium in the agricultural systems in the humid tropics. International Potash Institute, Berne, Switzerland, pp. 369-405.

Corrado, F., Quencez, P. and Tailliez, B. 1992. Boron deficiency in oil palm. Symptoms and corrections. *Oléagineux,* 47: 719-725.

Cunningham, S.D., Anderson, T.A. Schwab, A.P. and Hsu. F.C. 1996. Phytoremediation of soils contaminated with organic pollutants. *Adv. Agron.* 56: 55-114.

Datta, R.K 1991, Diseases and Pest of mulberry and their control. Central Sericulture Research and Training Centre, Mysore. P. 45.

Dekock, P.C., 1981. Iron nutrition under conditions of stress. *J. Plant Nutr.*, 3: 513-21.

Deotale, R.D. 2005. Macro and micronutrients deficiency symptoms in Nagpur orange. *J Soil and Crops.* 15: 284-286.

Dharmendran, J.P.D and Mahendran. P.P 2012. A book on sub surface drip irrigation for sugarcane, Kumbakonam, Tamil Nadu.

Dheebakaran, Ga and Jeganathan, R.2009. Crop nutrient interaction in relation to weather. In: Induced Nutrient Deficiency (eds). Ga. Dheebakaran, R. Jaganathan and De, Jegadeswari, TNAU, Coimbatore.

Dotaniya, M.L. and Meena, V.D. 2015. Rhizosphere Effect on Nutrient Availability in Soil and its Uptake by Plants: A Review. *Proc. Natl. Acad. Sci., India,* Sect. B. Biol. Sci. (Jan-Mar 2015) 85: 1-12.

Dow, D.I. and Roberts, S. 1982. Proposal: Critical nutrient ranges for crop diagnosis. *Agron.J.* 74: 401-403.

Dwivedi, B.S., Singh, V.K., Meena, M.C., Dey, A and Datta, S.P. 2016. Integrated Nutrient Management for Enhancing Nitrogen used Efficiency. *Indian J.Fert.* 12. 62-71.

Dybing, C. D., and Currier, H. B. 1961. Foliar penetration of chemicals. *Plant Physiol.* 36: 169–174.

Edmond, I.B. and Sefick, H.J. 1938. A description of certain nutrient deficiency symptoms of the Porto Rico sweetpotato. Proceedings of the American Society for Horticultural Science, 36: 544-549.

Eissenstat, D.M. and Volder, A. 2005. The Efficiency of nutrient acquisition over the Life of a Root. Ecological Studies, Vol. 181 H. BassiriRad (Ed.) Nutrient Acquisition by Plant An Ecological Perspective, Springer - Verlag Berlin Heidelberg.

Eskew, D. L., Welch, R. M. and Norvell, W. A. 1984. "Nickel in Higher Plants: Further Evidence for an Essential Role." *Plant Physiol.* 76: 691–693.

Fageria, N. K., Baligar, V. C. and Jones, C. A. 1997. Growth and mineral nutrition of field crops, 2nd edition. New York, USA.

Fageria, N. K., Baligar, V. C. and Clark, R. B. 2002. Micronutrients in crop production. *Adv. Agron.* 77: 185–268.

Fageria, N.K. 2009. The Use of Nutrients in Crop Plants. CRC Press Taylor and Francis Group, Newyork.

Fairhurst, T. 1998. Nutrient Deficiency Symptoms and Disorders in Oil Palm. Pocket Guide. Potash & Phosphate Institute, Singapore.

FAO, 1998. Guide to efficient plant nutrient management, Rome.

Fernandez. V., Sotiropoulos, T. and Brown, P. 2013. Foliar Fertilization, Scientific Principles and Field Practices, *International Fertilizer Industry Association (IFA),* Paris, France. 775: 90.

Finck, A. 1992. Fertilizers and their efficient use. In IFA World Fertilizer Manual, Paris, France, pp. 1-36.

Gade, R.M. Tanwar, R.K., Jeyakumar P and Kanwar, V. 2013. Leaf reddening and its management in cotton. NCIPM Technological Bulletin No.30. National Centre for Integrated Pest Management in cotton, India.

Gade, R.M. Tanwar, R.K., Jeyakumar, P. and Mane, S.S.2013. Nutritional management to combat leaf reddening in cotton, *J Plant Diseases Sci.* 8: 99-101.

Geethalakshmi,V., Christopher Lourduraj, A., John Joel, A. and Rajamanickam, K.*1993*. Nutrient Management in Groundnut. *Madras Agric. J.,* 80: 412-414.

George Thomas C. 2003. Forage Crops Production in the Tropic. Kalyani Publisher New Delhi.

Ghosh, S.P. 2000. Nutrient management in fruit crops. *Fert. News.* 45: 71-76.

Glass, A.D.M. 1989. Plant nutrition: An introduction to current concepts. Jones and Barlett Publishers, Boston M.A.

Gooding, M. J., and Davies. W. P. 1992. Foliar urea fertilization of cereals: A review. *Ferti. Res.* 32: 209–222.

Gopikumar. K. and Varghese. V. 2004. Sand culture studies of Teak (Tectona grandis) in relation to nutritional deficiency symptoms, growth and vigour. *J.Trop.For.sci.* 16: 46-61.

Gosnell, J.M and Long, A.C. 1969. A sulphur deficiency in sugarcane. Proc. South African Sugar Technologists Association, South Africa.

Grundon, N.J. Edwards, D.G, Takkar, P.N. Asher, C.J and Clark, R.B. 1987. Nutritional disorders of Grain sorghum, Australian centre for International Agricultural Research, Australia.

Gupta, V.K, and Sharma, S.K. 2005. Diseases of fruit crops. Kalyani Publishers. Ludhiana.

Hafeez, B., Khanif, Y.M., and Saleem, M. 2013. Role of zinc in plant nutrition. - A review. *Amer. J. Exptl Agri.* 3: 374-391.

Hamaza, S. Srinivasan, V and Dinesh, R. 2009. Nutrient diagnosis of cardamom garden in South India. *Indian J. agric.sci.* 79: 429-432.

Hanway, D.G. 1972. Advances in corn production - principles and practices, Kalyani Publishers, Ludhiana.

Hawkesford, M.J. and De Kok, L.J. 2006. Managing sulphur metabolism in plants. *Plant Cell and Environ.* 29: 382-395.

Haytova, D., 2013. A Review of Foliar fertilization of some vegetable crops. *Ann. Rev and Res.Biol.* 3: 455-465.

Hebblethwaite, P.D 1983. The Faba Beans- A basis for improvement. Betteworths, London.

Hegde, *D.M. 2000.* Nutrient management in oilseeds crops. *Fert. News.,* 45: 31-38 and 41.

Jain, G.L and Mahapatra I.C, 1987 Fertilizer Use for Intensive Maize Production, ICAR, New Delhi p. 77.

Johansen, C. 1990. Pigeon pea: Mineral Nutrition. Pages 209-231 in The Pigeon pea (Nene, Y.L., Hall, S.D., and Sheila, V.K., eds.). Wallingford, CAB International, UK.

Jones, B.J, Jr. 2001. Laboratory guide for conducting soil tests and Plant analysis. CRC Press London.

Jones, J.B. Jr. 1998. Plant nutrition Manual CRC Press London.

Jones, J.B. Jr., and Case, V.W. 1990. Sampling, handling, and analyzing plant tissue samples. In:R.L. Westerman (ed). Soil testing and plant analysis. 3rd ed. *Soil Sci. Soc. Am., Inc.* Madison, WI. pp. 389-447.

Kalbatia, A. and Pendias, H. 1994. Trace elements in soil and plants. 2nd ed. CRC Press London.

Kannan, S. 1980. Differences in iron stress response and iron uptake in some sorghum varieties. *J.Plant.Nutri.* 2: 347-358.

Kanniyan, S. 1993. Nitrogen fixing green manures. INSURF Progress Report. Tamil Nadu Agrl. Univ. Tamil Nadu. pp. 1-56.

Kanwar, J.S. and Randhawa, N.S. 1974. Micronutrient research in soils and plants in India, A Review, ICAR, New Delhi.

Karivaratharaju.T.V; Rathinavel. K. Bharathi .A, Raja.K and Baskaran.M 2008 Ready Reckoner for cotton cultivation and seed production. The SIMA Cotton Development and Research Association. Coimbatore. pp. 37-43.

Karthika, K.S., Priya, U.K. and Bhat, R. 2016. Nutrient deficiency disorders and their management in arecanut. *Indian Journal of Arecanut, Spices and Medicinal Plants Vol.* 18. 30-33.

Katyal, J.C. and Randhawa, N.S. 1983. Micronutrients. FAO Fertilizer and plant nutrition, Bulletin. No.7. FAO, Rome.

Katyal, J. C. 1981. Micronutrient Research in Pigeon pea. Pages 221-226. In Proceedings of the International Workshop on Pigeon peas, 15-19 Dec 1980, ICRISAT Center, India. Vol. 2. Patancheru.

Kohli, R.R. and Srivastava, A.K. 1997. Citrus nutrition - A retrospection. *Agri. Rev.* 18: 128-38.

Kolay. A.K. 2010. Soil fertility - Essential elements, Atlantic publishers. New Delhi. India. pp. 29-62.

Korikanthimath, V.S. 1994. Nutrition of cardamom. In: Advances in Horticulture.(Eds. K.L.Chanda), Malhotra Publishing House Ltd. New Delhi.

Kotur, S.C and Keshava murthy, S.V, 1999. Root activity patterns in fruits, Indian Hort. (January - March), p 31.

Kumar. S. and Nauriyal. J.P. and Cahoon. G.A. 1974. Survey of nutrient status of mango trees in Punjab. *Punjab Hort. J.* 14: 111-115.

Latha, M.R and Nadanassababady, T.2003. Foliar nutrition in crops - a review. *Agric. Rev.,* 24(3): 229-234.

Lenka, N.K., Biswas, A.K and Ramesh, K. 2015. Soil health enhancement, *Indian J. Fert.,* Vol. 11: 28-37.

Lindsay, W. L. 1974. "Role of chelation in Micronutrient Availability." In: The Plant Root and Its Environment, edited by E. E. Carson, 507"524. Charlottesville: University Press of Virginia.

Lindsay, W. L. and Norvell, W.A. 1978. Develoment of a Dtpa soil Test for Zinc, Iron, Manganese and Copper. *Soil Sci. Amer, J.* 42: 421-428.

Lopez, A.M. and Espinosa, J.M. 2000. Manual on the nutrition and Fertilization of Banana. Potash and phosphate Institute of Canada, Canada.

Lynch, J.P., and Deikman. J. 1998. Phosphorus in Plant Biology: Regulatory Roles in Molecular, Cellular, Organismic, and Ecosystem Processes, 401 pp. American Society of Plant Physiologists, Rockville, Maryland.

Maksymiec.W. 1997. Effect of Copper on cellular processes in higher plants. *Photosynthetica.* 34: 321-342.

Malavolta, E and da Cruz, V.F. 1971. A meaning for foliar diagnosis. In: Recent advances in Plant Nutrition, R.M. Samish (ed.). Vol.I. Gordon and Breach Science Publishers, Paris. pp. 1-14.

Malewar, G.U. 2005. Micronutrient stresses in soils and crops: Serious sickness and clinical approaches for sustainable agriculture. *J.Indian Soc. Soil Sci.* 53: 484-499.

Mapples,R.L., Thompson, W.R., and Varvil, J. 1988. Potassium deficiency in cotton. *Better Crops and Plant Food.* 73: 6-9.

Malhotra, S.K., Maheswarappa, H.P., Selvamani, V. and Chowdappa, P. 2017. Diagnosis and management of soil fertility constraints in coconut: A review. *Indian J. Agric.Sci.* 87: 711-726.

Marathe, R.A. and Shinde. Y.R. 2014. Identification of nutrient antagonism and deficiency symptoms in pomegranate. Souvenir, National Symposium cum exhibition on 'Pomegranate for Nutrition, Livelihood Security and Entrepreneurship Development', December 5-7, 2014, NRC on Pomegranate, Solapur, India.

Marathe, R.A., Babu. D.K and Shinde. Y.R. 2016. Soil and leaf nutritional constraints in major pomegranate growing states of India. *Agric. Sci. Digest.,* 36: 52-55.

Marboh, E.S., Singh, S.K., Pandey, S., Nath, V., Gupta, A.K. and Rongener, A. 2017. Fruit cracking in litchi (Litchi chinensis): An overview. *Indian J. Agric. Sci.* 87: 3-11.

Marschner, P. 2011. Mineral Nutrition of Higher Plants, 3rd Ed. Academic Press. New York.

Mengel, K. and Kirkby, E.A. 2001. Principles of Plant Nutrition. International Potash Insitute, Bern, Switzerland.

Mendel, R.R. and Schwarz, G, 1999. Molybdoenzymes and molybdenum cofactor in plants. *Crit. Rev. Plant Sci.* 18: 33–69.

Miyasaka. S.C. 2002. Nutrient Deficiencies and Excesses in Colacasia. Cooperative Extension Service, University of Hawai at Manoa.

Mohandass, S., Balamohan. T.N and Arjunan. G. 2000. Nutritional disorders in Crop Plants. Devi publications,Tiruchirappalli, Tamil Nadu.

Mortvedt, J.J. (ed.). 1991. Micronutrients in Agriculture. Second edition. SSSA Book Series. No. 4. Soil Sci.Soc.Amer., Madison, U.S.A.

Murthy, I.Y.L.N. 2016. Molybdenum in oilseed crops of India - A review. *Indian J. Fert.* 12: 40-50.

Natesan, S and Ranganathan , V. 1986. Sulphur nutrition in Plantation Crops. *Fert. News.* 31: 43-49.

Naidu, L.G.K., Ramamurthy, V., Sidhu, G.S. and Sarkar, D. 2011. Emerging deficiency of potassium in soils and crops of India, *Karnataka J.Agri.Sci.* 24: 12-19.

Naynard, D.N. 1979. Nutrition disorders of vegetable crops: A review. *J.Plant.Nutri.* 1: 1-23.

Norton, K.R. 1965. Boron deficiency in bananas. *Tropical Agri.* (Trinidad). 42: 361-365.

O' Sullivan, J.N. 2010. Yam Nutrition; Nutrient Disorder and Soil Fertility Management. Australian centre for International Agricultural Research, Canberra, Australia.

O.Sullivan, J.N., Asher, C.J. and Blamey, F.P.C.1997. Nutrient disorders of sweet potato. Australian Centre for International Agricultural Research, Canberra, Australia.

Pacheco, A.R. and Tailliez, B.J. 1986. Copper deficiency treatment in the oil palm nursery. *Oléagineux,* 41: 377-380.

Pasricha, N.S., Aulakh, M.S. Bahl, G.S. and Baddesha, M.S.1988. Fertilizer use research in oilseed crops. *Fert. News,* 33: 15-22.

Pathak, H.2010. Trend of fertility status of Indian soils. *Curr. Adv. Agric. Sci.* 2: 10-12.

Pathmanabhan, G. (ed) 2011. A hand book on mineral nutrition and diagnostic techniques for nutritional disorders of crops. Agrobios (India) Jodhpur.

Brown, P. H., Welch, R.M. and Cary, E.E 1987. Nickel: A Micronutrient Essential for Higher Plants. Plant Physiol. 85: 801-803.

Pattanayak, S.K. 2016. Biological Nitrogen Fixation - Status, Potential and Prospects in Supplementing Nitrogen Needs of the Crops. Indian *J.Fert.* 12: 94-103.

Petrazzini, L.L. 2014. Nutritional deficiency in crisp head lettuce grown in hydroponics. *Horticultura Brasileira* 32: 310-313.

Pillai, N.G., Mohankumar, B., Kabeerathumma, S. and Nair, P.G. 1986. Deficiency symptoms of micronutrients in sweet potato. *J.Root Crops* 12: 91-95.

Pilon-Smits, E.A. H., Quinn, C . F., Tapken, W., Malagoli, M. and Schiavon, M. 2009. Physiological functions of beneficial elements. *Curr.Opin. in Biotechnol.* 12: 267-274.

Plucknett. L.D. and Sprague. H.B. 1990. Detecting Mineral Nutrient deficiencies in Tropical and Temperate Crops. International Book Distributors. Dehra Dun, India.

Prakash, M., Balakrishnan. K and Rathinasamy. A. 2013. Nutritional Disorder in fruit crops. New India Publishing Agency. New Delhi. P. 178.

Prasad, M.S 2008. Maize Production Technology, Agrotech Publishing House. Udaipur. pp. 21-84.

Prasad, R., Shivay, Y.S. and Kumar, D. 2016. Interactions of zinc with other nutrients in soil and plants - A review. *Indian J.Fert.* 12: 16-26.

Purba, A.Y.L. and Turner, P.D. 1973. Severe boron deficiency in young oil palms in Sumatra. *Planter,* 49: 10-13.

Pushpadas, M.V. 2001. Zinc Nutrition of Coffee. Plant Horti Tech. 2: 64-66.

Rahman, F. and Sharma, P.C. 1974. Zinc as a fertilizer for tea. *Two and a Bud,* 21: 44-45.

Raj, G.B. and Patneick, M.C. 2008. Boron deficiency in coconut orchards of East Godavari districty of Andhra Pradesh. *Indian J. Coconut.* 7: 17-20.

Rajaratnam, J.A. 1972. 'Hook leaf' and 'fish-tail leaf': boron deficiency symptoms of the oil palm. *Planter,* 48: 120.

Rakkiyappan, P., 1995. Soil management (In sugarcane production manual, (Alexander, K.C. and Arulraj, S. eds). Sugarcane breeding Institute, Coimbatore.

Rakkiyappan, P., 2003. Management of problem soils. In (Eds. Balasundaram, N., Thiagarajan, R. & Rajula Chandaran). Advances in sugarcane production Technology. Sugarcane Breeding Institute, Coimbatore-641 007. pp. 80-87.

Ram. S. 1989. Factors associated with black tip and internal necrosis in mango and their control. *Acta Hort.* 231: 797-804.

Ramachandrappa B.K and Nanjappa, H.V. 2008. Fertigation Technology. Agrobios India, Jodhpur pp. 161-170.

Rashid, A., and Ryan, J. 2004. Micronutrient constraints to crop production in soils with Mediterranean-type characteristics: A review. *J.Plant.Nutr.* 27: 959-975.

Rattan, R.K. (2015) Mineral nutrition of plants. In Soil Science: An Introduction (R.K. Rattan, J.C. Katyal, B.S. Dwivedi, A.K. Sarkar, Tapas Bhattacharyya, J.C. Tarafdar and S.S. Kukal, Eds.) pp. 499-539. *Indian Society of Soil Science,* New Delhi.

Rattan, R.K., Datta, S.P. and Katyal, J.C. 2008. Micronutrient management-Research achievements and future challenges. *Indian J. Fert.* 4: 93-100.

Rengel, Z. 2015. Availability of Mn, Zn and Fe in the rhizosphere. *J. Soil Sci and Plant Nutr.* 15(2): 397-409.

Rengel, Z. and Marschner, P. 2005. Nutrient availability and management in the rhizosphere: exploiting genotypic differences. *New Phytologist* 168: 305-312.

Rodriguez, S.J., Cibes. H.R. and Ibanez, J.G. 1968. Deficiency symptoms displayed by Guava under greenhouse conditions. Technical Paper 44, University of Puerto Rico, Puerto Rico.

Roy, R.N. Finck, A., Blair, G.J. and Tandon, H.L.S. 2007. Plant nutrition for food security. Discovery Publishing House, New Delhi, p. 347.

Sanyal, S.K and Debnath, A. 2017. Chemistry and Dynamics of phosphorus in Indian soils in Relation to crop nutrition. *Indian J.Fert.* 13: 62-76.

Sanyal, S.K., Majumdar, K. and Singh, V.K. 2014. Nutrient management in Indian agriculture with special reference to nutrient mining - A relook J. Indian *Soc.Soil Sci.* 62: 261-275.

Sakal, R., Singh, A.P. Singh, R.B. and Bhogal, G.S.1996. Research Bulletin on 25 years of research on micro and secondary nutrients in *Soil & Crops of Bihar.* Dept, of Soil Sci., R.A.U., Pusa, Bihar. pp. 1-98.

Sandanandan, A.K., Hamza, S. and Srinivasan, V. 2000. Foliar nutrient diagnostic norms for optimizing cardamom production. In spices and Aromatic plants, Ind., Soc., for spices, IISR, Calicut, pp. 101-104.

Sawyer, J. 2004. Nutrient Deficiencies and Application Injuries in Field Crops, Department of Agronomy, Iowa State University. pp. 1-5.

Scaif, A., and M. Turner. 1983. Diagnosis of mineral disorders in plants. Vol. 2. Vegetables. Her majesty's Stationary Office. London.

Schactman, D, Reid, R.J. and Ayling, S.M. 1988. Phosphorus uptake by plants: from soil to cell. *Plant Physiol.* 116: 447-453.

Sekhon, B. S. 2003. "Chelates for Micronutrient Nutrition among Crops." Resonance 8: 46–53.

Sharma, C.P. 1996. Deficiency symptom and critical concentration of micronutrient in crop plants. Lucknow University, Lucknow, India.

Sharma, M.K. and Kumar. P. 2011. A Guide to Identifying and Managing Nutrient deficiencies in cereal crops. IBDC South Asia programme, India p. 49.

Sharma, R.M., Pandey, S.N. and Pandy, V. 2010. The Pear, Production, Post-Harvest Management and Protection. Ibdc Publishers, Lucknow pp. 261-75.

Shinde, B.N. and P.Kale. 1985. Boron for groundnut in dryland agriculture. *Fert. News.,* 30(51): 53-54.

Shivay, Y.S. and Kumar, D. 2008. A practical manual of field crops management. Division of Agronomy, Indian Agricultural Research Institute, New Delhi p. 113.

Shorrocks. M. 1964. Mineral Deficiencies in Hevea and Associated Cover Plants. Rubber Research Institute, Kuala Lumpur, Malaysia.

Shukla, A.K., Tiwari, P.K., Prakash, C., Patra, A.K., Meena, M.C., Singh, P., Tagore, G.C., and Rai, H.K. 2016. Current status of micronutrient deficiencies in soils and crop-specific recommendations for different agroclimatic zones of Madhya Pradesh. *Indian J.Fert.* 12: 26-35.

Shukla, A.R., Tiwari, P.K., Singh, P and Tiwari, K.N. 2017. Macro importance of Micronutrients in Indian Agriculture. *Indian Fmg.* 66; 67: 27-32.

Singh, G. 2013, Guava, Westville Publishing House, New Delhi, pp. 74-78.

Singh, N.P. 2013. A Guide to Identification and Management of Nutrient Deficiencies & Cultural Practices in Plantation Crops. Krishi Vigyan Kendra, ICAR Research Complex for Goa. p. 21.

Singh, S. and Naqvi, S.A.M.H. 2001. Citrus, International Book distributing co. Lucknow p. 187-198.

Singh, S.S. 2011. Soil Fertility and nutrient management. Kalyani Publishers, New Delhi.

Sivadasan, C.R., Nair, C.K., Srinivasan, K. and Mathews, K. 1991. Zinc requirement of cardamom. *J. Plantation Crops,* 18: 171-173.

Srinivasarao, C. H., Ganeshamurthy, A. N., and Ali, M. 2003. Nutritional Constraints in Pulse Production, Indian Institute of Pulses Research, Kanpur, India.

Srivastava, A.K. and Singh .S. 1998. Fertilizer use efficiency in Citrus. Bull No. 1 pp. 1-66.

Srivastava, A.K., Malhotra, S.K and Krishna kumar, N.K. 2015. Exploiting nutrient - microbe synergy in unlocking productivity potential of perennial fruits: A review. *Indian J. Agri. sci.* 85: 459-481.

Srivastava. R.S. 1998. Mango cultivation. International Book Distributing Co. Lucknow, India.

Stover, R. N. Simmonds. 1987. Bananas. 3rd edition. Longman, London.

Subramanian, K.S., Duraisami, V.P. Chitdeshwari, T., Govindaswamy, M. and Murugappan, V.2004. Visual Diagnostic kit (VDK) for identification and prescription of nutrient disorders in crop plants. TNAU Publication on Technology Release, Tamil Nadu Agricultural University, Coimbatore.

Sujatha, M.P. 2003. Diagnosis of Micronutrient Deficiencies in Teak Seedlings. Soil Science Division, Kerala Forest Research Institute, Kerala, India.

Sukla, R., Sharma, Y.K and Sukla, A.K 2014. Molecular mechanism of nutrient uptake in plants. *Int. J. Curr.Res. Aca. Rev.* 2: 142-154.

Sundara, B. 1998. Sugarcane cultivation. Vikas Publishing House, New Delhi. pp. 100-118.

Swarup, A.. and Ganeshmurthy, A.N. 1998. Emerging nutrient deficiencies under intensive cropping systems and remedial measures for sustainable high productivity. *Fert. News.* 43: 37-40 and 43-50.

Taiz, L. and Zeiger, E. 1998. Plant physiology, 356 pp. 2nd Ed. Sinauer Associates, Sunderland, MA.

Takkar, P.N. 1996. Micronutrient research and sustainable agricultural productivity in India. *J.Indian. Soc.Soil Sci.* 44: 562-581.

Takkar, P.N. and Randhawa, N.S. 1978. Micronutrients in Indian Agriculture. *Fert. News.* 23: 3-26.

Tandon. H.L.S. 1987. Fertilizer Recommendations for Horticultural Crops in India - A Guidebook. Repro Offset Printers. New Delhi.

Thampan, P.K. 1982. Handbook on coconut palm. Oxford and IBH publishing Co., New Delhi. p. 311.

Thorne, W., 1957. Zinc deficiency and its control. *Adv.Agron.* 9: 31-65.

Tondon,H.L.S. 1991. Sulphur Research and Agricultural Production in India. Fertilizer Development and Consultation Organization, New Delhi.

Turner, J., Lamber, M.J and Edwards, D.W. 1979. A guide to identifying nutritional and pathologic disorder of Pinus Radiata. Research Note No. 36. Forestry Commission of N.S.W., Sydney. p. 6-7.

Turner, P.D. 1981. Oil Palm Diseases and Disorders. Oxford University Press, Kuala Lumpur, p. 280.

Venkata Ram CS, 1968. Zinc deficiency in tea in Southern India. *Planter's Chron.* 63: 111-112.

Verma D.P. and Palani, M. 1997. Manuring of tea in South India (revised recommendations). In: Hand Book of Tea culture, UPASI Tea Research Institute, 1-33.

Verma, D.P. 1997. Sulphur nutrition in Tea the Planter's chron. (July) pp. 297-303.

Verma, D.P. and Sharma, V.S. 1995. Cost effective nutritional management for sustainable tea productivity in South India. *Fert News.* 40: 11-19.

Wahid P.A. 1984. Diagnosis and correction of Nutrient deficiencies in coconut palm *J. Plantation Crops* 12: 98-111.

Wallace, A and Lunt, O.R., 1960. Iron chlorosis in horticulture plants. A review. *Proc. Am. Soc. Hort. Sci.,* 75: 819-41.

Wanasuria, S. 1991. Copper deficiency of oil palm on sandy soils in Sumatra. *Better Crops International,* 7: 22-24.

Welch, R.M. 1995. Micronutrient nutrition of plants. *Critical.Rev.Plant Sci.* 14: 49-82.

Yadav, M.K and Solanki, V.K. 2015. Use of micronutrients in tropical and sub-tropical fruit crops: A review. *African J.Agri. Res.* 10: 416-22.

Yoshida, S. and Tanaka, A. 1969. Zinc deficiency of the rice plant in calcareous soils. *Soil Sci and Plant Nutri.* 15: 75-80.

Yoshida, S. 1981. Fundamentals of rice crops science. IRRI, Phillipines p. 269.

❑❑❑

Appendices

Appendix I. Symbols and relative atomic masses ("atomic weights") of important elements:

Symbol	Element	Relative atomic mass
Al	aluminium	27
B	boron	10.8
C	carsbon	12
Ca	calcium	40.1
Cl	chlorine	35.5
Co	cobalt	58.9
Cu	copper	63.5
Fe	iron	55.9
H	hydrogen	1
K	potassium	39.1
Mg	magnesium	24.3
Mn	manganese	54.9
Mo	molybdenum	95.9
N	nitrogen	14
Na	sodium	23
O	oxygen	16
P	phosphorus	31
S	sulphur	32.1
Si	silicon	28.1
Zn	zinc	65.4

* Atomic masses are based on carbon-12.

A value given in brackets denotes the mass number of the longest-lived or best-known isotope.

Appendix II. Absorption mechanism of different nutrients

Essential Elements	Source	Forms	Mechanism
C (Carbon)	Air	CO_2	Mostly through leaves by diffusion of gases
H_2 (Hydrogen)	Water	H^+, HOH	Mostly through roots by mass flow of soil solution
O_2 (Oxygen)	Air	O_2, OH^{-1}, CO_3^{-2}, SO_4^{-2}	Mostly through leaves by diffusion of gases
N_2 (Nitrogen)	Air/Fertilizer	NH_4^+, NO_3^{-1}	Mass flow
P_2O_5 (Phosphorus)	Soil/Fertilizer	$H_2PO_4^{-1}$, HPO_4^{-2}	Diffusion
K (Potassium)	Soil/Fertilizer	K^{+1}	Diffusion
Ca (Calcium)	Soil/Fertilizer	Ca^{+2}	Mass flow
Mg (Magnesium)	Soil/Fertilizer	Mg^{+2}	Diffusion
S (Sulphur)	Soil/Fertilizer	SO_4^{-2}	Mass flow
B (Boron)	Soil/Fertilizer	H_3BO_3, $H_2BO_3^{-2}$, $B(OH)_4^{-1}$	Mass flow
Cl (Chlorine)	Soil/Fertilizer	Cl^{-1}	Mass flow
Cu (Copper)	Soil/Fertilizer	Cu^{+2}	Diffusion
Fe (Iron)	Soil/Fertilizer	Fe^{+2}, Fe^{+3}	Diffusion
Mn (Manganese)	Soil/Fertilizer	Mn^{+2}	Diffusion
Mo (Molybdenum)	Soil/Fertilizer	MoO_4^{-2}	Mass flow
Zn (Zinc)	Soil/Fertilizer	Zn^{+2}	Root interception

Appendix III. Fertilizers and its relation with plant nutrient

1 kg. Nitrogen (N)	=	5 kg Ammonium Sulphate (20% N).
		or 2.2 kg. Urea (45%N).
		or 4 kg. Calcium Ammonium Nitrate (CAN, 25% N).
		or 4 kg. Ammonium chloride (25% N).
		or 3.84 kg. Ammonium Sulphate Nitrate (26% N)
1 kg. Phosphorus (P_2O_5)	=	6.250 kg Super Phosphate, (16% P_2O_5)
1 kg. Potassium (K_2O)	=	1.66. kg. Muriate of Potash (60% K_2O).

The following conversion factors may be used for determining the quantities of nutrients from given quantity of fertilizers:

Fertilizer material	Multiplied by	Corresponding quantity of nutrients
Ammonium Sulphate	0.200	Nitrogen
Urea	0.450	Nitrogen
Ammonium Sulphate nitrate	0.260	Nitrogen
Ammonium Chloride	0.250	Nitrogen
Ammonium Nitrate	0.330	Nitrogen
Sodium Nitrate	0.160	Nitrogen
Superphosphate (Single)	0.160	P_2O_5
Superphosphate (Double)	0.350	P_2O_5
Di-calcium Phosphate	0.350	P_2O_5
Muriate of Potash	0.600	K_2O
Sulphate of Potash	0.500	K_2O

Appendix IV. Compatibility of Fertilizers

	1	2	3	4	5	6	7	8	9	10	11	12
1. Muriate of Potash	R	R	R	O	O	O	O	O	R	R	R	R
2. Sulphate of Potash	R	R	R	O	O	O	R	O	R	R	R	R
3. Sulphate of Ammonia	R	R	R	R	O	O	X	O	R	R	X	R
4. Calcium Ammonium Nitrate	O	O	R	R	O	O	X	O	O	O	X	O
5. Potassium & Sodium Nitrate	O	O	O	O	R	O	O	O	O	O	R	O
6. Calcium Nitrate	O	O	O	O	O	R	O	O	X	X	O	X
7. Calcium Urea Cyanamide	O	R	X	X	O	O	R	O	X	X	R	X
8. Urea	O	O	O	O	O	O	O	O	O	O	O	O
9. Super Phosphate	R	R	R	O	O	X	X	O	R	R	X	R
10. Ammonium Phosphate	R	R	R	O	O	X	X	O	R	R	X	R
11. Basic Slag	R	R	X	X	R	O	R	O	X	X	R	X
12. Diammonium Phosphate	R	R	R	O	O	X	X	O	R	R	X	R

R: Fertilizers can be mixed

O: Fertilizers may be mixed before use

X: Fertilizers cannot me mixed.

Appendix V. Organic Manures and nutrient content

Manure	Percentage content		
	Nitrogen(N)	(P_2O_5)	Potash(K_2O)
Coir pith	1.20	1.20	1.20
Blood meal	10-12	1.2	1.0
Press mud	1-1.5	4-5	2-7
Bone meal			
1. Raw bone meal	3-4	20-25	-
2. Steamed bone meal	1-2	25-30	-
Fish meal	4-10	3.9	0.3-1.5
Animal refuse	0.3-0.4	0.1-0.2	0.1-0.3
Cattle dung, fresh	0.4-0.5	0.3-0.4	0.3-0.4
Horse dung, fresh	0.5 -0.5	0.4-0.6	0.3-1.0
Poultry manure, fresh	1.0-1.8	1.4-1.8	0.8-0.9
Sewage sludge, dry	2.0-3.5	1.0-5.0	0.2-0.5
Sewage sludge, activate dry	4.0-7.0	2.1-4.2	0.5-0.7
Cattle urine	0.9-1.2	trace	0.5-1.0
Horse urine	1.2-1.5	trace	1.3-1.5
Human urine	0.6-1.0	0.1-0.2	0.2-0.3
Sheep urine	1.5-1.7	trace	1.8-2.0
Ash, coal	0.73	0.45	0.53
Ash, household	0.5-1.9	1.6-4.2	2.3-12.0
Ash, wood	0.1-0.2	0.8-5.9	1.5-36.0
Rural compost, dry	0.5-1.0	0.4-0.8	0.8-1.2
Urban compost, dry	0.7-2.0	0.9-3.0	1.0-2.0
Farmyard manure, dry	0.4-1.5	0.3-0.9	0.3-1.9
Filter press cake	1.0-1.5	4.0-5.0	2.0-7.0
Rice hulls	0.3-0.5	0.2-0.5	0.3-0.5
Groundnut husks	1.6-1.8	0.3-0.5	1.1-1.7
Banana, dry	0.61	0.12	1.00
Cotton	0.44	0.10	0.66
Maize	0.42	1.57	1.65
Paddy	0.36	0. 08	0.71
Tobacco	1.12	0.84	0.80
Pigeon pea	1.10	0.58	1.28
Wheat	0.53	0.10	1.10

Sugarcane trash	0.35	0.10	0.60
Tobacco dust	1.10	0.31	0.93
Tree leaves, dry			
Calotropis gigantea	0.35	0.12	0.36
Careya arborea	1.67	0.40	2.20
Cassia ariculata	0.98	0.12	0.67
Dillenia pentagyna	1.34	0.50	3.20
Madhuca indica	1.66	0.50	2.00
Pongamia pinnata	3.69	2.41	2.42
Pterocarpus marsupium	1.97	0.40	2.90
Terimalia chebula	1.46	0.35	1.35
Terminalia paniculata	1.70	0.40	1.60
Terminalia tomentosa	1.39	0.40	1.80
Xylia dolabriformis	1.37	0.30	1.61
Green manures, fresh			
Cowpea (*vigna unguiculata*)	0.71	0.15	0.58
Sesbania aculeata	0.62	-	-
Cluster-bea *(cyamopsistetragonoloba)*	0.34	-	-
Horse-gram *(Dolichos biflorus)*	0.33	-	-
Mothbean	0.80	-	-
Green gram *(vigna radiate)*	0.72	0.18	0.53
Sunnhemp (*Crotalaria juncea*)	0.75	0.12	0.51
Blackgram (*vigna mungo*)	0.85	0.18	0.53
Non edible Oil Cakes			
Castor cake	4.3	1.8	1.3
Cotton cake	3.9	1.8	1.6
Karanj cake	3.9	0.9	1.2
Mahua cake	2.5	0.8	1.8
Neem cake	5.2	1.0	1.4
Safflower cake	4.9	1.4	1.2
Edible Oil cakes			
Coconut cake	3.0	1.9	1.8
Groundnut cake	7.3	1.5	1.3
Niger cake	4.7	1.8	1.3
Rape seed cake	5.2	1.8	1.2
Sesame cake	6.2	2.0	1.2

Appendix VI. Micronutrient content in organic manure(ppm)

Organic material	Fe	Mn	Cu	Zn	B
(a) Animal manures					
Cattle manure	690-1518	167-389	24-40	128-183	13-30
Farm Yard manure (FYM)	1135-3515	229-668	17-29	83-128	12-27
Goat manure	1449-2174	246-505	20-38	112-184	29-66
Pig manure	1020-1990	180-207	45-48	186-575	4-13
Poultry manure					
Broiler 1-2 litter	723-1565	213-421	27-40	166-271	15-27
Layer litter	1144-2215	287-450	22-38	182-329	12-23
(b) Other dead materials					
Biogas residue	1097-1644	402-616	37-53	300-412	22-46
Compost	2090-4064	201-505	11-25	75-169	12-20
Sewage sludge	2275-3322	100-287	7-11	69-177	3-5
(c) Green manures					
Acacia	870-994	78-92	7-9	54-61	46-49
Calapogonium	340-542	80-140	4-6	64-69	30-33
Centrosema	252-420	38-62	4-7	42-49	79-84
Gliricidia	212-450	85-156	4-5	57-70	41-69
Pueraria	117-280	82-120	4-6	64-70	20-24

Appendix VII. Sources of Plant Nutrient in Soil

Nutrient	Sources
N	Organic matter.
P	Apatite (Calcium Phosphate minerals), Fe, or Al, Phosphates, organic matter.
K	Micas, Feldspars.
Ca	Dolomite, Calcite, Apatite, Calcium Carbonate, Gypsum, Augite.
Mg	Dolomite, Muscovite, Biotite, Augite, Olivine, Hornblende.
Fe	Pyrites, Magnetite.
Mn	Manganite, Pyrolusite, Augite, Olivine, Hornblende.
B	Tourmaline
C	Carbamate.
Cu	Augite, Biotite, Chacopyrite, Olivine, Hornblende.
Zn	Augite, Biotite, Olivine, Hornblende, Sphalerite.
S	Gypsum, Pyrite, Organic matter.
Mo	Olivine.
Cl	Apatite.

Appendix VIII. Optimum pH range for availability of different nutrients

Nutrient	Optimum pH range
N	6.0-8.0
P	6.5-7.5
K	6.0-7.5
Ca	7.0-8.5
Mg	7.0-8.5
Fe	6.0 and below
Mn	5.0-6.5
B	5.0-7.0
Cu	5.0-7.0
Zn	5.0-7.0
S	6.0 and above
Mo	7.0 and above

Appendix IX. Lime requirement of acid soil

pH of soil buffer suspension	Lime required to bring the soil to desired pH (tons / acre)		
	pH 6.0	pH 6.4	pH 6.8
6.7	1.0	1.2	1.4
6.6	1.4	1.7	1.9
6.5	1.8	2.2	2.5
6.4	2.3	2.7	3.1
6.3	2.7	3.2	3.7
6.2	3.1	3.7	4.2
6.1	3.5	4.2	4.8
6.0	3.9	4.7	5.4
5.9	4.4	5.2	6.0
5.8	4.8	5.7	6.5
5.7	5.2	6.2	7.1
5.6	5.6	6.7	7.7
5.5	6.0	7.2	8.3
5.4	6.5	7.7	8.9
5.3	6.9	8.2	9.4
5.2	7.4	8.6	10.0
5.1	7.8	9.1	10.6
5.0	8.2	9.6	11.2
4.9	8.6	10.1	11.8
4.8	9.1	10.6	12.4

❑❑❑